高等教育"十三五"工科类全英教材

Fluid Mechanics in Civil Engineering

工程流体力学（土木类）

主编　尹小玲（YIN XIAOLING）　唐小南（TANG XIAONAN）
主审　【英】DONALD W KNIGHT

华南理工大学出版社
SOUTH CHINA UNIVERSITY OF TECHNOLOGY PRESS
·广州·

图书在版编目(CIP)数据

工程流体力学:土木类 = Fluid Mechanics in Civil Engineering:英文/尹小玲,唐小南主编. —广州:华南理工大学出版社,2019.1
高等教育"十三五"工科类全英教材
ISBN 978-7-5623-5474-1

Ⅰ.①工⋯ Ⅱ.①尹⋯ ②唐⋯ Ⅲ.①工程力学-流体力学-高等学校-教材-英文②土木工程-流体力学-高等学校-教材-英文 Ⅳ.①TB126 ②TU

中国版本图书馆 CIP 数据核字(2017)第 285685 号

Fluid Mechanics in Civil Engineering

工程流体力学(土木类)

尹小玲　唐小南　主编

出 版 人:卢家明
出版发行:华南理工大学出版社
　　　　　(广州五山华南理工大学17号楼,邮编510640)
　　　　　http://www.scutpress.com.cn E-mail:scutc13@scut.edu.cn
　　　　　营销部电话:020-87113487 87111048(传真)
策划编辑:赖淑华
责任编辑:骆　婷　庄　彦
印 刷 者:虎彩印艺股份有限公司
开　　本:787mm×1092mm　1/16　印张:20.5　字数:682千
版　　次:2017年12月第1版　2019年1月第2次印刷
定　　价:74.00元

版权所有　盗版必究　　印装差错　负责调换

Preface

This book is written for civil engineering students who are interested in learning some basic principle and application of fluid mechanics. We make effort to present the basic principles thoroughly to help understand physical circumstances, emphasizing physical concepts rather than mathematics. For the sake of the majority of students who work as civil engineers in future, emphasis of this book is more on engineering application of fundamental fluid mechanics. However, students should have background knowledge in statics and calculus.

This book is organized in two parts: basic theory and engineering application. The part of basic theory includes introduction, hydrostatics, hydrodynamics, flow friction and head loss. The part of engineering application includes orifice flow, nozzle flow, hydraulic calculation of long and short pipelines, pipe networks, steady open-channel flow, and seepage flow. The contents are feasibly arranged and suit for flexibly teaching hours.

This book is also organized to support the development of skills for problem solving. A brief summary of every chapter is given at the end of each chapter, followed by exercises, which include multiple-choice questions and problems. The answers to the selected problems are given at the end of the book.

We are most grateful to all who gave us support in the preparation of this book. We particularly thank Professor Donald W Knight (University of Birmingham, UK) to read the whole manuscript and give valuable comments. Also we would like to express our gratitude to Ms Bu Yu (former professor at South China University of Technology, China) who provided much of the book's framework and contents that was based heavily upon her earlier book titled Hydraulics.

Xiaonan Tang
(Xi'an Jiaotong-Liverpool University)
Xiaoling Yin
(South China University of Technology)
July 2017

Contents

Chapter 1 Introduction 1

1.1 What is about hydraulics? 1
1.2 Fluids and their properties 2
　1.2.1 The basic characteristic of fluid 2
　1.2.2 International system of units (SI) and engineering units 2
　1.2.3 The main physical properties of liquids 4
1.3 Forces acting on the fluid 14
　1.3.1 Surface force 14
　1.3.2 Body force 15
Multiple-choice questions (one option) 15
Problems 16

Chapter 2 Hydrostatics 18

2.1 Concept of hydrostatic pressure 18
　2.1.1 Definition of hydrostatic pressure 18
　2.1.2 Features of hydrostatic pressure 19
2.2 Hydrostatic differential equation and isobaric surface 21
　2.2.1 Differential equation of fluid in equilibrium 21
　2.2.2 Isobaric surface 23
2.3 Distribution of hydrostatic pressure under gravity 24
　2.3.1 Basic formula of hydrostatic pressure under gravity 24
　2.3.2 Absolute, relative and vacuum pressures 26
　2.3.3 Energy significance and geometric meaning of the basic hydrostatic pressure equation 30
2.4 The application of hydrostatics in measurement 32
　2.4.1 Piezometer 32
　2.4.2 Differential gauge 33
2.5 Total hydrostatic force acting on a plane surface 35
　2.5.1 Graphic method 36
　2.5.2 Analytical method 40
2.6 Total hydrostatic forces acting on curved surfaces 46
　2.6.1 Magnitude of total hydrostatic force on a curved surface 47
　2.6.2 Direction of total hydrostatic force 50

2.6.3 Acting point of total hydrostatic force ································ 50
2.7 Total hydrostatic force on a body, buoyancy, stability of a floating body ············ 54
  2.7.1 Total hydrostatic force acting on a body —Archimedes principle ················ 54
  2.7.2 Equilibrium of a sinking body, submerged body and floating body ·············· 55
Chapter summary ·· 56
Multiple-choice questions (one option) ··· 57
Problems ··· 58

Chapter 3 Basic equations of steady total flow ································ 67

3.1 Two methods for describing motion of fluid ·································· 67
  3.1.1 Lagrangian method and Eulerian method ································ 67
  3.1.2 Acceleration of particle: local, convective and total acceleration ················ 70
  3.1.3 Some basic concepts of fluid movement ································ 73
3.2 Continuity equation of steady total flow ······································ 79
3.3 Energy equation of steady total flow ··· 82
  3.3.1 Energy equation of steady streamtube flow of ideal fluid ··················· 82
  3.3.2 Energy equation of steady streamtube flow of real fluid ·················· 85
  3.3.3 Energy equation of steady total flow of real fluid ·························· 86
3.4 Momentum equation of steady total flow ··································· 104
  3.4.1 Derivation of the momentum equation ································· 104
  3.4.2 Conditions and tips in the application of the momentum equation ·············· 109
  3.4.3 Application examples of the momentum equation ························ 111
  3.4.4 Similarities and differences between the momentum equation and energy equation
   ··· 116
Chapter summary ··· 116
Multiple-choice questions (one option) ··· 117
Problems ·· 119

Chapter 4 Types of flow and head loss ··· 128

4.1 The classification of flow resistance and head loss ···························· 128
  4.1.1 The classification of flow resistance ······································ 128
  4.1.2 The classification of head losses ··· 131
  4.1.3 The superposition principle of head losses ································ 131
4.2 Two regimes of real fluid flow ·· 132
  4.2.1 Reynolds' experiment ·· 132
  4.2.2 The identification of laminar and turbulent flows ·························· 135
  4.2.3 The physical meaning of Reynolds number ······························· 136
4.3 The relationship between frictional head loss and shear stress of uniform flow ······ 138
  4.3.1 The relationship between frictional head loss and wall shear stress ············ 138

4.3.2　The relationship between frictional head loss and shear stress ……………… 139
4.3.3　The general calculation formula for frictional head loss ………………… 141
4.4　Laminar flow in circular pipes ………………………………………………… 143
　　4.4.1　The velocity distribution of laminar flow ………………………………… 143
　　4.4.2　The mean flow velocity of laminar flow ………………………………… 144
　　4.4.3　The flow rate of laminar flow ……………………………………………… 144
　　4.4.4　The frictional head loss of laminar flow ………………………………… 144
　　4.4.5　The kinetic correction coefficient of laminar flow ……………………… 145
4.5　The basic concepts of turbulent flow …………………………………………… 146
　　4.5.1　Developing process of turbulent flow …………………………………… 146
　　4.5.2　Fluctuation and time averaged motion of turbulent flow ……………… 148
　　4.5.3　The shear stress and Prandtl's theory of turbulent flow ……………… 152
　　4.5.4　The viscous sublayer and flow zone of turbulent flow ………………… 156
　　4.5.5　The velocity distribution of turbulent flow ……………………………… 159
4.6　Frictional head losses of turbulent flow ……………………………………… 164
　　4.6.1　Experiment of frictional resistance coefficient ………………………… 164
　　4.6.2　Frictional resistance coefficient of commercial pipes ………………… 169
　　4.6.3　Empirical formulae for frictional head loss ……………………………… 173
4.7　Local head loss …………………………………………………………………… 177
　　4.7.1　Local head loss of sudden expansion of pipe …………………………… 177
　　4.7.2　Local head loss coefficient ………………………………………………… 179
4.8　Basic concepts of boundary layer and flow resistance around an object …… 185
　　4.8.1　Basic concept of boundary layer …………………………………………… 185
　　4.8.2　Separation of boundary layer and flow resistance ……………………… 187
Chapter summary ………………………………………………………………………… 191
Multiple-choice questions (one option) ……………………………………………… 192
Problems …………………………………………………………………………………… 194

Chapter 5　Steady orifice, nozzle and pipe flow ………………………………… 197

5.1　Introduction ……………………………………………………………………… 197
5.2　Basic formulae for steady flow through orifice and nozzle …………………… 198
　　5.2.1　Steady flow through thin-wall orifice …………………………………… 198
　　5.2.2　Steady flow through nozzle ………………………………………………… 201
5.3　Steady flow in pressurized pipes ………………………………………………… 205
　　5.3.1　Hydraulic calculation of hydraulically short pipes ……………………… 205
　　5.3.2　Hydraulic calculation of hydraulically long pipes ……………………… 212
　　5.3.3　Hydraulic calculation for pipeline networks ……………………………… 221
Chapter summary ………………………………………………………………………… 226
Review questions ………………………………………………………………………… 227

Multiple-choice questions (one option) ······ 227
Problems ······ 229

Chapter 6 Steady flow in an open channel ······ 233

6.1 Geometry of open channel ······ 233
   6.1.1 Longitudinal bed slope of open channel ······ 233
   6.1.2 Cross-section of open channel ······ 234
   6.1.3 Geometrical parameters of flow cross-section ······ 234
   6.1.4 Prismatic and non-prismatic channel ······ 235
6.2 Uniform flow in open channel ······ 236
   6.2.1 Characteristics and conditions of uniform open-channel flow ······ 236
   6.2.2 Basic equations for uniform open-channel flow ······ 238
   6.2.3 Hydraulic calculation of uniform open-channel flow ······ 240
   6.2.4 The optimum hydraulic cross-section ······ 243
6.3 Steady non-uniform open-channel flow ······ 245
   6.3.1 Flow regime of open-channel flow ······ 245
   6.3.2 Specific energy ······ 248
   6.3.3 Critical depth ······ 249
   6.3.4 Critical bed slope ······ 250
   6.3.5 Hydraulic jump and hydraulic drop ······ 253
   6.3.6 Surface profile of gradually varied flow in prismatic open channel ······ 259
   6.3.7 Computation of surface profiles in steady gradually varied flow ······ 268
6.4 Weir flow and underflow of sluice gates ······ 271
   6.4.1 Types and basic formula of weir flow ······ 272
   6.4.2 Fundamental formula of underflow of a sluice gate ······ 278
Chapter summary ······ 280
Multiple-choice questions (one option) ······ 282
Problems ······ 283

Chapter 7 Seepage flow ······ 287

7.1 The phenomenon of seepage and the seepage model ······ 287
   7.1.1 Seepage phenomenon ······ 287
   7.1.2 State of water in soil ······ 288
   7.1.3 The characteristics of soil seepage ······ 289
   7.1.4 Seepage models ······ 289
7.2 The basic law of seepage flow ······ 290
   7.2.1 Darcy's Law ······ 290
   7.2.2 The limitations of Darcy's law ······ 291
   7.2.3 The coefficient of permeability ······ 292

7.3 Dupuit's formula of steady gradually varied seepage flow ... 293
   7.3.1 The velocity distribution in steady uniform and non-uniform seepage flows ... 293
   7.3.2 The basic differential equation and the seepage curve of steady gradually varied seepage flow ... 295
7.4 Seepage calculation of wells and catchment corridors ... 300
   7.4.1 Catchment corridors ... 300
   7.4.2 Fully penetrating open wells ... 300
   7.4.3 Fully penetrating artesian wells ... 302
   7.4.4 The drainage of large-diameter well and foundation ditch ... 303
   7.4.5 Well group ... 304
7.5 Graphical solution by drawing flow net ... 306
   7.5.1 Drawing of flow net for the planar confined seepage ... 307
   7.5.2 Seepage calculation by flow net ... 308
Chapter summary ... 310
Review questions ... 311
Multiple-choice questions (one option) ... 311
Problems ... 312

Answers to selected problems ... 314

References ... 317

Chapter 1   Introduction

1.1   What is about hydraulics?

Hydraulics is a branch of fluid mechanics that studies the engineering aspects of the behaviour of fluids, mainly water. It deals with the mechanical properties and laws of motion of fluids and their engineering applications. There are two major aspects of hydraulics, which differ from solid mechanics: one is the nature and properties of the liquid itself, which are very different from those of a solid; the other is that we are frequently concerned with the behaviour of a continuous stream of liquid rather than individual bodies or elements of known mass.

The basic principle of hydraulics includes two parts: hydrostatics and hydrodynamics. Hydrostatics is the study of fluids at rest, i.e. the mechanical characteristics of a stationary fluid. Hydrodynamics is the study of fluids in motion, i.e. the relationship between the force and motion of a moving fluid. Hydrostatics is the fundamental basis, whereas hydrodynamics describes the general principles of fluids in motion. However, it is extremely difficult to specify either the precise movement of a stream of fluid or that of individual particles within it. Therefore, it is often necessary to assume ideal, simplified conditions and patterns of the flow of fluid. The results so obtained for a basic analysis of the flow system may then be modified by introducing appropriate coefficients or factors determined experimentally in order to be applied in engineering practice. This approach has proved to be reasonably satisfactory so far.

Hydraulics is widely used in many engineering disciplines, for example, hydraulic engineering, civil engineering, water supply and distribution, port and shipping engineering, mechanical engineering, petroleum and chemical engineering, mining and metallurgical engineering, energy engineering, and environmental engineering. Therefore, hydraulics is an important fundamental subject required for many engineering programmes of universities, especially for civil engineering.

The pre-requisite knowledge for hydraulics is higher mathematics, physics, theoretical mechanics and materials mechanics.

1.2 Fluids and their properties

1.2.1 The basic characteristic of fluid

A substance has three states: solid, liquid and gas.

The molecules of a solid are usually close each other, and the attractive forces between the molecules are very large, so that a solid can retain its shape and volume. A solid can sustain certain degree of stretching, pressuring and shear forces. In contrast, the molecules of a liquid are less close so that their attractive forces are relatively smaller. A liquid can hardly bear stretching and resist tensile deformations, so the liquid will easily be deformed or flow if any small shear force acts on it. This property of flowing easily is defined as fluidity. Under certain pressure and temperature, a liquid cannot maintain a constant shape, but it will retain a definite volume for a given mass. If a liquid is filled into a container, a clearly defined free surface of liquid will be established.

Due to the molecules of a gas being much farther apart, the intermolecular force of a gas is much smaller than that of a liquid, so a gas has the similar characteristics of deformation as a liquid, i.e. easy change of its shape. In this respect, a liquid or a gas is called a fluid.

A study shows that 1 cm^3 of water contains 3.3×10^{23} molecules, which have the average distance apart of about 3×10^{-8} cm, and consequently liquid molecules each undertake complex microscopic motion.

Hydraulics is not concerned with the study of the micro-motion of liquid molecules, but their macroscopic mechanical movements as a whole. Therefore, the continuum concept of liquid is used in hydraulics, which assumes that no gap exists between the liquid particles that therefore they continuously fill the space occupied. Thus, the physical and motion properties of liquid particles are continuously distributed. It should be noted that the liquid particle refers to an infinitely small point inside the liquid with a corresponding mass. With the continuum concept of liquid, we can use mathematically continuous functions to study the liquid, and also meet the overall requirements of most practical engineering problems, because the distance between molecules is so small that it can be ignored when compared with the scale of flow in engineering problems.

Based on the continuum concept, liquids are generally considered to be uniform isotropic, i.e. the liquid is homogeneous, so the physical properties of each particle are the same in all directions. In a nutshell, the liquid in hydraulics is a free-flowing, incompressible, homogeneous isotropic continuum.

1.2.2 International system of units (SI) and engineering units

The international system of units (SI) was formally adopted for quantities and units in China, 1977. However, engineering units were used in some earlier published technical books and literature, and are still used by some senior technical professionals. In order to facilitate learning, it is necessary to clarify the differences and conversion relations between the two systems of units.

1.2.2.1 Dimensions and units

Dimension is a measure of a physical quantity, usually expressed in capital letters within square brackets, such as the length dimension in [L], the time dimension in [T], the mass dimension in [M], the force dimension in [F], and so on. The units of the same nature have the same dimension, such as year, month, day, hour, minute and second in [T]. However, the units of different natures have different dimensions, such as seconds, meters, Newton denoted by [T], [L], [F], respectively.

Unit is a specific measure of physical quantity values. Different physical quantities can have different units, and the same physical quantities may also have different units. For example, units of length can be used by centimetres, meters, kilometres, etc., whereas gravity units can be expressed in Newton, kN, tons, etc.

Dimensions can be divided into primary dimensions and derived dimensions.

Primary dimensions are independent, which cannot be deduced from other fundamental dimensions. This is to say that primary dimensions are not dependent on other fundamental dimensions. For example, length dimension [L], time dimension [T], and mass dimension [M] are independent of each other, so [M] can neither be derived from [L] and [T], nor [L] from [T] and [M], or [T] from [L] and [M].

Derived dimensions are the dimensions that can be derived from the primary dimensions. In mechanics, dimension of any physical quantity can be expressed as the index product of three primary dimensions. Any physical quantity x, may therefore have its dimension expressed as:

$$[x] = [L^\alpha T^\beta M^\gamma] \qquad (1-1)$$

where the x dimension depends on the indexes α, β, γ. When $\alpha \neq 0$, $\beta = \gamma = 0$, x is the quantity of geometry; when $\alpha \neq 0$, $\beta \neq 0$, $\gamma = 0$, it is the quantity of kinematics; when $\alpha \neq 0$, $\beta \neq 0$, $\gamma \neq 0$, it is the quantity of dynamics. When $\alpha = \beta = \gamma = 0$, that is $[x] = [L^0 T^0 M^0] = [1]$, x is a dimensionless quantity, called pure number.

1.2.2.2 The international system of units (SI) and system of engineering units

The difference between the two systems is in the choice of different fundamental dimensions, which leads to the different derived dimensions. Take Newton's second law $F = ma$ for example, which is often used in fluid mechanics.

International system of units (SI):

If the primary dimensions are [L], [T], [M], the derived dimension of a force is

$$[F] = \left[\frac{ML}{T^2}\right] = [MLT^{-2}]$$

If the units of length, time and mass are used by m, s, kg, respectively, then the unit of force is kg·m/s^2, i.e. 1 N (1 N = 1 kg·m/s^2).

Engineering units:

For the primary dimensions chosen as [L], [T], [F], the derived dimension of a mass is

$$[M] = \left[\frac{FT^2}{L}\right] = [FT^2L^{-1}]$$

If the units of length, time and force are used by m, s, kgf, respectively, then the engineering unit of mass is kgf·s²/m.

Note that, strictly speaking, kg and kgf are different: kg represents the mass, but kgf denotes the weight or force; however, the differences between the kg and kgf are not clearly identified in some books and materials.

The weight in the two units systems has the following basic relationship

$$1 \text{ kgf} = 9.8 \text{ N} \tag{1-2}$$

The basic relationship of mass conversion is

$$1 \text{ kgf} \cdot s^2/m = 9.8 \text{ kg} \tag{1-3}$$

i. e. 1 kgf force = 9.8 Newton and 1 engineering unit of mass = 9.8 kg.

1.2.3 The main physical properties of liquids

1.2.3.1 Inertia, mass and density

Inertia is a property of an object, by which it maintains its existing state of rest or uniform motion. Inertia is measured by mass. The greater the mass of an object, the greater is the inertia. When the motion state of a liquid changes as a result of external forces, due to the inertia of the liquid, the resistance to the external forces that attempts to maintain the original state of motion is called an inertia force. Let the mass of a liquid be m, the acceleration a, then the inertia force is

$$F = -ma \tag{1-4}$$

where the negative sign indicates that the direction of the inertia force of liquid is opposite to the direction of acceleration.

The mass of a liquid per unit volume is called the density, denoted by the symbol ρ.

For homogenous liquid

$$\rho = \frac{m}{V} \tag{1-5}$$

where m is the mass of liquid, and V is the volume of liquid.

For heterogeneous liquid

$$\rho = \lim_{\Delta V \to 0} \frac{\Delta m}{\Delta V} \tag{1-6}$$

In the international system of units (SI), the unit of density is kg/m³. In general, the density of water has little variation with temperature and pressure. Usually we take it as a constant. At the temperature of 4℃ and atmospheric pressure of 101.3 kPa, the maximum density of water is $\rho_w = 1000$ kg/m³.

The density of water at different temperatures is shown in Table 1-1.

Table 1-1 Values of the physical properties of water at different temperatures

Temperature $t/°C$	Specific weight $\gamma/\text{kN}\cdot\text{m}^{-3}$	Density $\rho/\text{kg}\cdot\text{m}^{-3}$	Kinetic viscosity coefficient μ $/10^{-3}\text{N}\cdot\text{s}\cdot\text{m}^{-3}$	Dynamic viscosity coefficient $\nu/10^{-6}\text{ m}^2\cdot\text{s}^{-1}$	Bulk modulus $K/10^9\text{ N}\cdot\text{m}^{-2}$	Surface tension coefficient $\sigma/\text{N}\cdot\text{m}^{-1}$
0	9.805	999.9	1.781	1.785	2.02	0.0756
5	9.807	1000.0	1.518	1.519	2.06	0.0749
10	9.804	999.7	1.307	1.306	2.10	0.0742
15	9.798	999.1	1.139	1.139	2.15	0.0735
20	9.789	998.2	1.002	1.003	2.18	0.0728
25	9.777	997.0	0.890	0.893	2.22	0.0720
30	9.764	995.7	0.798	0.800	2.25	0.0712
40	9.730	992.2	0.653	0.658	2.28	0.0696
50	9.689	988.0	0.547	0.553	2.29	0.0679
60	9.642	983.2	0.466	0.474	2.28	0.0662
70	9.589	977.8	0.404	0.413	2.25	0.0644
80	9.530	971.8	0.354	0.364	2.20	0.0626
90	9.466	965.3	0.315	0.326	2.14	0.0608
100	9.399	958.4	0.282	0.294	2.07	0.0589

1.2.3.2 Gravity, weight and specific weight

The attractive force between matter is called universal gravitation. Earth's gravitational force on an object is called gravity, or the weight, denoted as the symbol G. For the mass of a liquid m, the gravitational force is

$$G = mg \qquad (1-7)$$

where g is the acceleration of gravity, taken as 9.8 m/s².

The gravitational force per unit volume of fluid, or simply the weight per unit volume, is defined as specific weight, given by the symbol γ.

For homogeneous liquid

$$\gamma = \frac{G}{V} \qquad (1-8)$$

For non-homogeneous liquid

$$\gamma = \lim_{\Delta V \to 0} \frac{\Delta G}{\Delta V} \qquad (1-9)$$

In the international system of units, the unit of specific weight is N/m³ or kN/m³. The specific weight of water has little variation with temperature and pressure, generally regarded as constant. The specific weight of water (γ_w) is taken 9800 N/m³ at 4°C and 101 325 Pa (a standard atmospheric pressure).

From Equation (1-7): $G = mg$, by dividing the volume V on both sides of the equation, we have $\frac{G}{V} = \frac{m}{V}g$, which shows a relationship between specific weight and density:

$$\gamma = \rho g \qquad (1-10)$$

or
$$\rho = \gamma/g \qquad (1-11)$$

Specific weights of some common liquids are given in Table 1-2.

Table 1-2 γ values of common liquids (at a standard atmospheric pressure)

Liquid	Gasoline	Pure alcohol	Distilled water	Sea water	Mercury
Specific weight/N·m^{-3}	6 664 ~ 7 350	7 778.3	9 800	9 996 ~ 10 084	133 280
Temperature/°C	15	15	4	15	0

1.2.3.3 Viscosity and viscosity coefficient

Viscosity is of most unique and inherent physical properties of fluids. It is different from the other physics concepts learned in the past.

Fluid at rest cannot resist shearing forces and has shear deformation. However, when in motion, there is a relative motion between the liquid particles or motion layers, which will generate internal friction to resist the relative motion between the fluid particles or flow layers. The internal friction does work and dissipates mechanical energy. This feature is called viscosity of fluid.

Take an example of the motion of liquid in a wide and shallow flume. If the abscissa denotes the velocity u of the liquid, the vertical coordinate y represents the distance of the particle from the bottom of the flume, set measuring points at different depths, and then measure the average velocity of these points in the motion direction. Thus we can obtain the variation of velocity with depth, $u = u(y)$, shown in Fig. 1-1.

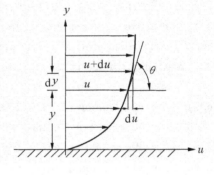

Fig. 1-1

The liquid in contact with the bottom of the flume has a zero velocity due to the adhesion of solid boundary, where no relative sliding motion exists. This is called the no-slip condition for the wall in fluid mechanics. The liquid above the bottom has different velocities, which are subject to the impact of retardation of bottom boundary. The closer to the bottom, the greater the impact of the retardation is, and the flow velocity is lower. However, the farther away from the bottom, the less the impact of the retardation is, so the flow velocity is higher.

The velocity of liquid on the free surface is slightly smaller than that of the liquid at the lower layers, which is due to the influence of air resistance. Assume that the velocity is u at the distance y to the bottom boundary, the velocity of liquid at the position of $y + \mathrm{d}y$ will be $u + \mathrm{d}u$, where $\mathrm{d}y$ is the distance between the two layers, and $\mathrm{d}u$ is the difference of their velocities. Due to the different velocities of flow at the two layers, relative motion exists. For a viscous liquid, a resistance takes place to such a relative motion: the liquid of the lower layer retards the liquid of the upper layer by a frictional force in the opposite direction of motion, trying to slow down the motion of the faster layer; the liquid on the upper layer takes the liquid of the lower forwards by a frictional force in the direction of flow, trying to speed up the velocity of the slower moving layer. The two frictional forces are of the same size but in opposite directions, and both of them resist the relative movement. Note that this type of frictional force is different from the external friction in physics, because it does not exist between the liquid and solid walls, but only in between the liquid particles or flow layers. So this is called the internal friction. Internal friction is an important reason for the formation of the flow resistance. In order to overcome the internal friction and maintain the movement of liquid, it is required to consume efficient mechanical energy. Therefore, the viscosity of the liquid is the source of the mechanical energy loss.

In 1686, Newton proposed the well tested and verified Newton's law of internal friction, described as follows: when liquid has parallel linear motion along a solid surface, internal frictional force is generated along the interface of two adjacent layers in relative motion. The internal friction is related to the property of the liquid, and proportional to both the gradient of velocity and the contact area of liquid, but is nothing to do with the normal pressure on the contact surface. Its mathematical expression is

$$F = \mu A \frac{du}{dy} \tag{1-12}$$

where μ—the dynamic viscosity coefficient of fluid, $N \cdot s / m^2$, i.e., $Pa \cdot s$;

A—the contact area of fluid, m^2;

du/dy—the velocity gradient, i.e., grad u, which is the velocity increment of flow per unit depth, s^{-1}.

Let τ represent the frictional force per unit area, i.e., the viscous shear stress, then

$$\tau = \frac{F}{A} = \mu \frac{du}{dy} \tag{1-13}$$

【Example 1 - 1】 A Newtonian fluid has the viscosity coefficient of $0.368 \ N \cdot s/m^2$, and its velocity distribution near the fixed wall is shown in Fig. 1 - 2. Determine the viscous shear stress at the wall (note that d is the height between the layer with the maximum velocity and the wall).

Solution: From Newton's law of internal friction, the viscous shear stress at the wall is

$$\tau_0 = \mu \frac{du}{dy}\bigg|_{y=0}$$

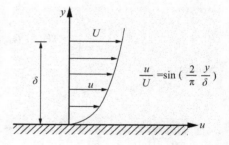

Fig. 1 - 2

where

$$\frac{du}{dy} = \frac{\pi}{2} \frac{U}{\delta} \cos\left(\frac{\pi}{2} \frac{y}{\delta}\right)$$

Then

$$\tau_0 = 0.368 \cdot \frac{\pi}{2} \frac{U}{\delta} = 0.578 \frac{U}{\delta}$$

For the fluid, the viscous force is towards the left, opposite to the direction of flow, and retards the flow; for the walls, the force acts in the right direction, consistent with the flow direction.

【Example 1 – 2】 A movable plate between two fixed parallel plates forms two gaps, which are filled with two different types of Newtonian fluid, as shown in Fig. 1 – 3a. If the moving plate moves at a speed of 4 m/s horizontally, find the shear stress on the fixed plates. Assume that the surface area of the plates is very large, ignore the marginal impact of the plates, and that the fluid between the plates has a linear distribution of velocity.

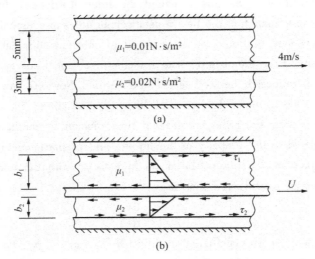

Fig. 1 – 3

Solution: The linear velocity distribution of the fluid between the plates is triangular as shown in Fig. 1 – 3b, where the velocity of the fluid at the fixed wall is zero, the velocity of the fluid at the moving plate is U. Therefore, for the fluid with the viscosity coefficient μ_1, the velocity gradient is U/b_1; for the fluid with μ_2, its velocity gradient is U/b_2.

By Newton's law of internal friction, and substituting the known conditions, the viscous shear stress is

$$\tau_1 = \mu_1 \frac{U}{b_1} = 0.01 \times \frac{4}{5 \times 10^{-3}} = 8 \text{ N}$$

Similarly,

$$\tau_2 = \mu_2 \frac{U}{b_2} = 0.02 \times \frac{4}{3 \times 10^{-3}} = 26.7 \text{ N}$$

The directions of the shear stress are shown in Fig. 1 – 3b.

【Example 1 – 3】 A rotating cylinder viscometer has a fixed outer cylinder, whereas the inner cylinder is rotated by a synchronous motor, filled with a liquid sample between the inner and outer cylinders, as shown in Fig. 1 – 4. The radius of the inner cylinder $r_1 = 1.93$ cm, with the height of $h = 7$ cm, the radius of the outer cylinder $r_2 = 2$ cm, the inner cylinder speed $n = 10$ r/min, and the torque on the rotating shaft is measured to be $M = 0.004$ N · m. Calculate the viscosity coefficient μ of the liquid.

Solution: Under the impact of the rotating inner cylinder, the liquid in the gap of the two cylinders will do circular motion. Because the gap is very small, a linear liquid velocity distribution is approximated. By ignoring the influence of both ends of the cylinder, the wall shear stress on the inner cylinder is

$$\tau = \mu \frac{du}{dy} = \mu \frac{\omega r_1}{\delta}$$

where δ is the thickness of the gap between the inner and outer cylinders, so $\delta = r_2 - r_1$; ω is the angular velocity of the inner cylinder rotation, and $\omega = 2\pi n/60$.

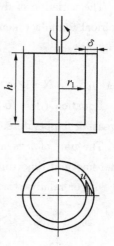

Torque: $\qquad M = \tau h r_1 \cdot 2\pi r_1$

Viscosity coefficient: $\mu = \dfrac{M\delta}{2\pi \omega r_1^3 h} = 0.952$ Pa·s

Regarding velocity gradient (du/dy), which is actually defined as the rate of shear deformation of liquid particle.

Fig. 1-4

As shown in Fig. 1-5, take the analysis on a rectangular element $ABCD$, which is taken from the two flow layers (1-1 and 2-2) similar to that in Fig. 1-1. After the period of time dt, the element is moved to a new location $A'B'C'D'$. Because the flow layer 2-2 is du faster than the layer 1-1,

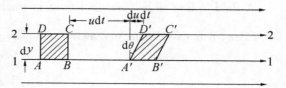

Fig. 1-5

$ABCD$ is not only translated at the distance of udt, but also deformed in shape into a parallelogram. Thus there is a shear deformation (or angular distortion), because the points D and C on the layer 2-2 have moved faster as a distance of $dudt$, both AD and BC have had the angular displacement $d\theta$, with a shear strain rate $d\theta/dt$. As both dt and $d\theta$ are small, $d\theta$ can be considered as

$$d\theta \approx \tan(d\theta) = \frac{dudt}{dy}$$

Then
$$\frac{du}{dy} = \frac{d\theta}{dt} \qquad (1-14)$$

Therefore, the Newton's internal friction Equation (1-13) can be expressed as

$$\tau = \mu \frac{d\theta}{dt} \qquad (1-15)$$

Thus, Newton's theory of fluid friction can be simplified as: viscous shear stress of fluid is proportional to the rate of shear deformation. However, in fluid mechanics the commonly used formula is Equation (1-13).

The viscosity of fluid can be measured by dynamic viscosity μ and kinematic viscosity ν. The unit of μ is Pa·s or kPa·s. Because Pa (N/m^2) includes the quantity of force N, a quantity of dynamics, μ is so called the dynamic viscosity or absolute viscosity. The greater μ, the stronger the viscosity of the fluid is.

Fluid Mechanics in Civil Engineering

The variation of dynamic viscosity of water with temperature can be described by Poiseuille empirical formula (named after J. L. M. Poiseuille):

$$\mu = \frac{0.00178}{1 + 0.0337t + 0.000221t^2} \quad (1-16)$$

where μ is in N·s/m², and t is in degree(℃).

Equation (1-16) shows that at a certain pressure, the viscosity of water decreases with increasing temperature.

The unit of ν is m²/s. Because ν does not contain the dimension of force but the length dimension and time dimension, it is a quantity of kinematics. Therefore, ν is called kinematic viscosity or relative viscosity. The relationship between μ and ν is

$$\mu = \rho\nu \quad (1-17)$$

or

$$\nu = \frac{\mu}{\rho} \quad (1-17')$$

A fluid in line with Newton's law of internal friction ($\tau = \mu \frac{du}{dy}$) is called Newtonian fluid. Otherwise, it is called non-Newtonian fluid. The shear stress of non-Newtonian fluid is expressed by

$$\tau = \tau_0 + K\left(\frac{du}{dy}\right)^\alpha \quad (1-18)$$

where τ is the initial viscous shear stress; du/dy is the velocity gradient; α is the index; K is a coefficient.

Fig. 1-6 shows the variation of several different fluid shear stresses (τ) with the velocity gradient (du/dy). A is a Newtonian fluid, line A is a straight line passing through the origin and has a fixed slope as the viscosity of the fluid does not change at the same temperature. B is an ideal Bingham fluid, which starts to deform with a constant strain rate after a certain value of shear stress; for example, mud, blood plasma, toothpaste. C is a pseudo-plastic fluid, whose viscosity decreases with increasing shear rate; for example, nylon, rubber, cellulose acetate solution, oil paint and paint. D is a dilatant fluid, in which viscosity increases with increasing shear rate, such as dough, thick starch paste, and so on.

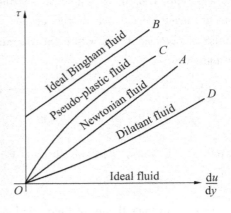

Fig. 1-6

When using Newton's law of internal friction, we should understand the range of its application. In nature and engineering, water and air are Newtonian fluid.

Real fluids, either liquid or gas, are viscous. The viscosity of real fluids makes it very difficult and complicated to study their motion. To simplify problems, the concept of an ideal fluid is introduced. The so-called ideal fluid refers to a fluid without viscosity, i. e., $\mu = 0$. In fact, viscosity and resistance always exist whenever a fluid is moving. The ideal fluid actually does not exist, so it is just a simplified mechanical model of fluid.

Since viscosity is not considered for the ideal fluid, the theoretical analysis of fluid motion is greatly simplified, thereby easily obtaining the theoretical solution. These theoretical solutions are applicable to the real flows in which the viscosity has little effect on the flow motion. For the fluids in which the viscosity cannot be ignored, the solutions for ideal flow, with some modification, can be applied to many practical flow problems, thus making the problems solved relatively easily.

The analysis of ideal flow is not only a basis for analysing complex fluid flows, but also an effective method for solving the problem of viscous fluid motion.

1.2.3.4 Compressibility and compressibility factor

A solid will be deformed under the action of an external force, and if the force does not exceed the elastic limit, when it is removed the solid will recover back to its original state. This property is called elasticity of solid.

Fluid cannot undertake tension, but can bear pressure. A fluid will reduce its volume under a pressure, in which an inherent internal force (elastic force) of the fluid is generated to balance the pressure. Once the pressure is removed, the fluid will remain its original state. This property is called compressibility or elasticity of fluid.

The compressibility of a fluid is expressed by a volume compressibility factor k or bulk modulus K. For a fluid of volume V and density ρ, when the pressure increases $\mathrm{d}p$, the corresponding reduction in the volume is $\mathrm{d}V$, and the corresponding density increases $\mathrm{d}\rho$. The ratio of the relative volume compression of the fluid ($\mathrm{d}V/V$) to the pressure is defined as the volume compressibility factor, denoted by k and expressed as

$$k = -\frac{\mathrm{d}V/V}{\mathrm{d}p} \quad (1-19)$$

where the negative sign is as a result of the positive pressure increment. This is to say that the volume increment $\mathrm{d}V = V_{\mathrm{final}} - V_{\mathrm{initial}} < 0$, when $\mathrm{d}p = p_{\mathrm{final}} - p_{\mathrm{initial}} > 0$. Therefore, the sign of $\mathrm{d}V$ and $\mathrm{d}p$ are always opposite, and the volume compressibility factor k is always a positive number. Thus, Equation (1-19) has a minus sign on the right. For the same increment of pressure, the greater k value, the easier is the fluid compressed. The unit of k is m^2/N.

When the liquid is compressed, its mass does not change, so

$$\mathrm{d}m = \mathrm{d}(\rho V) = \rho \mathrm{d}V + V \mathrm{d}\rho = 0$$

Then,
$$\frac{\mathrm{d}V}{V} = -\frac{\mathrm{d}\rho}{\rho}$$

Thus, the volume compressibility factor can be written as

$$k = \frac{1}{\rho}\frac{\mathrm{d}\rho}{\mathrm{d}p} \quad (1-20)$$

Fluid Mechanics in Civil Engineering

In engineering, similar to the solid, the bulk modulus K is used to represent the compressibility of fluid. The general definition of K is the ratio of the stress to strain, which is reciprocal to the volume compressibility factor k, i. e.

$$K = \frac{1}{k} = -\frac{dp}{\frac{dV}{V}} = \frac{dp}{\frac{d\rho}{\rho}} \qquad (1-21)$$

Equation (1 – 21) shows that the larger the value of K, the more difficult is a fluid compressed. $K = \infty$ means absolutely incompressible. The unit of K is N/m^2.

Different types of fluid have different k and K values. k and K of a fluid vary with temperature and pressure. The K values of water at different temperatures are shown in Table 1 – 1. Due to little changes, they are usually taken constant, usually $K = 2.1 \times 10^9$ N/m^2, i. e. the value at 10℃. In this case, an increment of one atmospheric pressure, i. e. $dp = 98000$ N/m^2, will result in $\left|\frac{dV}{V}\right| = \frac{d\rho}{K} = \frac{98\,000}{2.1 \times 10^9} \approx \frac{1}{20\,000}$. The relative compressive volume of the water is only 1/20 000. Therefore, under normal circumstances water is thought to be incompressible, approximately. Only in some specific cases, such as in a hydropower plant, a sudden closure of the inlet valve will have a sudden and rapid increase of pressure in the piping, where the water is compressed, so the compressibility of water has to be considered.

1. 2. 3. 5 Surface tension and surface tension coefficient

Fig. 1 – 7 is a schematic diagram of attractive force circle of a single liquid molecule. A molecule within the liquid is attracted equally in all four directions by the other modules nearby. Due to the symmetry, the left and right attractions balance each other; for the molecules at the free surface, the downward attraction of liquid is greater than the upward attraction of the liquid from the gas molecules, resulting in the unbalanced attraction,

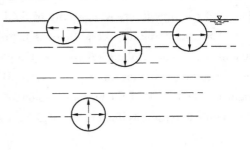

Fig. 1 – 7

which pulls the molecules at the free surface towards the liquid. This effect causes the liquid surface to be pulled into the liquid like a surface tension. The surface tension is generated on the liquid surface, which is in contact with the gas. But it can also be generated in the interfaces between a liquid and solid or between two different liquids. Surface tension does not exist inside the liquid, so it is a phenomenon of local force. Because surface tension is small, its impact on the macro motion of liquid is negligible. However, in a curved surface with a small radius of curvature, the surface tension must be taken into account when an additional pressure force caused by the surface tension is considerably high. For example, the surface tension of water is considered in the cases: the capillary phenomena in soil and rock, water droplets, and the broken thin tongue of water of large curvature. In the section of the commonly used piezometer tubes for measurement, their diameter has to consider the influence of surface tension.

The surface tension is measured by a surface tension coefficient σ, which is defined as the tension force of surface per unit length, in units of N/m. σ varies with the type of liquid and temperature. At 20℃, water $\sigma = 0.074$ N/m, and mercury $\sigma = 0.54$ N/m. The surface tension coefficient of water at different temperatures is given in Table 1-1.

In the following, take the capillary phenomenon in engineering measurement as an example to illustrate the influence of surface tension.

If a glass tube of d diameter, open in the downward end, is inserted into a container filled with mercury, the level of mercury inside the tube will be lower at a height of h than the level outside. If the same tube is inserted into a container filled with water, the level of water inside the tube is h' higher than the water level of the container. This phenomenon is called capillarity, and the height of the liquid column, h or h', is called capillary height.

Since the cohesive force of mercury is greater than the adhesion of the glass wall, the mercury surface inside the fine opening glass tube forms an upward convex meniscus. The cohesive force of water is less than the adhesion of the glass wall, so the water surface inside the tube is a downward convex surface of the meniscus, shown in Fig. 1-8.

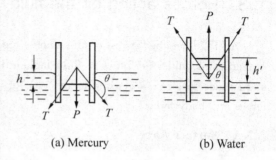

(a) Mercury   (b) Water

Fig. 1-8

Fig. 1-8a shows the upward convex meniscus of mercury. Its surface tension T, tangent to the meniscus, is directed downward. The additional pressure (P) is pointing downward, the same as the direction of gravity. In order to maintain the equilibrium of forces in vertical direction, the mercury inside the tube will fall in a capillary height h to offset the impact of the additional pressures generated by the surface tension. As shown in Fig. 1-8b for a convex meniscus of water, its surface tension T is upward and tangent to the meniscus, and the resulting additional pressure P is pointing upward, opposite to the direction of gravity, so the water column within the capillary rise of height h' is to maintain the static equilibrium vertically. Thus, based on the force equilibrium, the relationship between capillary height and diameter can be obtained:

$$2\pi r \cdot \sigma \cdot \cos\theta = \gamma h \pi r^2$$

Then
$$h = \frac{2\sigma\cos\theta}{\gamma r} = \frac{4\sigma\cos\theta}{\rho g d} \qquad (1-22)$$

where σ—surface tension coefficient;

ρ—liquid density;

g—acceleration of gravity;

d—capillary diameter;

θ—contact angle between the solid and the liquid surface; different fluids have different contact angles. The contact angle of mercury with the glass is about 140°, and the contact angle of water with the glass is 0° to 9°.

Equation (1 – 22) shows that for a given liquid and solid side wall the height of capillary is inversely proportional to the diameter of capillary.

For water at 20℃, $\rho = 9789$ N/m^3, $d = 0.073$ N/m, $\theta = 5°$, then

$$h = 30/d \ (d \text{ and } h \text{ in mm})$$

If the measured error is to be less than 3 mm, a tube of at least 1 cm diameter should be chosen.

In the above section, five types of physical properties of liquid were introduced: inertia, gravitational force, viscosity, compressibility and surface tension. Their impact on the movement of the liquid is different. In general, the inertial, gravitational, viscous forces are most important, whilst the elastic force and the surface tension have the impact only on some special water flow.

1.3 Forces acting on the fluid

All fluids, either at rest or on motion, are subjected to various forces. Based on the physical properties of fluid, the forces of fluid include inertial force, gravity, viscous force, elastic force and surface tension; according to the type of point of action, they can be divided into surface forces and body forces.

1.3.1 Surface force

Surface force acts on a surface of a liquid and is proportional to the surface area.

Surface force can be divided into pressure force (perpendicular to the plane of action) and shear force (parallel to the plane of action). The surface force can be expressed by the total force or the shear force per unit area (i.e. stress).

The average pressure force per unit area is called average pressure p:

$$p = \frac{P}{A} \tag{1-23}$$

where P is the total pressure; A is the area of surface which the pressure acts on.

The average shear stress per unit area is

$$\tau = \frac{F}{A} \tag{1-24}$$

where F is the total shear force; A is the area of surface which the total shear force acts on.

Based on the continuum concept of fluid, the concept of point stress can be obtained by taking its limit value.

Pressure at a point (point pressure):

$$p = \lim_{\Delta A \to 0} \frac{\Delta P}{\Delta A} \tag{1-25}$$

Shear stress of a point:

$$\tau = \lim_{\Delta A \to 0} \frac{\Delta F}{\Delta A} \tag{1-26}$$

The units of both pressure and shear stress are N/m^2, i.e. Pascal (Pa).

1.3.2 Body force

Body force is defined as a force acting on each particle of the liquid and is proportional to the mass of liquid, such as gravity, inertial force. In a homogeneous liquid, mass is proportional to the volume; therefore the body force is also known as volume force.

The body force can be measured by the total force or the force per unit mass that is commonly used and defined as the body force per unit mass of the liquid.

If the mass of a homogeneous liquid is m and the total body force acting on it is F, then the body force per unit mass f is

$$f = \frac{F}{m} \qquad (1-27)$$

where the bold font denotes the vectors. If the components of the force F in the coordinates x-, y-, z- axis are F_x, F_y, F_z, respectively, the corresponding components of the body force per unit mass f in the x-, y-, z-axis are denoted by X, Y, Z. That is

$$\left. \begin{aligned} X &= \frac{F_x}{m} \\ Y &= \frac{F_y}{m} \\ Z &= \frac{F_z}{m} \end{aligned} \right\} \qquad (1-28)$$

The gravity G is a body force. If the positive z-axis is taken vertically downward, the force of gravity per unit mass becomes

$$X = Y = 0, \quad Z = g = \frac{G}{m} \qquad (1-29)$$

Therefore, the body force per unit mass has the same dimension as acceleration g, i.e. $[L/T^2]$, in units of m/s^2.

Multiple-choice questions (one option)

1 - 1   Fluid at rest _____ shear stress.
  (A) can withstand   (B) can withstand very small
  (C) cannot sustain   (D) can hold (for viscous fluid)

1 - 2   Based on the assumption of a continuum concept, the liquid particles means ____.
  (A) the liquid molecules
  (B) the spatial geometry points inside the liquid
  (C) the solid particles inside the liquid
  (D) the small elements of liquid, which have a large number of molecules and an infinitely small volume

1 - 3   When the temperature rises, the viscosity of water _____.
  (A) becomes smaller   (B) becomes larger
  (C) does not change   (D) cannot be determined

1 − 4   When the temperature rises, the viscosity of air _____.
(A) becomes smaller (B) becomes larger
(C) does not change (D) cannot be determined

1 − 5   The relationship between dynamic viscosity coefficient and kinematic viscosity coefficient is _____.
(A) $\nu = \mu\rho$ (B) $\nu = \mu/\rho$ (C) $\nu = \rho/\mu$ (D) $\nu = \mu/p$

1 − 6   The unit of kinematic viscosity coefficient is _____.
(A) s/m^2 (B) m^2/s (C) $N \cdot s/m^2$ (D) $mN \cdot m^2/s$

1 − 7   The viscosity of fluid has no relationship with _____.
(A) molecular cohesion (B) molecular momentum exchange
(C) temperature (D) velocity gradient

1 − 8   Newton's law of internal friction is directly related to the factors of _____.
(A) velocity and shear stress (B) shear stress and shear deformation
(C) shear stress and shear strain rate (D) shear stress and pressure

1 − 9   Under _____ conditions, the compressibility of fluid is the rate of volume change per unit pressure change.
(A) isobaric (B) isothermal
(C) the same density (D) constant volume

1 − 10 _____ is a non-Newtonian fluid.
(A) Air (B) Water (C) Gasoline (D) Asphalt

1 − 11 Mass force acting on the liquid is comprised of _____.
(A) pressure force (B) frictional force
(C) shear stress force (D) gravitational force

1 − 12 The forces acting on the liquid surface are not _____.
(A) the pressure (B) the shear stress force
(C) the frictional force (D) the inertial force

1 − 13 Ideal fluid is defined as the fluid that is _____.
(A) incompressible
(B) of constant viscosity coefficient
(C) of no viscosity
(D) in line with Newton's law of internal friction

1 − 14 The force per unit mass is defined as the mass force acting on the fluid of per unit _____.
(A) area (B) volume (C) mass (D) weight

1 − 15 The SI unit of the force per unit mass is _____.
(A) N/m^2 (B) N/m^3 (C) m/s^2 (D) N

Problems

1 − 1   The water at 20℃ has the specific weight $\gamma = 9789$ N/m^3 and the dynamic viscosity

coefficient $\mu = 1.002 \times 10^{-3}$ N·s/m². Take the acceleration of gravity $g = 9.8$ m/s², determine the kinematic viscosity coefficient ν.

1-2 The bulk modulus of water K is 1.96×10^9 Pa. How much pressure change is needed for a volume deduction of 1%?

1-3 If $h = p/\gamma$, where γ is the specific weight of the liquid and p is the pressure with the dimension of $[F/L^2]$, what is the dimension of h?

1-4 For a velocity distribution of water flow given by $u = u_m \left(\dfrac{y}{H}\right)^{2/3}$, where u_m is the surface velocity of flow, H is the depth, and y is the distance from the wall, if the water temperature is 20℃, calculate the velocity gradient and shear stress at $y/H = 0.25$ and 0.5.

1-5 Fig. 1-9 shows that a flat plate horizontally moves on the oil. The velocity $u = 1$ m/s, the distance between the plate and the fixed boundary $\delta = 1$ mm, and the dynamic viscosity coefficient of oil $\mu = 1.15$ N·s/m². Assuming that the velocity of the oil driven by the plate is a linear distribution, determine the viscous resistance force on the plate per unit area.

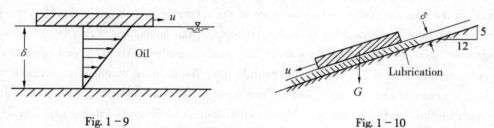

Fig. 1-9   Fig. 1-10

1-6 A piece of wood, which has the bottom area of 40 cm × 45 cm, the height of 1 cm, and the mass of 5 kg, moves downward along a lubricating inclined surface, as shown in Fig. 1-10. The wood velocity $u = 1$ m/s, the thickness of the oil $\delta = 1$ mm, and assume a linear velocity distribution in the oil layer. Calculate the dynamic viscosity coefficient of the oil.

1-7 If a liquid is held in a still tank, what is the body force of liquid per unit mass?

1-8 The velocity distribution is known as: (a) rectangular distribution; (b) triangular distribution; (c) parabolic distribution (horizontal symmetric axis), as shown in Fig. 1-11. Please draw the corresponding distribution of shear stress(τ) with depth(y).

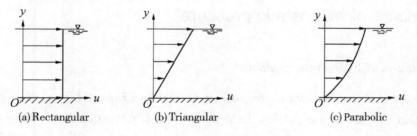

(a) Rectangular   (b) Triangular   (c) Parabolic

Fig. 1-11

Chapter 2 Hydrostatics

Hydrostatics is the study of the laws of fluid at rest and its application of engineering. Fluid equilibrium includes two cases: fluid at rest and in relative equilibrium. The difference between the two cases depends on the choice of the reference coordinate system. Take the Earth as the reference coordinate system, the fluid has no relative motion relative to the Earth, which is called static state. Statics is a special state of motion. By taking the coordinate system that is fixedly connected to a container containing liquid as the reference coordinate system, when the container does uniform motion along a straight line or uniform rotary motion around its central vertical axis, the liquid in the container is in motion in terms of the reference system of the Earth. However, the fluid or the fluid particles have no motion relative to the container, which is just like that passengers sitting in a moving train have no relative motion relative to the train carriage and other passengers, so this is known as the relative equilibrium. The common characteristic of both cases above is that forces on each fluid particle are balanced. At equilibrium state, there is no relative motion between fluid particles, where neither internal friction nor viscous effect exists. Therefore, in the study of hydrostatics problems, the real fluid and ideal fluid are the same, so there is no need to distinguish them.

At equilibrium state, the interactions between fluid particles or between the fluid particles and the solid boundaries are taken in the form of pressure. Hydrostatics is mainly concerned with the distribution law of hydrostatic pressure under static equilibrium conditions, the calculation of point pressure, and the computation of total hydrostatic forces on flat and curved planes.

Under certain conditions, the hydrodynamic pressure distribution of water flow will also follow the distribution law of hydrostatic pressure.

2.1 Concept of hydrostatic pressure

2.1.1 Definition of hydrostatic pressure

The concept of normal stress was previously mentioned in Chapter 1. In a quiescent fluid, a surface of action with the area of ΔA has the pressure force of ΔP acting on it. When the area is reduced to a point, the limiting value of average pressure $\Delta P/\Delta A$ is defined as the hydrostatic pressure at that point, expressed by p,

$$p = \lim_{\Delta A \to 0} \frac{\Delta P}{\Delta A} \qquad (2-1)$$

In the SI system, the unit of hydrostatic force (denoted by letter P) is N or kN, whereas the unit of hydrostatic pressure (denoted by letter p) is N/m² or kN/m², also Pa or kPa. P is force and p is stress, which should be distinguished clearly in calculations.

In materials mechanics, tensile stress denotes positive but compressive stress denotes negative. Fluids at rest only have compressive stress but not tensile stress; hence hydrostatic pressure always denotes positive.

2.1.2 Features of hydrostatic pressure

Hydrostatic pressure has two important features.

Feature 1: The direction of hydrostatic pressure is in line with the inward normal of the acting surface.

This feature can be proved by the method of reduction to absurdity as follows.

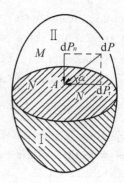

Fig. 2 – 1

Take an arbitrary portion of fluid M from a fluid at rest, as shown in Fig. 2 – 1. An arbitrary plane $N - N$ divides M into two parts of Ⅰ and Ⅱ; if Part Ⅰ is taken as free body, then the force acted by Part Ⅱ on Part Ⅰ is hydrostatic force. Take an arbitrary point A on the plane $N - N$, the hydrostatic force on the small area dA around A is dP. If dP is not perpendicular to dA, but intersects with dA at any angle α, then the force dP can be decomposed into a force dP_n that is perpendicular to dA and a force dP_τ that is parallel to dA. Obviously, dP_τ is a tangential force parallel to the surface. Since a static fluid cannot withstand any shear deformation, the fluid cannot remain static state and will move as long as dP_τ exists. This contradicts the premise of static fluid. It means that dP_τ cannot exist. In other words, the hydrostatic force dP and the corresponding hydrostatic pressure p must be perpendicular to the acting surface dA. Furthermore, if dP is not pointing in the direction of the inward normal but the outward normal of the surface, which means that the fluid is subject to a tensile force rather than a compressive force, then it is impossible for the fluid to remain in equilibrium. Thus, dP pointing the outward normal contradicts the premise of static fluid.

The above method proves that hydrostatic pressure must have the same direction of inward normal of the acting surface.

Feature 2: The hydrostatic pressures at a point are the same in all directions.

Consider the equilibrium of a small fluid element, in the form of a tetrahedron $Oabc$, as shown in Fig. 2 – 2. The three mutually perpendicular planes dA_x, dA_y, dA_z are perpendicular to coordinate axes x, y and z, respectively. The direction of the plane dA_n is arbitrary, and the length of the three mutually orthogonal edges is dx, dy and dz. External forces acting on the small tetrahedron are surface forces and body forces, which will be discussed in the following.

(1) Surface forces acting on the tetrahedron

$$P_x = p_x dA_x = p_x \cdot \frac{1}{2}dydz$$

$$P_y = p_y dA_y = p_y \cdot \frac{1}{2}dzdx$$

$$P_z = p_z dA_z = p_z \cdot \frac{1}{2}dxdy$$

$$P_n = p_n dA_n$$

where P_x, P_y, P_z and P_n are the surface pressure forces that are parallel to coordinate axes x, y, z and the normal direction of the plan dA_n, respectively; dA_x, dA_y and dA_z are the area of the three orthogonal surfaces of the tetrahedron; dA_n is the area of the slant surface of the tetrahedron.

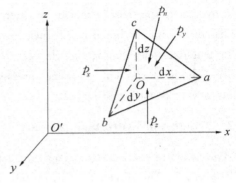

Fig. 2-2

(2) Body force acting on the tetrahedron

The volume of the tetrahedron is $\Delta V = \frac{1}{6}dxdydz$; assume that X, Y and Z are the projections of the body force per unit mass of the tetrahedral fluid element onto the three axes. The projections of the total body force onto the three axes are

$$F_x = \frac{1}{6}\rho X dxdydz$$

$$F_y = \frac{1}{6}\rho Y dxdydz$$

$$F_z = \frac{1}{6}\rho Z dxdydz$$

For the tetrahedron in equilibrium, the sum of the force projections onto the axes should be equal to zero, which is $\sum F_x = 0$, $\sum F_y = 0$ and $\sum F_z = 0$, thus

$$P_x - P_n \cos(n,x) + \frac{1}{6}\rho X dxdydz = 0$$

$$P_y - P_n \cos(n,y) + \frac{1}{6}\rho Y dxdydz = 0$$

$$P_z - P_n \cos(n,z) + \frac{1}{6}\rho Z dxdydz = 0$$

where $\cos(n,x)$, $\cos(n,y)$ and $\cos(n,z)$ are the cosine of the angle between the normal n of arbitrary slant plane and the axes x, y and z. Taking x direction as an example, we have

$$p_x \frac{1}{2}dydz - p_n dA_n \cos(n,x) + \frac{1}{6}\rho dxdydz = 0$$

where $dA_n \cos(n,x) = \frac{1}{2}dydz$, because the projection of the slant plane on the x-axis is dA_x.

When dx, dy and dz are approaching zero, which means that the tetrahedron tends to be one point, the component F_x of the body force is an infinitesimal of high order that is negligible. Thus,

$$p_x \frac{1}{2}dydz - p_n \frac{1}{2}dydz = 0$$

then
$$p_x = p_n$$
Similarly, in y direction, $p_y = p_n$; in z direction, $p_z = p_n$. Therefore,

$$p_x = p_y = p_z = p_n \tag{2-2}$$

Equation (2-2) shows: At any point in a fluid, the hydrostatic pressures are equal in all directions, i.e. $p_x = p_y = p_z = p_n$. Hydrostatic pressure is a scalar function of space coordinates, not a vector function, i.e. $p = p(x, y, z)$, and its magnitude has nothing to do with directions.

2.2 Hydrostatic differential equation and isobaric surface

2.2.1 Differential equation of fluid in equilibrium

In equilibrium fluid, take a small element in the form of a parallel hexahedron, with its edges parallel with the axes and the lengths being dx, dy and dz, as shown in Fig. 2-3.

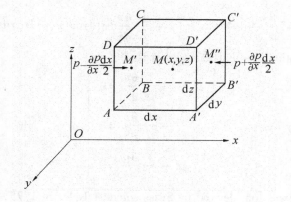

Fig. 2-3

The left, right, front, back, top and bottom faces of the hexahedron bear the surface pressures from the surrounding fluid. Since each face is small, the surface force can be considered as the product of the pressure at the centroid point of the face (regarded as the average pressure of the face) and the face area. Assume that the coordinate of the centroid M of the hexahedron is (x, y, z) and the pressure is p, and then the pressures at the centroids M' and M'' of the left and right faces $ABCD$ and $A'B'C'D'$ are $p - \frac{1}{2}\frac{\partial p}{\partial x}dx$ and $p + \frac{1}{2}\frac{\partial p}{\partial x}dx$, respectively. $\frac{\partial p}{\partial x}$ is the rate of change of pressure in the x-axis direction, and is called the pressure gradient. The distance between M' and M is $-\frac{dx}{2}$, the distance between M'' and M is $+\frac{dx}{2}$, and the differences of pressures at M' and M, and M'' and M are $-\frac{1}{2}\frac{\partial p}{\partial x}dx$ and $\frac{1}{2}\frac{\partial p}{\partial x}dx$, respectively. Thus the surface forces along the x axis acting on the faces $ABCD$ and $A'B'C'D'$ are $\left(p - \frac{1}{2}\frac{\partial p}{\partial x}dx\right)dydz$ and $\left(p + \frac{1}{2}\frac{\partial p}{\partial x}dx\right)dydz$, respectively.

Similarly, forces acting on the front and back faces along the y axis are $\left(p - \frac{1}{2}\frac{\partial p}{\partial y}dy\right)dxdz$ and $\left(p + \frac{1}{2}\frac{\partial p}{\partial y}dy\right)dxdz$, respectively. Forces acting on the top and bottom faces along the z axis are $\left(p - \frac{1}{2}\frac{\partial p}{\partial z}dz\right)dxdy$ and $\left(p + \frac{1}{2}\frac{\partial p}{\partial z}dz\right)dxdy$, respectively.

The body forces acting on the hexahedron are
Body force along the x axis: $X\rho\, dx\, dy\, dz$
Body force along the y axis: $Y\rho\, dx\, dy\, dz$
Body force along the z axis: $Z\rho\, dx\, dy\, dz$
where X, Y and Z are the components of unit body force in x, y and z axis, respectively.

Under the condition of static equilibrium, the sum of forces in each direction should be zero. Along the x axis:

$$\left(p - \frac{1}{2}\frac{\partial p}{\partial x}dx\right)dydz - \left(p + \frac{1}{2}\frac{\partial p}{\partial x}dx\right)dydz + X\rho dxdydz = 0$$

Divide the above equation by $\rho dxdydz$, we have

$$X - \frac{1}{\rho}\frac{\partial p}{\partial x} = 0$$

Similarly, along the y axis:
$$Y - \frac{1}{\rho}\frac{\partial p}{\partial y} = 0 \qquad (2-3)$$

Along the z axis:
$$Z - \frac{1}{\rho}\frac{\partial p}{\partial z} = 0$$

Equation $(2-3)$ was first derived by Swiss scientist Euler in 1775, and it is called Euler's hydrostatic equilibrium differential equations. Its physical meaning is that in equilibrium liquid, the change rate of the hydrostatic pressure in a certain direction is equal to the body force per unit volume in that direction.

Rewrite Equation $(2-3)$ to

$$\left.\begin{array}{l}\dfrac{\partial p}{\partial x} = \rho X \\[6pt] \dfrac{\partial p}{\partial y} = \rho Y \\[6pt] \dfrac{\partial p}{\partial z} = \rho Z\end{array}\right\} \qquad (2-3')$$

Multiply each equation by dx, dy and dz, respectively, and then add them together:

$$\frac{\partial p}{\partial x}dx + \frac{\partial p}{\partial y}dy + \frac{\partial p}{\partial z}dz = \rho(Xdx + Ydy + Zdz)$$

Since hydrostatic pressure is a scalar function of space coordinates, i. e.
$$p = p(x,y,z)$$
Then, the total differential dp of hydrostatic pressure becomes
$$dp = \rho(Xdx + Ydy + Zdz) \qquad (2-4)$$
where $Xdx + Ydy + Zdz$ can be regarded as the work $\boldsymbol{f} \cdot d\boldsymbol{l}$ done by unit body force $\boldsymbol{f}$ along a certain direction for a distance d$\boldsymbol{l}$. This can be explained as follows:

Mathematically, "Any vector in space can be expressed by a linear form of three basic unit vectors". As unit body force $\boldsymbol{f}$ and distance d$\boldsymbol{l}$ are vectors, they can be expressed as
$$\boldsymbol{f} = X\boldsymbol{i} + Y\boldsymbol{j} + z\boldsymbol{k}$$
and
$$d\boldsymbol{l} = dx\boldsymbol{i} + dy\boldsymbol{j} + dz\boldsymbol{k}$$
where $\{X, Y, Z\}$ and $\{dx, dy, dz\}$ are the components of $\boldsymbol{f}$ and d$\boldsymbol{l}$ on the $\{x, y, z\}$ coordinate axes, respectively; $\{\boldsymbol{i}, \boldsymbol{j}, \boldsymbol{k}\}$ are the unit vectors along the corresponding coordinate axes of $\{x, y, z\}$. Therefore,
$$\boldsymbol{f} \cdot d\boldsymbol{l} = (X\boldsymbol{i} + Y\boldsymbol{j} + Z\boldsymbol{k}) \cdot (dx\boldsymbol{i} + dy\boldsymbol{j} + dz\boldsymbol{k})$$
Because the scalar product of vectors $\boldsymbol{A}$ and $\boldsymbol{B}$ is equal to the product of their lengths and cosines of intersect angle θ, i. e. $\boldsymbol{A} \cdot \boldsymbol{B} = |\boldsymbol{A}||\boldsymbol{B}|\cos\theta$, the unit vectors have the following relationships:
$$i^2 = j^2 = k^2 = 1$$
$$\boldsymbol{i} \cdot \boldsymbol{j} = \boldsymbol{j} \cdot \boldsymbol{k} = \boldsymbol{k} \cdot \boldsymbol{i} = 0$$
Applying above relationships to $\boldsymbol{f} \cdot d\boldsymbol{l}$, we have
$$\boldsymbol{f} \cdot d\boldsymbol{l} = |\boldsymbol{f}||d\boldsymbol{l}|\cos\theta = Xdx + Ydy + Zdz \qquad (2-5)$$
Therefore, the physical meaning of the left side of Equation (2-5) is the work done by unit body force ($\boldsymbol{f}$) along a certain direction for a distance (d$\boldsymbol{l}$).

2.2.2 Isobaric surface

Definition: Isobaric surface (surface of equal pressure) is defined as the surface that consists of points with the same pressure in a continuous liquid of the same type. In other words, the pressure increment on an isobaric surface is zero, i. e. d$p = 0$.

Isobaric surface can be plane or curved. Different isobaric surfaces have different pressures.

Since d$p = 0$ on an isobaric surface, from Equation (2-4),
$$\rho(Xdx + Ydy + Zdz) = 0$$
As ρ is the density of liquid which cannot be zero, then
$$Xdx + Ydy + Zdz = 0$$
From Equation (2-5),
$$\boldsymbol{f} \cdot d\boldsymbol{l} = |\boldsymbol{f}||d\boldsymbol{l}|\cos\theta = 0$$
where neither $|\boldsymbol{f}|$ nor $|d\boldsymbol{l}|$ can be zero, so $\cos\theta = 0$, which means that the angle between the two vectors is 90°, thus indicating that the work done by the body force along an isobaric surface is zero, or that the isobaric surface is perpendicular to the body force.

If a liquid is at absolute rest, the body force acting on the liquid is gravity only. Gravity always points to the core of the Earth, so on a large scale, an isobaric surface is a surface perpendicular to the gravity everywhere; on a small scale, an isobaric surface is a horizontal plane.

The free surface of a liquid, which is in contact with the air, is exerted by the same atmospheric pressure. The free surface of equilibrium fluid is an isobaric surface, in other words, the free surface of a fluid at rest is horizontal.

For a static(body force is limited to gravity), continuous fluid of the same type, the isobaric surfaces are horizontal, or any horizontal plane is an isobaric surface.

Fig. 2-4 shows several connecting vessels in different boundary conditions, where some of the horizontal planes are isobaric surfaces, but some of them are not.

Using the concept of isobaric surfaces, we can obtain the pressure of an unknown point by pressure of a known point in the calculation.

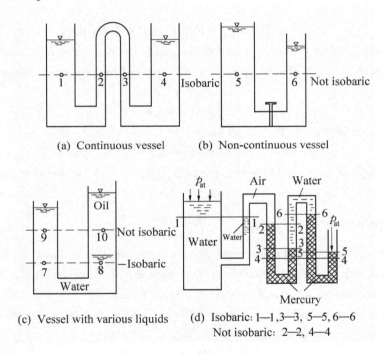

Fig. 2-4

2.3　Distribution of hydrostatic pressure under gravity

In practical engineering, the most hydrostatics problems are of fluid equilibrium with no motion relative to the Earth, which means that the body force is limited to gravity only.

2.3.1　Basic formula of hydrostatic pressure under gravity

Consider a static fluid under gravity as shown in Fig. 2-5, where the origin of rectangular coordinate system is set on the free surface, and the z axis points upward. As the body force is gravity only, there is no body force on the horizontal plane. Therefore, $X = Y = 0$, and $Z = -\dfrac{G}{m} = -g$, where the negative sign is used because the gravity is opposite to the positive direction of the z axis. Substituting above relationships into Equation (2-4), we have

$$dp = \rho(Xdx + Ydy + Zdz) = -\rho g dz$$

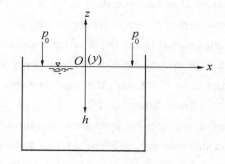

Fig. 2-5

In a homogeneous liquid, ρ is a constant, from $\gamma = \rho g$ in Equation (1-10),

$$dp = -\gamma dz$$

Integrating above equation,

$$p = -\gamma z + c_1 \tag{2-6}$$

Dividing by γ and rearranging,

$$z + \frac{p}{\gamma} = c \tag{2-7}$$

where

$$c = \frac{c_1}{\gamma}$$

Noting the boundary conditions at the free surface: $z = z_0 = 0$, and $p = p_0$, where p_0 is the free surface pressure, which may be equal, larger or smaller than the atmospheric pressure, so that the integration constant $c_1 = p_0$, and $c = \dfrac{p_0}{\gamma}$. Substituting into Equation (2-6) gives

$$p = p_0 + \gamma h \tag{2-8}$$

where $h = -z$, and h is the depth of that point below the free surface.

Equations (2-7) and (2-8) are the two basic forms of hydrostatics formula for describing the distribution of pressure of static liquid under gravity. These two equations are very important in hydrostatic pressure calculations.

Equation (2-7) is commonly used for theoretical analysis. In the equation z is the height of selected point to the reference plane; $\dfrac{p}{\gamma}$ is the pressure head. The physical meaning of Equation (2-7) is that: in a static fluid, although the height z and pressure head $\dfrac{p}{\gamma}$ at every point are different, their sums i.e. $(\dfrac{p}{\gamma} + z)$ are the same.

Equation (2-8) is the basis for calculating point hydrostatic pressure in practice. This equation indicates that: in a static liquid where the body force is limited to gravity only, the hydrostatic pressure at any point is equal to the sum of the free surface pressure (p_0) and the weight per unit area (γh) of the fluid column between that point and the free surface. Following the Pascal's law, which states that a change in pressure at any point in an enclosed fluid at rest is transmitted undiminished to all points in the fluid, the free surface pressure will equivalently be transferred to every point inside the fluid; the second part of the hydrostatic pressure is proportional to the depth of the point below the fluid surface, the greater the column height, i. e. the greater the fluid depth, the larger is the point hydrostatic pressure. All the points located at the same depth have the same pressure.

From Equation (2 – 8), we can also derive a relationship of the hydrostatic pressures between two points at different depths below the free surface.

Fig. 2 – 6 shows a closed container with static fluid. The specific weight of fluid is γ, and the surface pressure is p_0, which is larger than the atmospheric pressure (p_{at}). Take any two points 1 and 2 within the container, their depths are h_1 and h_2, respectively. From Equation (2 – 8), the pressures of the two points are

$$p_1 = p_0 + \gamma h_1$$
$$p_2 = p_0 + \gamma h_2$$

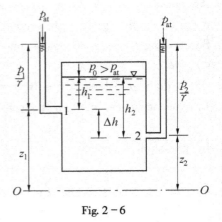

Fig. 2 – 6

Subtracting one from the other equation above,

$$p_2 - p_1 = \gamma(h_2 - h_1) = \gamma \Delta h$$

or
$$\left. \begin{array}{l} p_2 = p_1 + \gamma \Delta h \\ p_1 = p_2 - \gamma \Delta h \end{array} \right\} \quad (2-9)$$

Equation (2 – 9) shows that: if the pressure at a shallow point is known, the pressure at a deeper point is equal to the known pressure at the shallow point added by the weight per unit area of water column between the two points; if the pressure at a deep point is known, the pressure at a shallower point is equal to the known pressure at the deep point subtracted by the weight per unit area of water column between the two points. For gas, if the impact of its weight is ignored, the internal pressure of gas is equal everywhere.

2.3.2 Absolute, relative and vacuum pressures

The hydrostatic pressure at a point can have different values with reference to different zero pressures. Zero pressure usually has two definitions: absolute zero pressure and relative zero pressure. Therefore we have absolute pressure and relative pressure.

2.3.2.1 Absolute pressure

If the pressure of "perfect vacuum" (or absolute vacuum) with complete absence of any gas is referred as zero pressure, the pressure measured relative to this zero pressure is called absolute pressure, denoted by p_{abs}. This zero pressure is called absolute zero pressure.

A standard atmospheric pressure is the atmospheric pressure on sea level at 0℃, denoted by 1 p_{atm}. In terms of absolute pressure, 1 p_{atm} = 101 325 N/m². In engineering, for simplicity of calculation, engineering atmospheric pressure is used, denoted by p_{at}. In terms of absolute pressure, 1 $p_{at} = 98\,000$ N/m². Here we should note that although both p_{atm} and p_{at} represent the atmospheric pressure, they have different meanings and values. Unless specified, the engineering atmospheric pressure is used in this book.

2.3.2.2 Relative pressure

Since the atmospheric pressure acts on all the things on the Earth, removing such pressure will make calculations much simpler. Therefore, we introduce relative zero pressure and relative pressure.

If a local atmospheric pressure is referred as zero pressure, the pressure measured relative to this zero pressure is called relative pressure, also known as gauge pressure. Usually without the subscript, p refers to the relative pressure, and this zero pressure is called relative zero pressure.

The relationship between relative pressure (p_r) and absolute pressure (p_{abs}) is

$$p_r = p_{abs} - p_{at} \tag{2-10}$$

or

$$p_{abs} = p_{at} + p_r \tag{2-11}$$

2.3.2.3 Vacuum pressure

The difference between the relative zero pressure and absolute zero pressure is 1 local atmospheric pressure. Absolute pressures are always positive, but relative pressures can be either positive or negative depending on two situations:

If $p_{abs} > p_{at}$, $p_r > 0$, the relative pressure is positive;

If $p_{abs} < p_{at}$, $p_r < 0$, the relative pressure is negative.

When an absolute pressure is below the atmospheric pressure, its relative pressure is negative, and its absolute value is called vacuum or suction pressure, denoted by p_v.

$$p_v = p_{at} - p_{abs} = |p_{abs} - p_{at}| = |p_r| \tag{2-12}$$

It is worth noting that a vacuum in fluid mechanics is referred to the pressure whose absolute pressure is smaller than the atmospheric pressure. It is different from the concept of a vacuum in physics, which means a free space where the average free path of gas molecules exceeds the size of container.

The relationship between absolute pressure, relative pressure and vacuum pressure is shown in Fig. 2-7.

The vertical coordinate represents the value of pressure; the reference plane of relative pressure is one atmospheric pressure (98 000 N/m²) above the reference plane of absolute pressure. Therefore, the absolute pressure is always positive, and the relative pressure can be positive or negative. When the absolute pressure is greater than the atmospheric pressure, the relative pressure is positive; when the absolute pressure is less than the atmospheric pressure, the relative pressure is negative. Since the vacuum pressure cannot be negative, it equals the

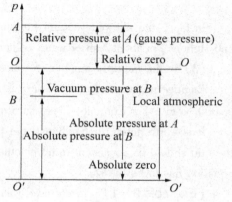

Fig. 2-7

absolute value of gauge pressure (which is negative).

In theory, when an absolute pressure is zero, its vacuum value reaches the maximum, i.e. $p_v = p_{at}$. However, in practical engineering, the liquid cannot reach the state of "absolute vacuum", because when the pressure of the free surface drops down to the vapour pressure, the liquid will rapidly be vaporized so that the pressure will not longer decline. Therefore, the actual maximum vacuum of a liquid will be less than the difference between the local atmospheric pressure and the saturated vapour pressure of the liquid.

2.3.2.4 Expression of pressure

Pressure can be expressed by the following three methods.

(1) By stress

This method is originated from the definition of pressure, expressed by the force per unit area, such as N/m^2, KN/m^2 or Pa, kPa.

(2) By height of liquid column, or head

If the free surface pressure p_o is the atmospheric pressure, the relative pressure is zero, $p_o = 0$. Then, Equation (2-8) is reduced to $p = \gamma h$, which can be written as

$$h = \frac{p}{\gamma} \qquad (2-13)$$

Equation (2-13) shows that the hydrostatic pressure at any point can be converted to a liquid column height of any specific weight, also called pressure head. In hospital, blood pressure is measured in mmHg (mercury), despite in kPa elsewhere nowadays. This is typically a practical application of Equation (2-8).

In terms of water column height or head, 1 engineering atmospheric pressure is

$$h = \frac{p_{at}}{\gamma_w} = \frac{98\,000}{9\,800} = 10 \text{ m in water column}$$

The pressure on a diver is proportional to the diving depth. An increase of every 10m diving will increase the pressure by 98 000 Pa (1 engineering atmospheric pressure).

Theoretically, the maximum vacuum pressure is

$$h_v = \frac{p_v}{\gamma_w} = \frac{98\,000}{9\,800} = 10 \text{ m in water column}$$

In fact, the allowed vacuum pressure for in the construction machinery, e.g. pump sumps and siphons, is about 7 m of water column.

(3) By the multiple number of atmospheric pressures

Engineering atmospheric pressure is often used as the unit of pressure in engineering. $1p_{at} = 98\,000$ N/m^2. For example, a pressure of 147 kN/m^2 can be expressed as $1.5\,p_{at}$.

It is worth noting that although the pressure can be expressed by the above three methods, the unit should be consistent in all calculations. In other words, the dimension of each term in an equation has to be the same.

[Example 2-1] In a closed water tank (see Fig. 2-8), the absolute pressure p_0 on the free surface is 78 kN/m^2. Calculate the absolute hydrostatic pressure, relative hydrostatic pressure and vacuum pressure at point *C* that is located at the depth (h) of 1.5 m below the water surface.

Solution:

$p_{abs} = p_0 + \gamma_w h = 78 + 9.8 \times 1.5 = 92.7 \text{ kN/m}^2$

$p_r = p_{abs} - p_{at} = 92.7 - 98 = -5.3 \text{ kN/m}^2$

$p_v = |p_r| = -5.3 \text{ kN/m}^2$

$h_v = \dfrac{p_v}{\gamma_w} = \dfrac{5.3}{9.8} = 0.54$ m in water column

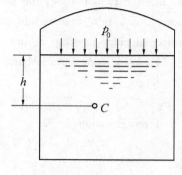

【Example 2-2】 In the same conditions as the above example, what is the submerged depth h of point *C* if the relative hydrostatic pressure p_r is 8.4 kN/m²?

Solution:

$p_r = p_{abs} - p_{at} = p_0 + \gamma_w h - p_{at}$

$h = \dfrac{p_r - p_0 - p_{at}}{\gamma_w}$

Fig. 2-8

Replacing by the numbers,

$$h = \dfrac{8.4 - 78 + 98}{9.8} = 2.898 \text{ m}$$

【Example 2-3】 Fig. 2-9 shows an oil tank with a horizontal bottom and an inclined side wall of 30°. The submerged length (l) of the side wall is 5 m, the pressure on the free surface $p_0 = p_{at} = 98$ kN/m², and the specific weight of oil is 7.8 kN/m². How much is the pressure on the bottom of the tank?

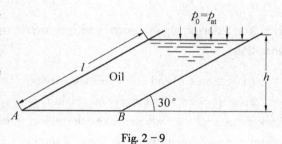

Fig. 2-9

Solution: The key to this problem is that the liquid column height in static equilibrium equation under gravity is the height in the vertical direction. Hence, the vertical height h is

$$h = l \cdot \sin 30° = 5 \times \dfrac{1}{2} = 2.5 \text{ m}$$

Because the bottom is also subject to the atmospheric pressure, the actual hydrostatic pressure on the bottom is a relative pressure, which is

$$p_r = \gamma_{oil} h = 7.8 \times 2.5 = 19.5 \text{ kN/m}^2$$

【Example 2-4】 In order to measure the vapor pressure in the boiler, a large-scale double-mercury manometer is used, as shown in Fig. 2-10. The various elevations are: $\nabla_1 = 2.3$ m, $\nabla_2 = 1.2$ m, $\nabla_3 = 2.5$ m, $\nabla_4 = 1.4$ m, and $\nabla_5 = 3.5$ m. What is the value of the relative pressure on the free surface of the pipe?

Solution: The key to the problem is about the application of the concept of isobaric surface and Equations (2-8) & (2-9), for calculating the pressure at the unknown point from the known point pressure. In Fig. 2-10 there are three isobaric surfaces: the horizontal plane passing through ∇_2 into the right U-tube; the horizontal plane passing through ∇_3 into the middle inverted U-tube; the horizontal plane passing ∇_4 into the left U-tube.

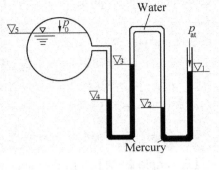

Fig. 2-10

From the concept of isobaric surface and Equation (2-8),
$$p_{\nabla_2} = p_{\nabla_1} + \gamma_m(\nabla_1 - \nabla_2) = \gamma_m(\nabla_1 - \nabla_2)$$
From Equation (2-9),
$$p_{\nabla_3} = p_{\nabla_2} - \gamma_w(\nabla_3 - \nabla_2)$$
and
$$p_{\nabla_4} = p_{\nabla_3} + \gamma_m(\nabla_3 - \nabla_4)$$
$$p_{\nabla_5} = p_{\nabla_4} - \gamma_w(\nabla_5 - \nabla_4)$$

Combining above equations gives:
$$p_{\nabla_5} = p_0 = \gamma_m(\nabla_1 - \nabla_2) - \gamma_w(\nabla_3 - \nabla_2) + \gamma_m(\nabla_3 - \nabla_4) - \gamma_w(\nabla_5 - \nabla_4)$$
where γ_w, γ_m are the specific weight of water and mercury, respectively.

Substituting in the numbers:
$$p_0 = 133.28 \times (1.1 + 1.1) - 9.8 \times (1.3 + 2.1) = 259.9 \text{ kN/m}^2$$

2.3.3 Energy significance and geometric meaning of the basic hydrostatic pressure equation

(1) Elevation head and elevation energy

From physics, if an object of weight G is removed from the reference plane to a height z, the work done is $G \cdot z$, which makes the object gain the elevation energy of $G \cdot z$. The elevation energy per unit weight is $G \cdot z/G = z$. Therefore, z denotes the elevation potential energy of a unit weight object relative to a reference plane, or simply called elevation energy. In hydraulics we often use "head" instead of height, so z is also called elevation head.

(2) Pressure head and pressure energy

Every point within a static liquid has pressure potential energy. If we make an opening on the sidewall of a closed container and place the hand on the opening, we can feel pressured. If a slender glass tube is connected to the opening, due to the pressure of the liquid, the liquid in the glass tube will rise. The height raised is equivalent to the pressure at the opening divided by the specific weight of the liquid, i.e. $h = \dfrac{p}{\gamma}$. The pressure energy is converted to the elevation energy, as shown in Fig. 2-11. For a liquid with weight G rising for height h, the elevation energy gained is $G \cdot h$. For

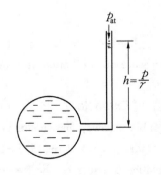

Fig. 2-11

liquid with unit weight, the elevation energy is $G \cdot h/G = h = \frac{p}{\gamma}$. Hence, $\frac{p}{\gamma}$ (based on the relative pressure at the opening of the container) stands for the pressure energy of a liquid with unit weight, also known as the piezometric height. In hydraulics it is called the pressure head.

(3) Piezometric head and potential energy

A small open tube, whose one end is connected to a container containing liquid while the other end is connected to the atmosphere, is called a piezometer. The sum of elevation energy and pressure energy of liquid per unit weight is $z + \frac{p}{\gamma}$, called the piezometric head. Since z represents the elevation energy of liquid per unit weight, and $\frac{p}{\gamma}$ is the pressure energy per unit weight, $z + \frac{p}{\gamma}$ represents the potential energy of liquid per unit weight, or unit potential energy for short.

The energy significance of the basic hydrostatic pressure equation $z + \frac{p}{\gamma} = c$ shows that, in static liquid, the potential energy of a liquid per unit weight conserves. Due to the conservation of potential energy, elevation potential energy and pressure potential energy can be converted to each other: if the elevation energy of liquid at one point is larger, then its pressure energy is smaller; if the elevation energy at one point is smaller, then its pressure energy is larger. The total potential energy remains unchanged.

The piezometric height (pressure head) $\frac{p}{\gamma}$ and piezometric head $z + \frac{p}{\gamma}$ are two different concepts. Their difference is: the piezometric height $\frac{p}{\gamma} = h$ depends on both the pressure at the measuring point and the specific weight of the liquid in the piezometer tube, regardless of the location of the reference plane; the piezometric head $z + \frac{p}{\gamma}$ also includes the elevation head, so it is related with the location of the reference plane. Thus, the piezometric head changes with the reference plane.

The geometric meaning of the basic hydrostatic pressure equation can be seen from Fig. 2-12. z_1 of point 1 is larger than z_2 of point 2, but $\frac{p_1}{\gamma} < \frac{p_2}{\gamma}$, nonetheless $z_1 + \frac{p_1}{\gamma} = z_2 + \frac{p_2}{\gamma}$, i. e. the piezometric heads at points 1, 2 are constant. The surface of the piezometer is leveled with the free surface of the container.

In a closed container, the surface pressure of a liquid may be greater or less than the atmospheric pressure, so the level of liquid in the piezometer can be higher or lower than the surface of the container, but the piezometric head of different points remains constant, like points A and B in Fig. 2-13. If the surface of a piezometer is higher than the measuring point, as $z + \frac{p}{\gamma} > z$, the pressure height $\frac{p}{\gamma}$ is positive; otherwise, as $z + \frac{p}{\gamma} < z$, the pressure height is negative.

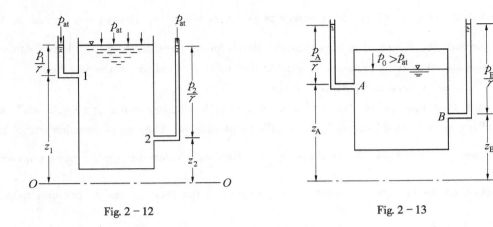

Fig. 2 – 12   Fig. 2 – 13

2.4 The application of hydrostatics in measurement

2.4.1 Piezometer

Piezometers are widely used for measurement of point pressure or free surface level. In a typical piezometer as shown in Fig. 2 – 11, the point pressure can be obtained from $p = \gamma h$ by reading the column height h. Like a piezometer, a communicating tube (containing a homogeneous liquid) is used to connect with a large container, as the level of liquid in the communicating tube is the same as the free surface of the large container, and thus by reading the level of liquid in the tube we can obtain the level of liquid in the large container. Equipment like an oil tank, boiler and electric thermos needs the use of a communicating tube to measure the free surface of liquid in a large container, which is usually hard to see.

In practical measurement, in order to improve measurement accuracy, an inclined piezometer is often used, as shown in Fig. 2 – 14a. In such a case, for the pressure p with a vertical column height of h, the segment length of observation l is larger than h, because $l = \dfrac{h}{\sin\alpha}$, where α is the angle of inclination and $\sin\alpha$ is less than 1. Another way to improve the precision of measurement is to use another liquid (specific weight γ'), which is light and immiscible with water, in a piezometer. Since $\gamma' < \gamma$, a larger liquid column height h can be obtained. Also we can use an inclined piezometer with light liquid to further improve the observation accuracy.

When the pressure to be measured is relatively large, the water column height is too high to read, so we can use a U-shaped mercury piezometer, as shown in Fig. 2 – 14b. The isobaric surface 1 – 1 is the interface between the two liquids. The water pressure at point A can be obtained by applying Equation (2 – 9) for the isobaric surface:

$$p_A = \gamma_m h_m - \gamma_w a \tag{2-14}$$

where γ_m and γ_w are the specific weight of mercury and water, respectively.

【Example 2 – 5】 In order to measure the water level in a water tank, a U-shaped mercury

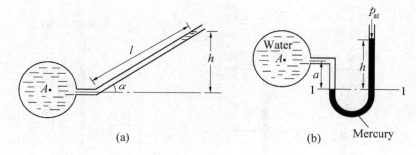

(a)   (b)

Fig. 2 – 14

piezometer is amounted on the ground with its left limb being connected with the tower through a hosepipe, as shown in Fig. 2 – 15. The mercury elevation in the left limb of the piezometer is measured as $\nabla_1 = 61.00$ m, the difference of mercury levels in the piezometer is $h_1 = 86$ cm. Find the elevation of water surface ∇_2 of the tank.

Solution: The interface between the mercury and the water in the left limb of the U-shaped piezometer is an isobaric surface. In terms of relative pressure, the atmospheric pressure acting on the free water surface of the tank $p_{at} = 0$. Hence

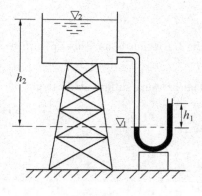

Fig. 2 – 15

$$\gamma_m h_1 = \gamma_w h_2$$

$$h_2 = \frac{\gamma_m h_1}{\gamma_w} = \frac{133.28 \times 0.86}{9.8} = 11.7 \text{ m}$$

The elevation of water surface in the water tank is

$$\nabla_2 = \nabla_1 + h_2 = 61 + 11.7 = 72.7 \text{ m}$$

2.4.2 Differential gauge

Piezometer is used to measure pressure at one point, while a differential gauge is used to measure the pressure difference between two points. Connect the piezometer which is connected to two measuring points, so the pressure difference can be obtained from the column difference of liquid between the two limbs of the piezometer. When the pressure difference between two points is large, a U-shaped mercury differential gauge can be used, as shown in Fig. 2 – 16a. Use the U-shaped mercury differential gauge to measure the pressure difference between points A and B, when the left and right containers contain two fluids (liquid or gas), whose specific weights are γ_A and γ_B respectively.

Since the specific weight of mercury $\gamma_m = 133\,280$ N/m^3, which is larger than any other liquid or gas, the liquid column h representing the pressure difference will be smaller. A U-shaped mercury differential gauge is usually amounted like the one in Fig. 2 – 16a, at the plane

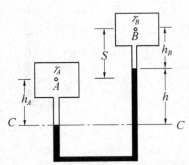 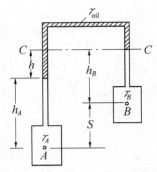

(a) U-shaped mercury differential gauge (b) Inverted U-shaped differential gauge

Fig. 2 – 16

$C-C$, which is an isobaric surface,

$$p_A + \gamma_A h_A = p_B + \gamma_B h_B + \gamma_m h$$

The pressure difference is then

$$p_A - p_B = \gamma_B h_B + \gamma_m h - \gamma_A h_A$$

as

$$h_A + S = h_B + h$$
$$h_A = h_B + h - S$$

So

$$p_A - p_B = \gamma_B h_B + \gamma_m h - \gamma_A(h_B + h - S) = (\gamma_m - \gamma_A)h + (\gamma_B - \gamma_A)h_B + \gamma_A S$$

When $\gamma_A = \gamma_B = \gamma$,

$$p_A - p_B = (\gamma_m - \gamma)h + \gamma S$$

When $\gamma_A = \gamma_B = \gamma$ and A, B are on the same elevation ($S = 0$),

$$p_A - p_B = (\gamma_m - \gamma)h \qquad (2-15)$$

Note that we often make a common mistake, like $p_A - p_B = \gamma_m h$.

If the pressure difference between the points is small, an inverted U-shaped differential gauge containing light liquid can be used, as shown in Fig. 2 – 16b. The interface $C-C$ of the two liquids is an isobaric surface, where the pressure equation is

$$p_A - \gamma_A h_A - \gamma_{oil} h = p_B - \gamma_B h_B$$

So

$$p_B - p_A = \gamma_B h_B - \gamma_A h_A - \gamma_{oil} h$$

and

$$h_B + S = h + h_A$$
$$h_A = h_B + S - h$$

Hence

$$p_B - p_A = \gamma_B h_B - \gamma_A(h_B + S - h) - \gamma_{oil} h = (\gamma_A - \gamma_{oil})h + (\gamma_B - \gamma_A)h_B - \gamma_A S$$

When $\gamma_A = \gamma_B = \gamma$,

$$p_B - p_A = (\gamma - \gamma_{oil})h - \gamma S$$

When $\gamma_A = \gamma_B = \gamma$ and $S = 0$,

$$p_B - p_A = (\gamma - \gamma_{\text{oil}})h \qquad (2-16)$$

where γ_{oil} is the specific weight of light liquid.

If the light liquid is a gas in a differential gauge, the weight of gas is negligible and the specific weight of gas is approximately considered as zero, so there is no pressure difference in the gas column.

When $\gamma_A = \gamma_B = \gamma$ but $S \neq 0$, it can be shown that $\dfrac{\gamma_m - \gamma}{\gamma}h$ is the difference of piezometric head between points A and B, not the difference of pressure head.

【Example 2-6】 To measure the pressure difference between points 1, 2 on a pressurized water pipe, a U-shaped mercury differential gauge is connected to the two cross-sections, as shown in Fig. 2-17. The elevation difference between the two points is $\Delta z = 1.2$m. If the deflection of mercury is measured as $h = 100$ mm, find the pressure difference between points 1 and 2.

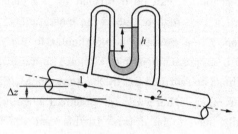

Fig. 2-17

Solution: According to the principle of a differential gauge, the piezometric head difference between points 1 and 2 is

$$\left(z_1 + \frac{p_1}{\gamma}\right) - \left(z_2 + \frac{p_2}{\gamma}\right) = \frac{\gamma_m \gamma}{\gamma}h$$

Hence,

$$\frac{p_1 - p_2}{\gamma} = -\Delta z + \left(\frac{\gamma_m}{\gamma} - 1\right)h$$

Substituting the known values into the above equation gives

$$\frac{p_1 - p_2}{\gamma} = -1.2 + (13.6 - 1) \times 0.1 = 0.06 \text{ m}$$

$$p_1 - p_2 = 0.06 \times 9800 = 588 \text{ N}$$

2.5 Total hydrostatic force acting on a plane surface

Water pressure force acting on the whole pressured surface of a body in a static state is called total hydrostatic force. Total hydrostatic force is one of important forces on a hydraulic structure. It is an important basis for designing gate structures, selecting hoisting equipment and checking the stability of water retaining buildings.

Fluid Mechanics in Civil Engineering

In fact, the calculation of total hydrostatic force on a plane is to obtain the resultant force of parallel distributed force on the pressured plane. From the distribution of the known point hydrostatic pressure, the magnitude, direction and acting point of total hydrostatic force on an entire surface can be obtained.

There are two methods to obtain the total hydrostatic force on a plane: a graphic method and an analytical method. The graphic method is suitable for a rectangular plane whose bottom edge is parallel to the water surface; the analytical method suits for any type of plane.

2.5.1 Graphic method

2.5.1.1 Hydrostatic pressure distribution diagram

The graphic method may be used to obtain a total hydrostatic force on a rectangular plane based on a hydrostatic pressure distribution diagram.

Hydrostatic pressure distribution can be described by a geometric diagram with a series of parallel arrow lines, for which the length of line is the magnitude of hydrostatic pressure and the arrow indicates the direction of the pressure perpendicular to the plane. This added diagram, which is comprised of such lines perpendicular to a plane, is the hydrostatic pressure distribution diagram. Because the buildings are surrounded by the atmosphere, the atmospheric pressure in all directions cancels out each other. Therefore, in practical engineering, only relative pressure distribution diagrams actually need to be drawn. In this case the basic hydrostatic equation is simplified as $p = \gamma h$, which is the linear distribution of hydrostatic pressure with depth h. As long as the pressures of two points along the depth on a plane are known, a pressure distribution diagram along the entire depth can be drawn.

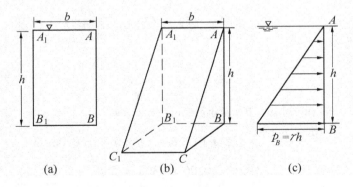

Fig. 2 - 18

Fig. 2 - 18a represents a vertical rectangular flat gate with water at one side, for which the area of the gate below water is ABB_1A_1 in a depth of h and a width of b. The 3D diagram of the hydrostatic pressure distribution of the gate is a prism of $ABCC_1B_1A_1$, as shown in Fig. 2 - 18b. Because the flat gate is a rectangular plane, the hydrostatic pressure distribution diagrams of each cross-section along BB_1 are the same, so the hydrostatic pressure distribution diagram only needs to be shown as in Fig. 2 - 18c. Generally a hydrostatic pressure distribution diagram is a two-dimensional diagram.

In different boundary conditions, hydrostatic pressure distribution diagrams of a structure are different, but they must satisfy the following conditions.

(1) The direction of hydrostatic pressure has to be the same as the direction of the inner normal of the acting surface;

(2) The magnitude of point pressure is independent of the direction of the acting surface;

(3) The magnitude of point pressure is determined by $p = \gamma h$.

Fig. 2 - 19 shows hydrostatic pressure distribution diagrams of various hydraulic structures for different boundary conditions.

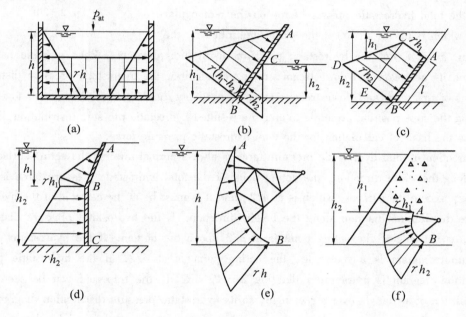

Fig. 2 - 19

2.5.1.2 Total hydrostatic pressure force on a rectangular plane by the graphic method

All perpendiculars to a plane must be parallel, so the hydrostatic pressure on a rectangular plane is parallel to each other. The total hydrostatic pressure force P is actually the integral of pressure force dP over a small area dA on the plane. Therefore, the total hydrostatic pressure force on a rectangular plane is actually the resultant force of parallel force system. This can be described by a mathematical expression:

$$P = \int_A dP = \int_A p dA$$

If the acting surface is a rectangle with height a and width b, and parallel to the axes, as shown in Fig. 2 - 20, then $dA = dxdy$, so the above equation with double integral is

$$P = \int_0^b \int_0^a p dx dy = \int_0^b dx \int_0^a p dy$$

Fluid Mechanics in Civil Engineering

where $\int_0^a p\,dy$ is the area of the hydrostatic pressure distribution diagram, represented by A_p, $\int_0^b dx$ is the width of acting surface b. So we have

$$P = \int_A dP = \int_A p\,dA = \int_V dV = A_p \cdot b \quad (2-17)$$

where V is the volume of hydrostatic pressure distribution diagram acting on the rectangular plane.

Hence, for a rectangular plane, as long as we draw the hydrostatic pressure distribution diagram correctly, we can obtain the total hydrostatic pressure force on the rectangular plane, which is the width (b) of the plane multiplied by the

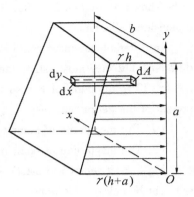

Fig. 2-20

geometric area (A_p) of the hydrostatic pressure distribution diagram. When a plane has water acting on its upstream and downstream sides, we can draw the hydrostatic pressure distribution diagrams of upstream and downstream water, respectively, and then combine them together by following the superposition principle to obtain a resulted hydrostatic pressure distribution diagram, which is the basis of calculation for the total hydrostatic pressure force.

Direction of total hydrostatic pressure force is always normal towards the acting surface.

Along the width direction, the acting point of the total hydrostatic pressure force is on the symmetry axis of the surface, which is at $b/2$, and it is must be at the centroid of the hydrostatic pressure distribution diagram along the height direction. If the hydrostatic pressure distribution diagram is a triangle, the acting point is at $a/3$ above the bottom. If the hydrostatic pressure distribution diagram is a rectangle, the acting point is at $a/2$. If the hydrostatic pressure distribution diagram is a trapezoid like Fig. 2-19 c & d, the trapezoid can be seen as the superposition of a triangle over a rectangle, so its hydrostatic pressure distribution diagram is the sum of area of both sections. The center position of pressure can be found by the mechanical principle of "the moment of resultant force to a point equals the algebraic sum of moments of each component force to that point". For example, the hydrostatic pressure distribution diagram (A_p) in Fig. 2-19c can be seen as the sum of Triangle CDE (A_{p1}) and Rectangle CABE (A_{p2}), so P_1 (pressure force for the triangle) acts at $\dfrac{l}{3}$ away from point B, and P_2 (pressure force for the rectangle) acts at $\dfrac{l}{2}$ away from point B. From "a resultant moment equals the algebraic sum of component moments",

$$P_1 y_{P_1} + P_2 y_{P_2} = P y_P$$

where $P_1 = A_{P_1} \cdot b$, $P_2 = A_{P_2} \cdot b$, $P = P_1 + P_2$, $y_{P_1} = l/3$, $y_{P_2} = l/2$,
Then,

$$y_P = \frac{P_1 y_{P_1} + P_2 y_{P_2}}{P} = \frac{l}{3} \frac{(2h_1 + h_2)}{(h_1 + h_2)}$$

where l is the length of the inclined surface, h_1, h_2 are the pressure heads of the upper and lower part of the trapezoidal hydrostatic pressure distribution diagram. For some complex hydrostatic pressure distribution diagrams, their coordinates of acting points of total hydrostatic pressure forces can be seen in Table 2-1.

Table 2-1 Total hydrostatic forces acting on rectangular planes

Diagram	Formulas
	$P = \dfrac{1}{2}\gamma H^2 b$ $h_D = \dfrac{2}{3}H$
	$P = \dfrac{1}{2}\gamma(H_1^{\ 2} - H_2^{\ 2})b$ $h_D = \dfrac{1}{3}\left(2H_1 - \dfrac{H_2^{\ 2}}{H_1 + H_2}\right)$
	$P = \dfrac{\dfrac{1}{2}\gamma H^2 b}{\sin\alpha}$ $h_D = \dfrac{2}{3}H$
	$P = \dfrac{1}{2}\gamma(H_1^{\ 2} - H_2^{\ 2})\dfrac{b}{\sin\alpha}$ $h_D = \dfrac{1}{3}\left(2H_1 - \dfrac{H_2^{\ 2}}{H_1 + H_2}\right)$
	$P = \dfrac{1}{2}\gamma(2H - l)lb$ $h_D = H - \dfrac{1}{3} \cdot \dfrac{3H - 2l}{2H - l}$
	$P = \gamma(H_1 - H_2)lb$ $h_D = H_1 - \dfrac{1}{2}l$
	$P = \gamma\dfrac{(2H_1 - l\sin\alpha)}{2}lb$ $h_D = H_1 - \dfrac{l\sin\alpha}{3} \cdot \dfrac{3H_1 - 2l\sin\alpha}{2H_1 - l\sin\alpha}$
	$P = \gamma(H_1 - H_2)lb$ $h_D = H_1 - 0.5l\sin\alpha$

2.5.2 Analytical method

When a surface is in arbitrary shape, such as a plane with no symmetry axis, unequal width along the depth, or irregular shape, we cannot use the hydrostatic pressure diagram to find the total hydrostatic force, but the analytical method can be used to determine the magnitude, direction and acting point of total hydrostatic force.

2.5.2.1 Magnitude of total hydrostatic force

Fig. 2 - 21 shows an arbitrary plane that is inclined at an angle α to the water surface. The plane $A'B'$ has water pressure on one side while the other side is the atmosphere. The pressure at the water surface is atmospheric, $p_0 = p_{at}$. The submerged area is A, the centroid of the submerged area is C, and the depth at the centroid is h_C.

For convenience of analysis, take the intersection OE between the plane and the water surface as the x axis (Ox), and the axis perpendicular to the Ox on the left as the y axis (Oy). Rotate the xOy plane to make it coincide with the paper, as shown in Fig. 2 - 21.

Consider a small element dA in the area A, for which the depth of its centroid below the water is h, its ordinate is y, and the hydrostatic pressure is $p = \gamma h$, so the hydrostatic force acting on the centroid is

$$dP = p \cdot dA = \gamma h dA = \gamma y \sin\alpha dA$$

Since the hydrostatic force dP on any small element dA has the same direction, the total hydrostatic force P on the whole area A equals the sum of hydrostatic force dP on each small element dA. Hence,

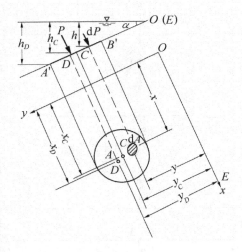

Fig. 2 - 21

$$P = \int_A dP = \int_A p dA = \int_A \gamma h dA = \int_A \gamma y \sin\alpha dA = \gamma \sin\alpha \int_A y dA \qquad (2-18)$$

where $\int_A y dA$ is the static moment (area moment) of each small area to a certain axis. According to the area moment theorem in materials mechanics, "the sum of static moment (area moment) of every small area to a certain axis is equal to the static moment (area moment) of the total area to the same axis," which is

$$\int_A y dA = y_C \cdot A$$

where y_C is the distance from the centroid of total area A to the Ox axis. Thus Equation (2-18) can be written as

$$P = \gamma \sin\alpha y_C \cdot A = \gamma h_C A = p_C \cdot A \qquad (2-19)$$

where $h_C = y_C \sin\alpha$ is the submerged depth of the centroid, and γh_C is the pressure on the centroid, equivalent to the average pressure p_C on the acting surface.

Equation (2−19) is a general equation for calculating total hydrostatic force on an arbitrary plane, regardless of the shape and position of the plane. It indicates that the total hydrostatic force acting on a plane of arbitrary shape is equal to the product of the area of plane and the hydrostatic force at the centroid of plane. It is worth noting that when a plane is partially submerged in water, the area and its centroid refer to the submerged part of plane.

2.5.2.2 Direction of total hydrostatic force

The direction of total hydrostatic force P always points to the acting surface.

2.5.2.3 Acting point of total hydrostatic force

The acting point of total hydrostatic force P is called the center of pressure, represented by D. Since the hydrostatic pressure is proportional to the depth, the greater the water depth, the greater the hydrostatic pressure will be, so the center of pressure D is generally below the centroid C of the plane, i.e. $y_D > y_C$.

The position of the center of pressure can be determined based on "the moment of resultant force to any axis equals the algebraic sum of moments of each component force to that axis". In order to determine y_D, we can take moment to the Ox axis.

Component moment of any component force dP to the Ox axis:
$$dM = dP \cdot y = \gamma y \sin\alpha dA \cdot y = \gamma \sin\alpha y^2 dA$$

The sum of moment of the hydrostatic pressure dP on each small area to the Ox axis:
$$\int_A dM = \int_A y dP = \int_A \gamma \sin\alpha y^2 dA = \gamma \sin\alpha \int_A y^2 dA \qquad (2-20)$$

where $\int_A y^2 dA$ is the inertia moment of area A to the Ox axis, denoted by I_{x0}. From the parallel-axis theorem of moment of inertia in materials mechanics, "the moment of inertia of an arbitrary plane to any axis is equal to the sum of the moment of inertia to the centroid axis of the plane (that is parallel to the axis) and the area multiplied by the square of the distance between the two axes", which can be expressed by
$$I_{x0} = I_{xC} + y_C^2 A \qquad (2-21)$$

where I_{x0} is the moment of inertia of area A to the cx axis that passes through the centroid C; y_C is the distance between axes Cx and Ox.

Substituting Equation (2−21) into Equation (2−20) gives
$$\int_A dM = \int_A y dP = \gamma \sin\alpha I_{x0} = \gamma \sin\alpha (I_{xC} + y_C^2 A) \qquad (2-22)$$

Moment of the total force P with respect to the Ox axis is
$$P \cdot y_D = \gamma h_C \cdot A \cdot y_D = \gamma \sin\alpha y_C A y_D \qquad (2-23)$$

Recall the relationship that the resultant moment equals the algebraic sum of moments of component forces, which means Equation (2−23) equals Equation (2−22), thus
$$\gamma \sin\alpha y_C A y_D = \gamma \sin\alpha (I_{xC} + y_C^2 A)$$

Then,
$$y_D = y_C + \left(\frac{I_{xC}}{y_C A}\right) \qquad (2-24)$$

or
$$e = y_D - y_C = \frac{I_{xC}}{y_C A}$$

where e is always larger than zero, so $y_D > y_C$.

As shown in Fig. 2-22,
$$y^* = y + y_C, \quad x^* = x + x_C$$

$$I_{x^*0} = I_{y^*} = \int_A y^{*2} dA = \int_A (y + y_C)^2 dA$$
$$= \int_A (y^2 + 2y_C y + y_C^2) dA$$
$$= \int_A y^2 dA + 2y_C \int_A y dA + \int_A y_C^2 dA$$
$$= I_{xC} + 2y_C \cdot A \cdot \zeta_C + y_C^2 A$$
$$= I_{xC} + A y_C^2$$

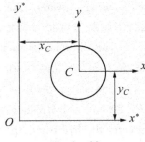

Fig. 2-22

where $\zeta_C = \int_A y dA$. Because the origin O of the xCy coordinate system is the centroid C of area A, the coordinate of centroid C in the xCy coordinate system is zero, $\zeta_C = 0$. Equation (2-21) is proved.

In Equation (2-24), y_D is not related with the inclined angle α; when $\alpha = 90°$, the surface is placed vertically, so $h = y$, $h_D = y_D$, $h_C = y_C$, and Equation (2-24) becomes $h_D = h_C + \frac{I_C}{h_C A}$.

Take moment of the hydrostatic force to the Oy axis:
$$P \cdot x_D = \int_A xp dA$$

Substituting $p = \gamma y \sin\alpha$ into above equation gives
$$P \cdot x_D = \gamma \sin\alpha \int_A xy dA \qquad (2-25)$$

Let $I_{xy} = \int_A xy dA$, named the product of inertia of the plane $A'B'$ with respect to the Ox and Oy axes. Substituting I_{xy} into Equation (2-25),
$$x_D = \frac{\gamma \sin\alpha I_{xy}}{\gamma \sin\alpha y_C A} = \frac{I_{xy}}{y_C A} \qquad (2-26)$$

Once y_D and x_D are obtained from Equations (2-24) and (2-26) respectively, the position of center of pressure is determined. Most surfaces in engineering have a symmetry axis parallel to the Oy axis, its pressure point D must be located on this axis of symmetry, so x_D is not needed to calculate.

If the pressure on a free surface is not equal to the atmospheric pressure, the above conclusions still hold true with the Ox axis relocated to the position of atmospheric pressure.

Table 2-2 shows the total hydrostatic forces on inclined planes and the displaced distances of center of pressure from the water surfaces.

Chapter 2 Hydrostatics

Table 2-2 Total hydrostatic forces acting on inclined planes

Figure	Total hydrostatic force P	Slant distance ζ_D from pressure center D to water surface
	$\dfrac{1}{2}lb(2\zeta_e+l)\gamma\sin\alpha$	$\zeta_e+\dfrac{(3\zeta_e+2l)l}{3(2\zeta_e+l)}$
	$\dfrac{1}{6}l[3\zeta_e(B+b)+l(B+2b)]\gamma\sin\alpha$	$\zeta_e+\dfrac{[2(B+2b)\zeta_e+(B+3b)l]l}{6(B+b)\zeta_e+2(B+2b)l}$
	$\dfrac{1}{6}lB(3\zeta_e+l)\gamma\sin\alpha$	$\zeta_e+\dfrac{(2\zeta_e+l)l}{2(3\zeta_e+l)}$
	$\dfrac{\pi}{8}d^2(2\zeta_e+d)\gamma\sin\alpha$	$\zeta_e+\dfrac{d(8\zeta_e+5d)}{8(2\zeta_e+d)}$
	$\dfrac{1}{24}d^2(3\pi\zeta_e+2d)\gamma\sin\alpha$	$\zeta_e+\dfrac{d(32\zeta_e+3\pi d)}{16(3\pi\zeta_e+2d)}$

Notes: ζ_e is the distance from the top edge (point) to the free surface that is in contact with the atmosphere.

【Example 2-7】 A rectangular flat gate is tilted in the inlet of a spillway tunnel (see Fig. 2-23). The tilt angle α is 45°, the width of gate b is 3 m, the gate length l is 4.5 m, the top edge of the gate is at 8 m below the water surface, and the gate weighs 300 N. Find: How much is the total hydrostatic pressure force P acting on the gate? Where is the acting point of the total hydrostatic pressure? How much tension T is required to pull the gate along the tilted surface (the friction coefficient between the gate and the gate groove $f=0.25$)?

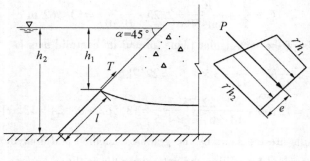

Fig. 2-23

Fluid Mechanics in Civil Engineering

Solution:

(1) Determining the total hydrostatic pressure force P and its acting point by hydrostatic pressure distribution diagram.

Because the gate is completely under the water, the diagram of pressure distribution is a trapezoid, where the hydrostatic pressure at the upper end of the gate is

$$p_1 = \gamma h_1 = 9.8 \times 8 = 78.4 \text{ kN/m}^2$$

At the lower end the hydrostatic pressure is

$$p_2 = \gamma h_2 = \gamma(h_1 + l\sin 45°) = 9.8 \times (8 + 4.5 \times 0.707) = 109.6 \text{ kN/m}^2$$

The area of the pressure distribution diagram is

$$A_P = \frac{1}{2}(\gamma h_1 + \gamma h_2) \cdot l = \frac{1}{2}(78.4 + 109.6) \times 4.5 = 423 \text{ kN/m}^2$$

The total hydrostatic pressure force is

$$P = A_p \cdot b = 423 \times 3 = 1269 \text{ kN}$$

The inclined distance between the acting point of total hydrostatic pressure and the bottom of the gate is

$$e = \frac{l(2h_1 + h_2)}{3(h_1 + h_2)} = \frac{4.5 \times (2 \times 8 + 8 + 4.5 \times 0.707)}{3 \times (8 + 8 + 4.5 \times 0.707)} = 2.13 \text{ m}$$

The inclined distance between the total pressure force P and the water surface is

$$l_D = (l + \frac{h_1}{\sin 45°}) - e = (4.5 + \frac{8}{0.707}) - 2.13 = 15.82 - 2.13 = 13.69 \text{ m}$$

(2) Determining P and l_D using the analytical method

$$P = p_C \cdot A = \gamma h_C \cdot bl$$

$$h_C = h_1 + \frac{l}{2}\sin 45° = 8 + \frac{4.5}{2} \times 0.707 = 9.59 \text{ m}$$

$$P = 9.8 \times 9.59 \times 3 \times 4.5 = 1269 \text{ kN}$$

The inclined distance between P and the water surface is

$$l_D = l_C + \frac{I_C}{l_C A}$$

$$l_C = \frac{l}{2} + \frac{h_1}{\sin 45°} = 2.25 + \frac{8}{0.707} = 13.562 \text{ m}$$

The moment of inertia of the rectangular plane around its centroid axis is

$$I_C = \frac{l}{12}bl^3 = \frac{1}{12} \times 3 \times 4.5^3 = 22.78 \text{ m}^4$$

$$l_D = 13.56 + \frac{22.78}{13.56 + 3 \times 4.5} = 13.562 + 0.124 = 13.69 \text{ m}$$

It shows that the results are the same by the graphical method and analytical method.

(3) Tension T required by pulling the gate along the inclined surface

Tension T that pulls the gate along the inclined surface needs to overcome the friction of the gate against its slot and the component force of the gate weight along the inclined surface.

$$T = (G\sin 45° + P) \times f + G\cos 45°$$
$$= 300 \times 0.707 \times 0.25 + 1269 \times 0.25 + 300 \times 0.707$$
$$= 53 + 317.3 + 212.1 = 582.4 \text{ kN}$$

【Example 2 – 8】 Fig. 2 – 24 shows a rectangular flap gate, which will be opened automatically by the hydrostatic pressure forces. When the upstream water level exceeds the operating level H, the gate will open automatically clockwise around the axis. If the gate weight and friction are neglected, find the location of the junction.

Solution:

First we analyze the working condition of the automatic flap gate: when the gate weight and friction are neglected, the total hydrostatic force P is the only force to generate a moment on the shaft. Assume the shaft is located at a height of a above the base. When the acting point of P is below the level of the shaft, i. e. $H_e < a$, its moment with respect to the shaft is counter-clockwise, so the gate will be closed. When the acting point of P is above the level of the

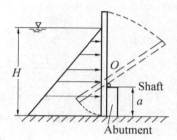

Fig. 2 – 24

shaft, i. e. $H_e > a$, its moment to the shaft is clockwise, so the gate will be opened automatically. When the acting point of P is exactly on the shaft, i. e. $H_e = a$, its moment with respect to the shaft is zero, so the gate remains upright but is in a critical state. Since the hydrostatic pressure diagram for the gate is a right-angled triangle, the shaft should be located at $1/3H$ above the base.

【Example 2 – 9】 Fig. 2 – 25 shows a flashboard inserted vertically in a trapezoidal channel, which has a bottom width (b) of 0.5 m and the cross-sectional side angle (θ) of 60°. If the water depth is 0.6 m, find the total hydrostatic force on the flashboard and the pressure center.

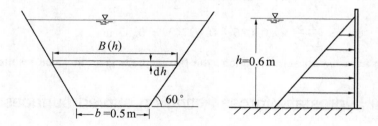

Fig. 2 – 25

Solution:

Since the width of flashboard varies with the depth, the graphic method is not suitable and the analytical method should be used instead. Note that the problem can also be solved by immediate integration for the parallel force distribution.

(1) By analytical method

Based on the equations of P and ζ_D in Table 2-2:

$$P = \frac{1}{6}l[3\zeta_e(B+b) + l(B+2b)]\gamma\sin\alpha$$

$$\zeta_D = \zeta_e + \frac{[2(B+2b)\zeta_e + (B+3b)l]l}{6(B+b)\zeta_e + 2(B+2b)l}$$

where $b = 0.5$ m, $B = b + 2 \times 0.6 \times \cot 60° = 0.5 + 0.693 = 1.193$ m, $l = h = 0.6$ m, $\zeta_e = 0$, $\alpha = 90°$.

Thus,

$$P = \frac{1}{6} \times 0.6 \times [0.6 \times (1.193 + 2 \times 0.5)] \times 9.8 = 1.29 \text{ kN}$$

$$\zeta_D = \frac{(1.193 + 3 \times 0.5) \times 0.6 \times 0.6}{2(1.193 + 2 \times 0.5) \times 0.6} = 0.37 \text{ m}$$

(2) By immediate integration

The flashboard width $B(h)$ at a depth of h is

$$B(h) = 0.5 + 2(0.6 - h)\cot 60°$$

The total hydrostatic pressure force on the flashboard is

$$P = \int_A dP = \int_A p dA = \int_0^h p \cdot B(h) \cdot dh$$

$$= \int_0^{0.6} \gamma h[0.5 + 2(0.6 - h)\cot 60°] dh$$

$$= \gamma\left[(0.5 + 1.2\cot 60°)\frac{h^2}{2} - \frac{2}{3}h^3\cot 60°\right]_0^{0.6}$$

$$= 9.8 \times (0.215 - 0.083) = 1.29 \text{ kN}$$

The location of pressure center is

$$h_D = \frac{1}{P}\int_0^h dP \cdot h = \frac{1}{P}\int_0^{0.6} \gamma h^2[0.5 + 2 \times (0.6 - h)\cot 60°] dh$$

$$= \frac{\gamma}{P}\left[(0.5 + 1.2\cot 60°)\frac{h^2}{2} - \frac{2}{3}h^3\cot 60°\right]_0^{0.6}$$

$$= \frac{9.8}{1.29} \times (0.086 - 0.0374) = 0.37 \text{ m}$$

As can be seen from the above calculation, the two methods give the same results.

2.6 Total hydrostatic forces acting on curved surfaces

In practical engineering, hydrostatic forces act on not only flat surfaces but also curved surfaces, such as curved gates, arch dams, circular sewers, culvert inlets, curved piers, side piers, and so on. Most curved surfaces in engineering are two-dimensional, such as cylindrical surfaces with horizontal generatrix. Therefore, this book focuses are on the calculation of total hydrostatic forces on two-dimensional curved surfaces.

The direction of hydrostatic pressure acting at a point on a curved surface is always along the direction of the inner normal of the acting curved surface, and its magnitude is proportional to the depth of the point, which is $p = \gamma h$ when the relative pressure reference is used. Fig. 2 – 19 e & f show the hydrostatic pressure distribution diagram of a curved surface AB. From Fig. 2 – 19 e & f, it can be seen that: normals of each point on the curved surfaces are neither parallel to each other nor necessarily intersecting; therefore the resultant force cannot be obtained by using immediate integration, like that for the plane. Instead, the calculation of total hydrostatic force on curved surfaces can be obtained by the so-called method of "combination after decomposition", which is described as follows:

First decompose a non-parallel force dP into a horizontal component force dP_x (parallel to the x-axis) and a vertical component force dP_z (parallel to the z-axis), as shown in Fig. 2 – 26. As both dP_x and dP_z are parallel force system, the algebraic sum P_x (the horizontal component) and P_y (the vertical component) of a total hydrostatic force can be found by immediate integration. Finally combining the resulting P_x and P_y gives the total hydrostatic force P on a curved surface.

2.6.1 Magnitude of total hydrostatic force on a curved surface

Suppose a two-directional curved surface with a horizontal generatrix of length b, which has hydrostatic pressure on its left side, see Fig. 2 – 26. Based on the "left-handed" three-dimensional coordinate, the xOz plane is on the paper, the xOy plane is on the water surface, and the y-axis is perpendicular to the paper and parallel to the horizontal generatrix of the curved surface, so the projection of the curved surface on the xOz plane is the curve AB. Take a small element dA on the AB with unit width, the centroid of dA is at h below the waster surface, and the hydrostatic pressure at the centroid is $p = \gamma h$. The hydrostatic pressure force acting on the small element dA is $dP = pdA = \gamma h dA$, and its direction points to the inner normal of dA at an angle of α with the horizontal. dP can be decomposed into a horizontal component force dP_x and a vertical component force dP_z.

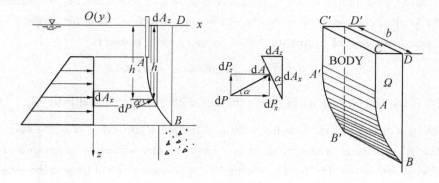

Fig. 2 – 26

2.6.1.1 Horizontal component force of the total hydrostatic force

In Fig. 2 – 26, the component force of dP in horizontal direction is

$$dP_x = dP\cos\alpha = \gamma h dA \cos\alpha \qquad (2-27)$$

Because the angle between the small curved surface dA and vertical plane is also α, $dA\cos\alpha$

can be seen as the projection dA_x of the curved surface dA on the vertical plane yOz, where the subscript x indicates the normal of the projection that is parallel to the x-axis. The horizontal component force on the entire curved surface AB is

$$P_x = \int_{A_x} dP_x = \int_{A_x} \gamma h dA \cos\alpha = \gamma \int_{A_x} h dA_x \qquad (2-28)$$

where $\int_{A_x} h dA_x$ is the area moment of the vertical projection plane with respect to the horizontal axis Oy. If h_C represents the submerged depth of the centroid C of the vertical projection plane, then the area moment is

$$\int_{A_x} h dA_x = h_{Cx} A_x$$

Substituting it into Equation (2-28) gives

$$P_x = \gamma h_{Cx} A_x = p_{Cx} A_x \qquad (2-29)$$

where A_x is the projection area of the curved surface AB on the yOz plane.

Equation (2-29) shows that the horizontal component force P_x of total hydrostatic force P acting on a curved surface equals the hydrostatic force acting on the projection plane A_x of the curved surface on the yOz plane. Therefore, calculating the horizontal component force of total hydrostatic force on a curved surface is converted to calculating the total hydrostatic force on the vertical plane A_x, which can be obtained by either the graphic method or analytical method, described previously. The acting line of horizontal component force P_x should pass through the pressure center of plane A_x.

2.6.1.2 Vertical component force of the total hydrostatic force

In Fig. 2-26, the component force of dP in vertical direction is

$$dP_z = dP\sin\alpha = \gamma h dA \sin\alpha \qquad (2-30)$$

where $dA\sin\alpha$ can be seen as the projection dA_z of the curved surface dA on the horizontal plane of xOy. The subscript z indicates the normal direction of the projection that is parallel with the z-axis. The vertical component force on the entire curved surface AB is the algebraic sum of the vertical component forces of hydrostatic forces on every small element:

$$P_z = \int_{A_z} dP_z = \int_{A_z} \gamma h dA \sin\alpha = \gamma \int_{A_z} h dA_z \qquad (2-31)$$

where $h dA_z$ is the volume of a small body with unit width, height h, and base dA_z. $\int_{A_z} h dA_z$ is the sum of many small bodies, which is the volume of the prism $ABCDA'B'C'D'$ between the curved surface $ABB'A'$ and its projection $CDC'D'$ on the free water surface (or extension of free water surface), as shown in Fig. 2-26. The volume of the prism multiplied by γ is the weight of water in the prism. Prism of $ABCDA'B'C'D'$ is called pressure prism, represented by V. Hence,

$$P_z = \gamma V \qquad (2-32)$$

Equation (2-32) indicates that the vertical component force P_z of the total hydrostatic force P acting on a curved surface equals the water weight in pressure prism.

The projection of pressure prism $ABCDA'B'C'D'$ on the xOz plane is comprised of the curve AB, its projection CD on the water surface or extension of water surface, and the vertical lines passing through A and B. If we define this area by Ω, the relationship between Ω and pressure prism V is

$$V = b \cdot \Omega \qquad (2-33)$$

where b is the width of the two-directional curved surface.

From the above analysis, it can be seen: if we want to find the vertical component force P_z of the total hydrostatic force P acting on a two-directional curved surface, we have to find the pressure prism V first. To determine V, we have to find its bottom area Ω as well. Once Ω is known, the vertical component force P_z can be obtained from Equation (2-33).

Usually, we call the bottom area Ω as pressure prism for simplicity. Next, we will introduce some general principles for drawing a pressure prism, followed by how to determine the direction of the vertical component force P_z and how to draw the pressure prism on a complex curved surface.

(1) How to draw pressure prism

Based on the side of a curved surface that is in contact with water, we determine the acting surface of a hydrostatic pressure, which can be used to construct the pressure prism. This so called "prism" should be comprised of six faces: top and bottom, left and right, front and back.

Top and bottom faces are the projection of the curved surface on water surface or extension of water surface and the curved surface itself, respectively. It should be noted that the "free water surface" is a horizontal plane where the relative pressure is zero.

Left and right faces are the vertical surfaces made through the left edge and the right edge of the curved surface, respectively.

Front and back faces are the vertical surfaces made through the front edge and the back edge of the curved surface, respectively.

The body surrounded by above six faces is the pressure prism. For a two-directional curved surface, the pressure prism is a prismoid. As the area of faces along the width is the same, only the surface areas of the prism at top and bottom, left and right are needed.

(2) How to determine the direction of the vertical component force P_z

First, the direction of total hydrostatic pressure force P on a single two-directional curved surface based on "direction of hydrostatic pressure is always consistent with the direction of the inner normal of the acting surface". Then decompose the total hydrostatic pressure force P into the horizontal component force P_x and the vertical component force P_z, which may point vertically upward or downward depending on the shape and the side of the curved surface that are in contact with water.

(3) For complex curved surfaces, how to draw the pressure prism and determine the direction of the vertical component force P_z

Take the S-curved surface in Fig. 2-27 as an example to show how to determine the pressure prism and the direction of the vertical component force P_z.

① Divide the curved surface into several segments, in which each segment can be used to construct a pressure prism. So we need to make a number of vertical planes tangent with the complex curved surface at the tangent points C and D, which divide the curved surface into three segments: AC, CD and DB.

② Draw the pressure prism for each segment.

③ Determine the direction of vertical component force P_{zi} in AC, CD and DB and express them in different symbols, lines or colours. For example, the horizontal lines represent upward vertical component forces, and the vertical lines represent downward vertical component forces.

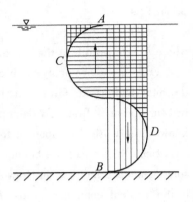

④ Superposition: The sections that have two kinds of lines indicate to have both upward and downward vertical component forces, and these two forces will cancel out each other, so the pressure prisms after superposition are those with only one kind of line:

Fig. 2 – 27

$$P_z = \sum_{i=1}^{n} P_{z_i} = P_{z_1} - P_{z_2}$$

⑤ If both sides of the curved surface have water, we draw the pressure prisms for each side separately and then superpose them.

Once the bottom size of prism Ω is known, Ω multiplying by b gives the volume of the pressure prism V, so $P_z = \gamma V = \gamma \cdot \Omega \cdot b$.

2.6.1.3 Total hydrostatic force

After obtaining the horizontal component forces P_x and vertical component forces P_z by "decomposition", we can combine these two force vectors into the total hydrostatic force by "combination".

$$P = \sqrt{P_x^2 + P_z^2} \qquad (2-34)$$

2.6.2 Direction of total hydrostatic force

Direction of total hydrostatic force is determined by the angle θ between P and the horizontal plane:

$$\theta = \arctan \frac{P_z}{P_x} \qquad (2-35)$$

Note that an angle between P and the vertical plane is not recommended to use.

2.6.3 Acting point of total hydrostatic force

For a circular cylindrical surface, since the hydrostatic forces at every point are perpendicular to the surface and point to the center, so the resultant force (total hydrostatic force) also points to the center of the circle O. Draw a line through the center at an angle of θ with the horizontal, which will intersect at m with the cylindrical surface. This intersection point m is the center of pressure, as shown in Fig. 2 – 28a.

If the curved surface is a non-circular cylindrical surface, the pressure center can be found by drawing a diagram: the horizontal component force P_x of total hydrostatic force passes through the pressure center of the projection plane A_x, the vertical component force P_z of total hydrostatic force passes through the centroid of the pressure prism, extend the acting lines of P_x and P_z to intersect at a point n, and then draw a line through the intersection n at an angle of θ with the horizontal, so this line will intersect with the cylindrical surface at point m, which is the pressure center, as shown in Fig. 2 – 28b.

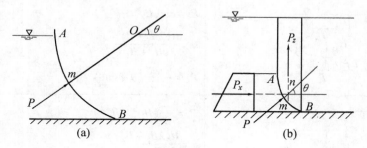

Fig. 2 – 28

In engineering sometimes we have 3D curved surfaces, such as the water eliminators in ball valves, cone valves, etc. The above methods for two-directional curved surfaces can also be applied to three-directional curved surfaces. What is different is that: for three-directional curved surfaces, apart from the projection on the yOz plane, we also have a projection on the xOz plane, on which there are the two component forces of not only P_x in x-axis direction but also P_y in y-axis direction. Similar to P_x, P_y equals the total pressure force on the projection plane A_y of the curved surface on the xOz plane. The vertical component force of a three-directional curved surface also equals the water weight of pressure prism. Here the pressure prism is comprised of the curved surface, its projection on the free water surface (or extension of the free water surface), and the vertical generatrix (not a plane) from the circumference of curved surface to the free water surface. Thus,

$$\left. \begin{array}{l} P_x = \gamma h_{C_x} A_x \\ P_y = \gamma h_{C_y} A_y \\ P_z = \gamma V \end{array} \right\} \qquad (2-36)$$

where h_{C_x} and h_{C_y} are the submerged depths of projection area A_x and A_y respectively.

$$P = \sqrt{P_x^2 + P_y^2 + P_z^2} \qquad (2-37)$$

The direction of the resultant force is represented by the cosines of angles α, β, γ respectively.

$$\cos\alpha = \frac{P_x}{P}, \quad \cos\beta = \frac{P_y}{P}, \quad \cos\gamma = \frac{P_z}{P} \qquad (2-38)$$

Table 2 – 3 lists the calculation equations for hydrostatic forces on several horizontal cylindrical surfaces.

Table 2-3 Total hydrostatic forces acting on horizontal cylindrical surfaces

Figure	Formulas
(First row)	$P_x = \dfrac{\gamma}{2}(H_2^2 - H_1^2)$ $h = \dfrac{H(2H_1 + H_2)}{3(H_1 + H_2)}$ $P_z = -\dfrac{\gamma R}{2}[(H_1 + H_2)(\cos\theta_1 - \cos\theta_2) + R(\theta - \sin\theta)]$
(Second row)	$P_x = \dfrac{\gamma}{2}(H_2^2 - H_1^2)$ $h = \dfrac{H(2H_1 + H_2)}{3(H_1 + H_2)}$ $P_z = -\dfrac{\gamma R^2}{2}(\theta - \sin\theta)$
(Third row)	$P_x = \dfrac{\gamma}{2}(H_2^2 - H_1^2)$ $h = \dfrac{H(2H_1 + H_2)}{3(H_1 + H_2)}$ $P_z = -\gamma R[H_1(1 - \cos\theta) + \dfrac{R}{4}(2\theta - \sin 2\theta)]$
(Fourth row)	$P_x = \dfrac{\gamma}{2}(H_2^2 - H_1^2)$ $h = \dfrac{H(2H_1 + H_2)}{3(H_1 + H_2)}$ $P_z = \gamma R[H_2(1 - \cos\theta) - \dfrac{R}{4}(2\theta - \sin 2\theta)]$

The method for calculating the total hydrostatic force of curved surfaces by decomposing it into horizontal force and vertical force, is also often used for slant plane.

【Example 2-10】 A thin-walled pipe with diameter d has a pressure head of H. Because H is much larger than d, it can be assumed that the pipe is acted by uniform hydrostatic pressure $p = \gamma H$, as shown in Fig. 2-29a. If the tensile strength allowed for the pipe is $[\sigma]$, what is the thickness(δ) of the pipe wall?

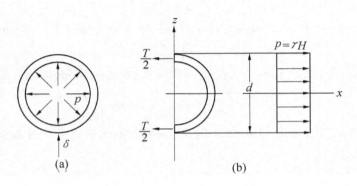

Fig. 2-29

Solution:

This problem is regarded as the calculation of circle tension. In order to analyze the relationship between the tension of pipe and hydrostatic force, take a pipe section of a unit length along the pipe axis and cut the pipe along the diameter into half, as shown in Fig. 2-29b. The tensile resistance of the pipe is

$$T = 2[\sigma]\delta \cdot 1$$

Take an analysis of the hydrostatic force, which is regarded as the curved surface force. Consider an arbitrary small area on the curved surface, since its force dP points to the curved surface perpendicularly, its vertical component force offsets each other, so only horizontal hydrostatic forces remain. The projection area of the half pipe on the yOz plane $A_x = d \times 1$, and from Equation (2-29) the horizontal hydrostatic component force on curve surface is

$$P = p_c A_x = \gamma H d$$

The static equilibrium equation in the x-axis direction

$$T = P$$

For the safety, T should be large than P

$$2[\sigma]\delta \geqslant \gamma H d$$

$$\delta \geqslant \frac{\gamma H d}{2[\sigma]}$$

【Example 2-11】 A horizontal cylinder has water only on its left side and the water levels with the top of cylinder, as shown in Fig. 2-30a, where diameter $d = 3$ m and $\theta = 45°$. Find the horizontal component force P_x and vertical component force P_z of the total hydrostatic force acting on the cylinder.

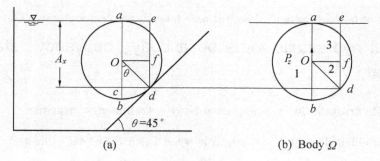

(a)　　　　　　　　(b) Body Ω

Fig. 2-30

Solution:

Consider a cylinder with unit width $b = 1$ m. The curved surface of abd is acting by the water force. Note that on the arched surface below cd, the horizontal component forces on two sides cancel out each other, so the projection of this surface on the yOz plane $A_x = \overline{ac} \times b$. Consider the relationship of geometry in Fig. 2-30:

Fluid Mechanics in Civil Engineering

$$\angle cOd = \theta = 45°$$

$$\overline{Oc} = \overline{Od} \cdot \cos\angle cOd = \frac{d}{2} \cdot \cos 45° = 1.5 \times 0.707 = 1.06 \text{ m}$$

$$\overline{ac} = \frac{d}{2} + \overline{Oc} = 1.5 + 1.06 = 2.56 \text{ m}$$

$$\overline{ae} = \overline{cd} = \overline{Od}\sin 45° = 1.5 \times 0.707 = 1.06 \text{ m}$$

$$P_x = \gamma h_{C_x} A_x = \gamma \times \frac{1}{2}\overline{ac} \times \overline{ac} \cdot b$$

$$= 9.8 \times \frac{2.56}{2} \times 2.56 \times 1 = 32.11 \text{ kN}$$

The pressure prism is *abdea*, which consists of three parts:

$$V = V_1 + V_2 + V_3 = (\Omega_1 + \Omega_2 + \Omega_3) \cdot b$$

$$= (\Omega_{225° \text{ sector}} + \Omega_{\triangle odf} + \Omega_{\text{rectangle } aefo}) \cdot b$$

$$= \frac{225}{360}\pi \gamma^2 + \frac{1}{2}\overline{ae} \cdot \overline{Oc} + \overline{ae} \cdot \overline{aO}$$

$$= \frac{225}{360}\pi \cdot 1.5^2 + \frac{1}{2} \times 1.06 \times 1.06 + 1.06 \times 1.5$$

$$= 4.42 + 0.56 + 1.59 = 6.57 \text{ m}^3$$

$$P_z = \gamma V = 9.8 \times 6.57 = 64.4 \text{ kN}$$

$$P = \sqrt{P_x^2 + P_z^2} = \sqrt{32.11^2 + 64.4^2} = 71.96 \text{ kN}$$

$$\alpha = \arctan\frac{P_z}{P_x} = 63.5°$$

The acting line of total hydrostatic force will pass through the center of circle O.

2.7 Total hydrostatic force on a body, buoyancy, stability of a floating body

2.7.1 Total hydrostatic force acting on a body—Archimedes principle

Any object either floating or submerged in water is acted by total hydrostatic force, which is equal to the sum of the hydrostatic forces on the surface of the object.

Fig. 2-31 shows an arbitrary shaped object submerged in static liquid. The total hydrostatic force P on the object can be decomposed into three component forces: horizontal component forces P_x, P_y and vertical component force P_z.

By the "left-hand" rule, of the coordinate system Ox axis, Oy axis and Oz axis, the xOy plane coincides with the free water surface. To find the horizontal component forces, we make a generatrix that is parallel with the horizontal circumscribed cylinder of Ox axis and tangent with the object, so the tangent points make a closed curve *abdc*, and the curve *abdc* divides the object into left and right two parts, so the horizontal component force P_x should be the algebraic sum of the horizontal component force P_{x1} on the left surface and the horizontal component force P_{x2} on the right surface. The projections of left and right parts on the yOz plane A_x are the same, so the pressures γh_{C_x} on their centroids are the same; hence $|P_{x_1}| = \gamma h_{C_x} A_x = |P_{x_2}|$ but P_{x1} and P_{x2} are in opposite directions, which leads to $P_x = P_{x_1} - P_{x_2} = 0$ after superposition. Similarly, we make a generatrix that is parallel with the horizontal circumscribed cylinder of Oy axis and tangent with the object, so the tangent points make a closed curve that divides the object into front and back two parts. Because their projections on the xOz plane A_x are the same, their centroids coincide, and the pressures γh_{C_y} on the centroids are the same, P_{y1} and P_{y2} are equal in magnitude but opposite in direction. Thus, $P_y = P_{y_1} - P_{y_2} = \gamma h_{C_y} A_y - \gamma h_{C_y} A_y = 0$ after superposition.

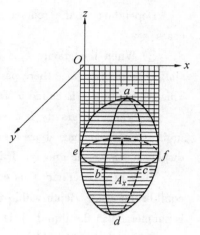

Fig. 2-31

The vertical component force is not equal to zero, because the hydrostatic pressure changes with depth. Similar to the above analysis, draw the vertical circumscribed cylinder of the object, so the curve $bcfe$ (comprised of tangent points) divides the object into upper and lower parts. The vertical component force P_z is the algebraic sum of the vertical component forces on the upper surface and lower surface. The pressure prism on the upper curved surface is V_1, $P_{z_1} = \gamma V_1$, with direction downward, as shown in Fig. 2-31; the pressure prism on the lower curved surface is V_2, $P_{z_2} = \gamma V_2$, with direction upward. The resultant force after superposition $P_z = \gamma(V_2 - V_1) = \gamma V$, with direction upward.

$$P = \sqrt{P_x^2 + P_y^2 + P_z^2} = P_z$$
$$P = \gamma V \qquad (2-39)$$

Equation (2-39) indicates that the total hydrostatic force of an object submerged in liquid is a vertically upward force, called buoyancy, whose magnitude is equal to the weight of the displaced liquid. The acting point of buoyancy is at the centroid of the submerged object, which is called the center of buoyancy. In 250 BC, the Greek scientist Archimedes published this principle, that is, the buoyancy of an object in static water is equal to the weight of the same volume of water displaced. Thus, this principle is called Archimedes principle.

In the initial proof of Archimedes principle, the object is assumed to be submerged in water completely, but the principle is also applicable to objects that are partially submerged in the water.

2.7.2 Equilibrium of a sinking body, submerged body and floating body

All objects submerged or floating in liquid are subjected to two forces: gravity G and buoyancy P_z. The gravity acts vertically downwards through the center of gravity, while the buoyancy acts vertically upwards through the center of buoyancy.

Depending on the relative magnitude of the gravity G and buoyancy P_z, there are three cases:

① When the gravity G is larger than the buoyancy, P_z, i. e. $G > P_z$, an object will be sinking to the bottom, where part of its weight will be supported by the bottom of the container. This type of object is called a sinking body.

② When the gravity G is smaller than the buoyancy, P_z, i. e. $G < P_z$, an object will rise until part of it emerges above the liquid, and an equilibrium is achieved when the volume of liquid displaced equals the gravity. This type of object is called a floating body.

③ When the gravity G is equal to the buoyancy, P_z, i. e. $G = P_z$, the object maintains in equilibrium at any depth within the liquid depending on the initial condition. Because the object is submerged in the liquid, it is called a submerged body.

When a sinking body reaches the bottom, the equilibrium is achieved—no more sinking, rising, or rotating. But for the balance of a floating body and a submerged body, two conditions must be satisfied (see Fig. 2 – 32):

$\sum F = 0$, which means the gravity G equals the buoyancy P_z;

$\sum m = 0$, which means the gravity G and buoyancy P_z are on the same vertical line.

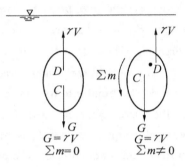

Fig. 2 – 32

If only $G = P_z$ is satisfied but the centers of gravity and buoyancy are not on the same vertical line, a couple will be generated and the object will rotate, as shown in Fig. 2 – 32.

Chapter summary

This chapter has described the characteristic of stress, the distribution law of hydrostatic pressure in static liquid, and the calculation of total hydrostatic force on surface, in addition to the pressure.

1. In static liquid, only normal stress (pressure) exists. Its magnitude does not depend on the orientation of the surface, but it is a continuous function of spatial coordinate, $p = p(x, y, z)$. In the static homogeneous fluid under gravity, an isobaric surface is a horizontal plane.

2. Hydrostatic pressure distribution under gravity has two forms of expression: Equation (2 – 7), $z + \dfrac{p}{\gamma} = c$ and Equation (2 – 8), $p = p_0 + \gamma h$. The two equations are both called the basic hydrostatic equation.

3. With regards to different reference pressures, pressures are defined as absolute pressures, relative pressures and vacuum pressures. Their relationships are described by Equations (2 – 11) and (2 – 12), which are

$$p_{abs} = p_{at} + p_r$$
$$p_v = p_{at} - p_{abs} = |p_r|$$

A pressure can be expressed by three ways: stress, column height of liquid, and number of atmospheric pressure.

4. The total hydrostatic force acting on a plane can be calculated by graphic method and analytical method. The graphic method only applies to the rectanglar surfaces, whose bottom edge is parallel to the liquid surface. The acting point of total hydrostatic force (center of pressure) is lower than the centroid of acting surface.

5. To calculate the total hydrostatic force on a two-directional curved surface, first we calculate the horizontal component force P_x and vertical component force P_z, where the horizontal component force P_x can be obtained from the projection area A_x of the curved surface on the vertical plane using the method for the flat plane; the vertical component force P_z can be obtained from the liquid weight in pressure prism. Then we calculate the resultant force of P_x and P_y, which is the total hydrostatic force P. If a two-directional curved surface is a cylindrical surface, the resultant force will point to the center of circle.

Multiple-choice questions (one option)

2-1 In static fluid, the magnitude of pressure on any point is not related to the _____.
 (A) direction of acting surface (B) position of the point
 (C) type of the fluid (D) gravitational acceleration

2-2 The basic hydrostatic pressure equation under gravity is _____.
 (A) $dp = -\rho dz$ (B) $dp = -\rho g dz$ (C) $dp = -g dz$ (D) $dp = \rho g dz$

2-3 In static liquid, _____ can exist.
 (A) normal stress
 (B) normal stress and tensile stress
 (C) normal stress and shear stress
 (D) normal stress, tensile stress and shear stress

2-4 Euler differential equation is _____.
 (A) $dp = -\rho(Xdx + Ydy + Zdz)$ (B) $dp = \rho(Xdx + Ydy + Zdz)$
 (C) $dp = -g(Xdx + Ydy + Zdz)$ (D) $dp = g(Xdx + Ydy + Zdz)$

2-5 $z + \dfrac{p}{\gamma} = c$ indicates that in static liquid, _____ of all points is equal.
 (A) piezometric height (B) elevation height
 (C) piezometric head (D) elevation head

2-6 Use a U-shaped mercury manometer to measure the pressure difference between points A and B in a horizontal pipe, and the reflection difference of mercury is 40 mm, so the pressure difference is _____ kPa.
 (A) 5.34 (B) 4.94 (C) 0.392 (D) 3.92

2-7. The relative pressure at a depth of 0.6 m below a static oil surface (the oil surface is

Fluid Mechanics in Civil Engineering

open to the atmosphere) is _____ kPa (the density of oil is 800 kg/m³).
(A)0.8　　　　　(B)0.6　　　　　(C)0.48　　　　　(D)4.7

2-8　Assuming the atmospheric pressure $p_0 = 10^5 \text{N/m}^3$, if the vacuum pressure at a certain point is 0.48×10^5 Pa, the relative pressure at this point is _____ Pa.
(A)0.52×10^5　　(B)1.48×10^5　　(C)0.48×10^5　　(D)10^5

2-9　For a tilted plate, the relationship between the submerged depth of its centroid h_C and that of the center of hydrostatic pressure h_d is h_C _____ h_d.
(A) >　　　　　(B) <　　　　　(C) =　　　　　(D) undetermined

2-10　In a pressure prism, there _____.
(A) must be filled with liquid　　(B) must be no liquid
(C) should be liquid, at least partly　　(D) could be liquid, or could be no liquid

2-11　The zero-pressure reference of relative pressure is _____.
(A) absolute vacuum　　(B) 1 standard atmospheric pressure
(C) the local atmospheric pressure　　(D) the pressure on the liquid surface

2-12　There is a vertically placed rectangular flat gate, and the water depth in front of it is 3 m, so the distance from the acting point of total hydrostatic pressure force to the water surface is _____.
(A)1.0 m　　(B)1.5 m　　(C)2.0 m　　(D)2.5 m

2-13　The buoyancy on a submerged body in a liquid is _____.
(A) proportional to the density of the submerged body
(B) proportional to the density of the liquid
(C) inversely proportional to the density of the submerged body
(D) inversely proportional to the density of the liquid

2-14　The relationship among absolute pressure p_{abs}, relative pressure p_r, vacuum pressure p_v and local atmospheric pressure p_a is _____.
(A) $p_{abs} = p_r + p_v$　　　　(B) $p_r = p_{abs} + p_a$
(C) $p_v = p_a - p_{abs}$　　　　(D) $p_a = p_r - p_{abs}$

2-15　When a liquid rotates within a container at the same angular velocity, the resultant force of the gravity and inertia force on the free surface is always _____ with the liquid surface.
(A) orthogonal　　(B) skew　　(C) tangent　　(D) undetermined

Problems

2-1　Fig. 2-33 shows a water supply system, in which $p_1 = 1.5 \times 10^6$ Pa, $p_2 = 4 \times 10^4$ Pa, $z_1 = z_2 = 0.5$ m, and $z_3 = 2.3$ m. The valve is closed, and find the pressure difference between points A and C, C and D, B and D.

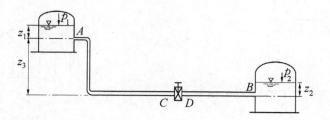

Fig. 2-33

2-2 A load G of 8341 N is added on the lid of a container that is filled with water. If the lid is fully sealed with the container, find the relative hydrostatic pressure at points A, B, C and D (see Fig. 2-34).

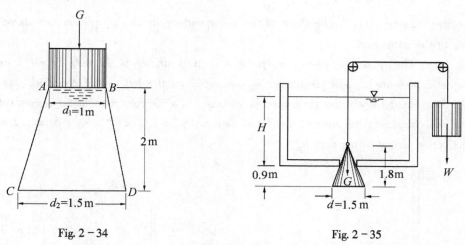

Fig. 2-34 Fig. 2-35

2-3 A conical automatic control valve is used for the circular hole at the bottom of a water tank, and the conical valve is suspended on pulley by a wire, which is tied with a metal weight W of 12 000 N at the other end of wire, as shown in Fig. 2-35. The weight of valve is 300 N. If the friction of the pulley is neglected, find how much of water depth H will be so that the valve will open automatically. ($V_{cone} = \dfrac{1}{3} \times$ bottom area $\times$ height)

2-4 A water tank is shown in Fig. 2-36. The bottom of the tank is 3 m above the ground, the water depth in the tank is 2 m, and the absolute pressure p_0 on the free surface is 68 600 Pa (0.7 engineering atmospheric pressure). ① Take the ground as a reference datum, what is the unit elevation energy and unit pressure energy at points A and B? ② Take the water surface of tank as a reference datum, what is the unit elevation energy and unit pressure energy at points A and B? Indicate the answers on the figure.

2-5 Fig. 2-37 shows a closed container with two piezometers connected on both sides. The top of the right piezometer is sealed, the water surface in it is 2 m higher than the water surface in the container, and the pressure p_0 on the water surface in the piezometer is 87.8 kN/m². The left piezometer is open to the atmosphere. Find ① the absolute pressure of the surface pressure p_C

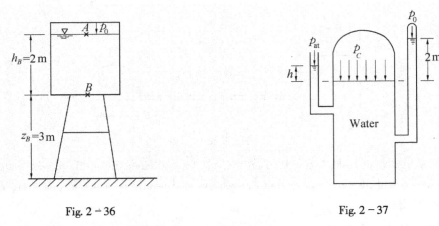

Fig. 2 - 36   Fig. 2 - 37

inside the container, ② the height h of the water surface in the left piezometer above the water surface in the container.

2 - 6   The relative pressure on point A in a container is 29 400 Pa (0.3 engineering atmospheric pressure). If a piezometer is connected on the left of the container, as shown in Fig. 2 - 38, how long will the glass tube be at least? If a mercury manometer is connected on the right of the container, the specific weight of mercury $\gamma_m = 133.28 \text{ kN/m}^3$, $h' = 0.2$ m, what is the deflection height of mercury h_m?

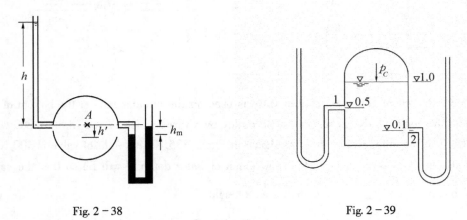

Fig. 2 - 38   Fig. 2 - 39

2 - 7   Fig. 2 - 39 shows a container containing water, where the water elevation is 1.0 m, and the installation elevation of two piezometers are 0.5 m and 0.1 m respectively. The local atmospheric pressure is $p_{at} = 98\,000 \text{ N/m}^2$. If the water column height $\dfrac{p_C}{\gamma} = 6$ m (absolute pressure), what is the elevation of water surfaces in the piezometers? What are the absolute pressure, relative pressure, pressure head and piezometric head at points 1 and 2? What is the degree of vacuum?

2 - 8   Fig. 2 - 40 shows a cylindrical barrel containing a light oil and a heavy oil. The specific weight of the light oil γ_1 is 6.2 kN/m^3, and the specific weight of the heavy oil γ_2 is 8.9 kN/m^3. If the weights of two oils are equal, find ① what are the depths (h_1 and h_2) of two oils?

② what height will the oil surfaces in the piezometers rise to?

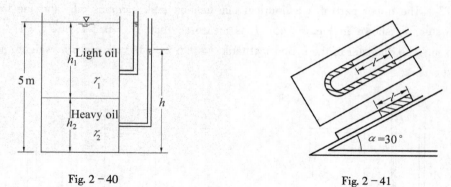

Fig. 2−40   Fig. 2−41

2−9 Fig. 2−41 shows an inclined manometer for air pressure difference measurement.

(1) When water is inside the manometer, the inclined angle $\alpha = 30°$, and the difference of the water surface readings is $l = 25$ cm, what is the air pressure difference at the two tube ends?

(2) When alcohol is inside the manometer (the specific weight of alcohol $\gamma = 7.85$ kN/m^3), and the inclined angle and the air pressure difference at the two pipe ends are the same as the case in (1), what is the difference of the water surface readings l?

2−10 A vacuometer for measuring the pressure at point A is shown in Fig. 2−42. $z = 1$ m, $h = 1.5$ m, and the local atmospheric pressure is $p_{at} = 98$ kN/m^2. Find the absolute pressure, relative pressure and degree of vacuum of point A (the weight of air may be neglected).

2−11 A type of three U-shaped mercury manometers in series is used to measure the pressure in a highly pressured water pipe, and on the top of the manometers is filled with water, as shown in Fig. 2−43. If the pressure at point M is equal to the atmospheric pressure, all the mercury surfaces are above the $O-O$ plane. When the reading on the right limb of the last manometer $h = 10$ mm, find the pressure at point M.

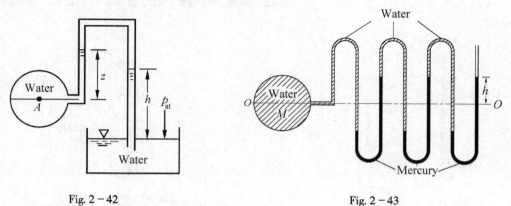

Fig. 2−42   Fig. 2−43

2−12 Point A on the top of a boiler is connected with a U-shaped manometer, and point B at the bottom of the boiler is connected with a piezometer, as shown in Fig. 2−44. The top of the manometer is closed. If the absolute pressure of p_0 is 9 800 N/m^2, the height difference (h_2) of mercury in the manometer is 0.7 m, and the local atmospheric pressure is 98 kN/m^2, find the vapour pressure p in the boiler and the height h_1 in the piezometer.

2-13 The upper part of a cylindrical clarification tank contains oil, and the lower part contains water, as shown in Fig. 2-45. It is measured that $\nabla_3 = 0.5$ m, $\nabla_2 = 1.4$ m, and $\nabla_1 = 1.6$ m. The diameter of the cylindrical tank $d = 0.6$ m. Find the specific weight and weight of oil in the tank.

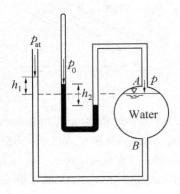

Fig. 2-44

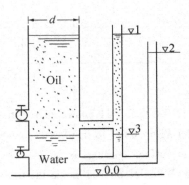

Fig. 2-45

2-14 For a container shown in Fig. 2-46, at point A is a vacuum gauge, whose reading is 29.32 kN/m^2, and $\gamma_{oil} = 7.8$ kN/m^3, find the liquid column height h_1 and h_2 of piezometers E and F, and the value of h in the left U-shaped piezometer. (unit of elevation: m)

2-15 Two containers contain the same type of liquid (specific weight $\gamma_A = \gamma_B = 11.1$ kN/m^3), and their centroids A and B are on the same elevation as shown in Fig. 2-47. A U-shaped manometer is used to measure the pressure difference between A and B (the manometer contains oil with the specific weight $\gamma_{oil} = 8.5$ kN/m^3). At point A, a mercury piezometer is installed as well. For $h_1 = 0.2$ m, $h_2 = 0.04$ m, and $l = 0.05$ m, find ① what is the pressure difference between points A and B? ② will a vacuum happen between points A and B? If so, what is its value?

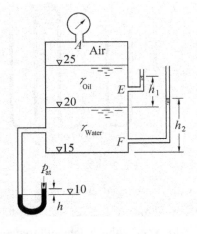

Fig. 2-46

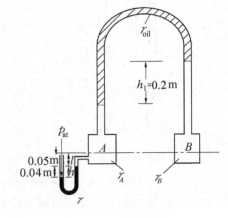

Fig. 2-47

2-16 In Fig. 2-48, draw the hydrostatic pressure distribution diagrams on the side walls in all the cases.

2-17 In Fig. 2-49, draw the hydrostatic pressure distribution diagrams on the surfaces labelled with letters in every figure.

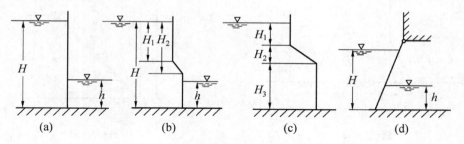

Fig. 2-48

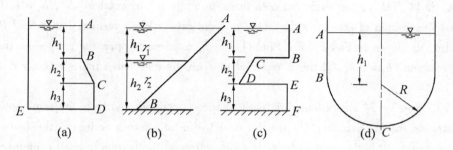

Fig. 2-49

2-18 In a container, the lower part of it contains water, while the upper part contains oil, as shown in Fig. 2-50. $h_1 = 1$ m, $h_2 = 2$ m, and the specific weight of oil $\gamma_{oil} = 7.84$ kN/m^3. Find the force per unit width acting on the side wall AB and its acting position.

2-19 In the pressure test of a pipeline, the reading of pressure gauge M is 980 kPa (10 engineering atmospheric pressure), and the diameter of pipe $d = 1$ m. Find the total hydrostatic force acting on the blind flange shown in Fig. 2-51.

2-20 Fig. 2-52 shows a concrete gravity dam. In order to check the stability of the dam, calculate the horizontal component force P_x and vertical component force P_z on the dam of 1 m length in the cases of with and without water downstream.

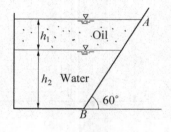

Fig. 2-50

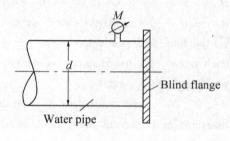

Fig. 2-51

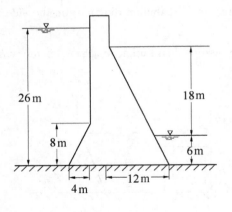

Fig. 2 – 52   Fig. 2 – 53

2 – 21 A 3 m wide rectangular flat gate AB is hinged at A. The weight of the gate (uniform thickness) is 14 700 N, the angle between the gate and the horizontal is 45°, $h_1 = 1.0$ m, and $h_2 = 2$ m. The friction of gate is neglectable, and the gate is lifted vertically at its end B through a steel wire, as shown in Fig. 2 – 53. Find ① the T_1 required to open the gate when there is no water downstream ($h_s = 0$); ② the T required to open the gate when there is water downstream ($h_s = h_2$).

2 – 22 Fig. 2 – 54 shows four differently shaped containers that have the same bottom areas A and water depths h. Explain: ① are the total hydrostatic forces acting on the bottoms of all containers equal? ② is the total hydrostatic force acting on the bottom of each container equal to the counterforce N on the container from the ground (neglecting the weight of container)?

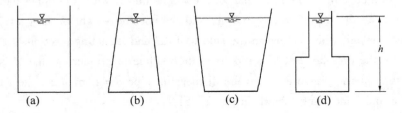

Fig. 2 – 54

2 – 23 Fig. 2 – 55 shows a flat gate, on the back of the gate lay three beams with each having equal force on. If the upstream water height $H = 4$ m, find ① the total hydrostatic force per unit width acting on each beam; ② the distance of each beam from the water surface (y_1, y_2, y_3).

2 – 24 Draw the hydrostatic pressure distribution diagrams of horizontal component forces on the two-dimensional curved surfaces in Fig. 2 – 56 and the pressure prisms on the vertical.

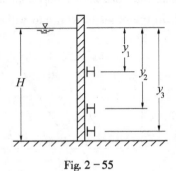

Fig. 2 – 55

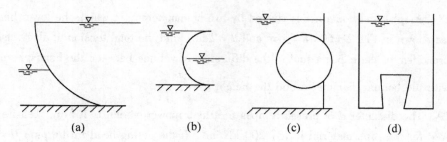

Fig. 2-56

2-25 Draw the pressure prisms for the curved surfaces in Fig. 2-57.

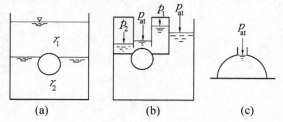

Fig. 2-57

2-26 A curved surface ABCD is comprised of three semi-arcs, as shown in Fig. 2-58. $R_1 = 0.5$ m, $R_2 = 1$ m, $R_3 = 1.5$ m, and the width b of the curved surface is 2 m. Find the horizontal component force and the vertical component force on the curved surface, as well as the direction of the vertical pressure force.

2-27 Fig. 2-59 shows a vertically placed rectangular plane with water on both sides. Draw a straight line through point C to divide the plane into a trapezoid and a triangle. If the total hydrostatic forces on the two areas are expected to be equal, what should the value of x be?

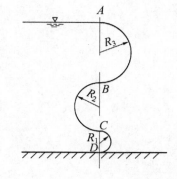

Fig. 2-58

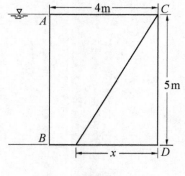

Fig. 2-59

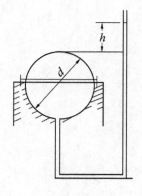

Fig. 2-60

2-28 A spherical container is riveted by two hemispheres, in which the lower hemisphere is fixed, as shown in Fig. 2-60. $h = 1$ m and $d = 2$ m. Find the total tension of all rivets. (Hint: the pressure prism of the cap is equal to the difference of volume between the hemisphere and the cylinder with the bottom diameter d and the height $h + \dfrac{d}{2}$.)

2-29 The diameter d of pressure pipes in a hydropower station is 1.5 m. Tensile strength $[\sigma]$ allowed for the pipe material is 137 200 kN/m². If the acting head in the pipe $H = 150$ m, design the thickness δ required for the pipe wall.

Chapter 3 Basic equations of steady total flow

In Chapter 2 the basic principle of hydrostatics and its application were introduced. However, in many practical engineering problems, the fluid is usually in motion, while a static fluid is only a special state of fluid motion. Only through a thorough analysis of the fluid in motion, a general principle of the law of motion of the liquid can be obtained. Hydrodynamics is a kind of science that studies the basic principles of fluid in motion and their applications in engineering. It covers the kinematics and dynamics of fluid. Due to the various physical properties of a fluid and the complexity of fluid boundary conditions, the motion of a fluid is much complex than the motion of a solid.

In fluid mechanics, the various physical quantities (such as velocity, acceleration, dynamic pressure of fluid) that represent the state of the movement are known as the movement parameters. The basic task of hydrodynamics is to study the change of these movement parameters with time and space, and establish the relationships between these movement parameters, which then will be used to solve the practical problems in engineering projects.

When a fluid is in mechanical motion, it still follows some common principles in physics and mechanics such as the law of conservation in mass, energy and momentum. In the mechanical motion of a fluid, the applied equations are usually the continuity equation, energy equation and momentum equation. These three equations are the important theoretical basis for the study of a real fluid flow. It is necessary to understand the physical meaning of these equations, the conditions which they are based on, and the range of their application.

3.1 Two methods for describing motion of fluid

3.1.1 Lagrangian method and Eulerian method

3.1.1.1 Lagrangian method

Lagrangian method is also known as a particle system method. It expresses the form and way of fluid movement by focusing on the movement of interior particles. The entire flow process of fluid can be obtained through the description of the process of each particle motion. This means that the trajectory of each particle in a fluid should be identified and then through it, the entire motion of the fluid can be fully understood.

In order to study the trajectory of motion of every fluid particle, these particles have to be distinguished. If a liquid particle is at position $A(a, b, c)$ in the Cartesian coordinate system at time t_0, then this particle is marked by its coordinates, also called its initial coordinates. When the particle moves to position $B(x, y, z)$ at another time t, its coordinate is called the coordinate of motion. This coordinate can be expressed as the function of t and the initial coordinates of this particular particle:

$$\left. \begin{array}{l} x = x(a,b,c,t) \\ y = y(a,b,c,t) \\ z = z(a,b,c,t) \end{array} \right\} \quad (3-1)$$

where a, b, c and t are called the Lagrangian variables.

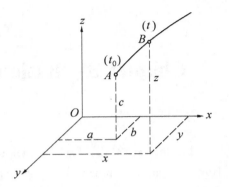

Fig. 3-1

If (a, b, c) is constant and t is variable, the position of a particle at any time can be obtained. If t is constant and (a, b, c) is variable, the space distribution of different particles at a certain time can be obtained. If both t and (a, b, c) are variable, then the trajectory of any fluid particle at any time can be obtained.

To obtain the velocity of a fluid particle at any time, Equation $(3-1)$ is taken by partial differentiation with respect to time t, and thus the components of velocity of the particle in x, y, z direction are

$$\left. \begin{array}{l} u_x = \dfrac{\partial x}{\partial t} = \dfrac{\partial x(a,b,c,t)}{\partial t} \\ u_y = \dfrac{\partial y}{\partial t} = \dfrac{\partial y(a,b,c,t)}{\partial t} \\ u_z = \dfrac{\partial z}{\partial t} = \dfrac{\partial z(a,b,c,t)}{\partial t} \end{array} \right\} \quad (3-2)$$

Taking the partial differentiation of Equation $(3-2)$ with respect to time gives the acceleration of the particle at x, y, z direction

$$\left. \begin{array}{l} a_x = \dfrac{\partial u_x}{\partial t} = \dfrac{\partial^2 x(a,b,c,t)}{\partial t^2} \\ a_y = \dfrac{\partial u_y}{\partial t} = \dfrac{\partial^2 y(a,b,c,t)}{\partial t^2} \\ a_z = \dfrac{\partial u_z}{\partial t} = \dfrac{\partial^2 z(a,b,c,t)}{\partial t^2} \end{array} \right\} \quad (3-3)$$

The physical meaning of the Lagrangian method is easy to understand but the method has difficulties in application in two aspects: to distinguish and track the particles technologically; to express complicated equations of the trajectory mathematically. Especially in turbulent motion, the flow is much difficult to describe by this method. In practical engineering, it is not necessary to know the motion of every particle of fluid. Therefore, the Lagrangian method is rarely used in fluid mechanics of engineering, but it is often used in some special situations such as in the study of jet flow and waves.

3.1.1.2 Eulerian method

Eulerian method is also known as a flow-field method. It is a method that describes the motion of fluid by focusing on the flow field. Eulerian method is not used to study the motion process of each particle, but to find the flow characteristics of particles passing through each fixed space point when the space is filled with fluid. The Eulerian method concentrates on observing and analyzing the motion situation of fluid particles through an arbitrary coordinate (x, y, z) at any time t. However, the Eulerian method cannot directly show where the particles come from and where they will go when reaching a point.

At each fixed point in a flow field, the changes of its movement elements with time are observed, which can describe the whole fluid motion. For example, in the space of fluid flow, the fluid particles at every fixed space point have a certain velocity of fluid at a time, and the combinations of the velocity of fluid at all the fixed space points constitute a velocity field of flow. If the velocity field of flow at every moment is known, then the entire motion state and process of fluid can be obtained. Similarly, in the space full of fluid flow, the pressure at every fixed space point at a certain time constitutes a pressure field, and the pressure field at different time can describe the change of pressure in the flow.

In the Cartesian coordinate system, each movement element is a function of coordinate (x, y, z) and t. The velocity field of flow can be described as
$$u = u(x, y, z, t)$$
The projection of velocity u on the coordinate axes can be expressed as
$$\left.\begin{array}{l} u_x = u_x(x,y,z,t) \\ u_y = u_y(x,y,z,t) \\ u_z = u_z(x,y,z,t) \end{array}\right\} \quad (3-4)$$
The pressure field can be expressed as
$$p = p(x,y,z,t) \quad (3-5)$$
where x, y and z are called the Euler variables.

If x, y and z in Equation (3-4) or (3-5) are constant and t is variable, then the changes of flow velocity and pressure at certain fixed points with time can be obtained. Otherwise, if t is constant and (x, y, z) is variable, then at the same instant the distributions of flow velocity and pressure with different space points in a flow field can be obtained. If t, x, y and z are all variables, then the flow velocity or pressure field at any time and any space point can be obtained.

For one-dimensional flow, the flow velocity or pressure is a function of the position (l) and time (t) if the coordinates are selected along the flow direction l:
$$u = u(l,t) \quad (3-6)$$
$$p = p(l,t) \quad (3-7)$$

In practical engineering, we usually want to know the motion of fluid at some particular position rather than the trajectory of fluid particles. Therefore, the Eulerian method is important in the study of fluid motion.

3.1.2 Acceleration of particle: local, convective and total acceleration

In this section, acceleration is described by Eulerian method. For a certain particle of liquid, the acceleration is defined as the ratio of change in the velocity of the liquid particle to the change in time, i.e. $\frac{du}{dt}$. The Eulerian method focuses on the motion elements at certain fixed space points. For a certain space point, the coordinates x, y and z are constant. Therefore, for the flow velocity u at a fixed point of space, its change rate with time is only the partial change rate of the flow velocity, $\frac{\partial u}{\partial t}$, not the total derivative of the flow velocity $\frac{du}{dt}$. In order to calculate the total derivative of the flow velocity with respect to time at a certain liquid particle, $\frac{du}{dt}$, the change of particle in motion should be considered through observing its trajectory over the time of dt. The coordinates of the trajectory of the moving liquid particles are the functions of time t,

$$x = x(t), \quad y = y(t), \quad z = z(t)$$

Therefore, the time t has two meanings: it is not only a variable among the four variables (x, y, z, t) of a multivariate function u, but also an independent variable of x, y and z. According to full derivative formula of composite function in the higher mathematics, acceleration a is the full derivative of velocity u with respective to time t. This can be expressed as follows:

$$a = \frac{du}{dt} = \frac{du(x,y,z,t)}{dt} = \frac{\partial u}{\partial t} + \frac{\partial u}{\partial x}\frac{dx}{dt} + \frac{\partial u}{\partial y}\frac{dy}{dt} + \frac{\partial u}{\partial z}\frac{dz}{dt}$$

Due to

$$\frac{dx}{dt} = u_x, \quad \frac{dy}{dt} = u_y, \quad \frac{dz}{dt} = u_z$$

thus

$$a = \frac{du}{dt} = \frac{\partial u}{\partial t} + u_x \frac{\partial u}{\partial x} + u_y \frac{\partial u}{\partial y} + u_z \frac{\partial u}{\partial z} \tag{3-8}$$

The components of the acceleration a of a particle in x, y and z direction are

$$\left.\begin{aligned} a_x &= \frac{du_x}{dt} = \frac{\partial u_x}{\partial t} + u_x \frac{\partial u_x}{\partial x} + u_y \frac{\partial u_x}{\partial y} + u_z \frac{\partial u_x}{\partial z} \\ a_y &= \frac{du_y}{dt} = \frac{\partial u_y}{\partial t} + u_x \frac{\partial u_y}{\partial x} + u_y \frac{\partial u_y}{\partial y} + u_z \frac{\partial u_y}{\partial z} \\ a_z &= \frac{du_z}{dt} = \frac{\partial u_z}{\partial t} + u_x \frac{\partial u_z}{\partial x} + u_y \frac{\partial u_z}{\partial y} + u_z \frac{\partial u_z}{\partial z} \end{aligned}\right\} \tag{3-9}$$

which can be written in the form of a tensor

$$a = \frac{du}{dt} = \frac{\partial u}{\partial t} + (u \cdot \nabla) u \tag{3-9'}$$

where the Hamilton operator (∇) is

$$\nabla = \frac{\partial}{\partial x} + \frac{\partial}{\partial y} + \frac{\partial}{\partial z}$$

For one-dimensional flow, Equation (3-9') can be written as

Chapter 3 Basic equations of steady total flow

$$a = \frac{du}{dt} = \frac{\partial u}{\partial t} + \frac{\partial u}{\partial l}\frac{dl}{dt}$$

Because $\frac{dl}{dt} = u$, then

$$a = \frac{du}{dt} = \frac{\partial u}{\partial t} + u\frac{\partial u}{\partial l} \qquad (3-10)$$

In Equations (3-9) and (3-10), $\frac{du}{dt}$ and $\frac{\partial u}{\partial t}$ are fundamentally different. For the full derivative $\frac{du}{dt}$, x, y and z are the functions of t. For the partial derivative $\frac{\partial u}{\partial t}$, x, y and z are taken as constant as they are coordinates at the fixed point in space. In terms of physical meaning, $\frac{\partial u}{\partial t}$ means the partial derivative of the liquid particle with time at a fixed point (x, y, z) and this is called local acceleration or position acceleration. Since the liquid is in motion, at different time the particles through a fixed point in space are not the same. Therefore, $\frac{\partial u}{\partial t}$ only expresses the rate of change of velocity with time at a fixed space point. It is not the change rate of velocity of the same liquid particle with time. $\frac{du}{dt}$ means the rate of change of velocity with time for the same liquid particle following a fixed trajectory at $x = x(t)$, $y = y(t)$ and $z = z(t)$. It is called total acceleration in fluid mechanics. The total acceleration includes not only the change of velocity of a fixed observed point with time but also the change of the velocity caused by the change of particle's position. The term $(u \cdot \nabla)u$ on the right side of Equation (3-9') means the change of velocity per unit time due to the change of the fluid particle's position. It is also called convective acceleration or spatial acceleration.

Two terms on the right hand of Equation (3-9') may not exist at the same time. If a physical quantity does not change with time, then $\frac{\partial u}{\partial t}$ is zero. If a physical quantity does not change with position, then $(u \cdot \nabla)u$ is zero.

Local acceleration plus convective acceleration is the total acceleration. In order to further understand the total acceleration, local acceleration and convective acceleration, a well-known example is taken to explain this in the following section. If the physical quantity that we study is temperature T instead of acceleration a, as temperature varies with season and region, let t as time and l as distance, the function expression of T is

$$T = T(t, l)$$

Similarly to Equation (3-9), we have

$$\frac{dT}{dt} = \frac{\partial T}{\partial t} + u\frac{\partial T}{\partial l}$$

If a traveller travels from Guangzhou to Beijing by train in an early summer, what is the change of temperature that he/she feels? The temperature change in one day can be expressed by $\frac{dT}{dt}$(℃/d), which consists of two parts: one is the $\frac{\partial T}{\partial t}$ that is the daily increasing rate of temperature (℃/d) in the early summer; the other is the temperature difference due to the change of location. If the velocity of the train is at 2000 km a day, i.e. $u = 2000$ km/d, the distance between Beijing and Guangzhou is 2000 km, and the temperature difference of two cities is 4℃, then $\frac{\partial T}{\partial l} = -4℃/2000$ km. When the traveller has arrived at Beijing, the temperature change he/she feels is

$$\frac{dT}{dt} = \frac{\partial T}{\partial t} + u\frac{\partial T}{\partial l}$$

$$= 1℃/d + 2000 \text{ km/d} \times \frac{-4℃}{2000 \text{ km}}$$

$$= -3℃/d$$

【Example 3-1】 In a flow field, the velocity increases uniformly with both time and distance. The distance between A and B is 2 m and C is at the middle, shown in Fig. 3-2. Given $t = 0$, $u_A = 1$ m/s, $u_B = 2$ m/s; $t = 5$ s, $u_A = 4$ m/s, $u_B = 8$ m/s. What is the expression of the acceleration at point C? And calculate the acceleration at C when $t = 0$ and $t = 5$ s.

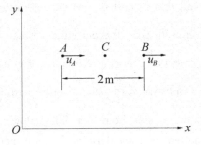

Fig. 3-2

Solution: In order to calculate the acceleration of flow at point C, Equation (3-10) is used as

$$a_C = \frac{\partial u}{\partial t}\bigg|_C + u_C \frac{\partial u}{\partial l}\bigg|_C$$

where the subscript C denotes the values of different physical quantities at point C. In this case, the finite difference of a variable is used to replace its partial differential, i.e. $\frac{\partial u}{\partial t}\bigg|_C$ replaced by $\frac{\Delta u}{\Delta t}\bigg|_C$, and $\frac{\partial u}{\partial l}\bigg|_C$ replaced by $\frac{\Delta u}{\Delta l}\bigg|_C$.

By interpolation or the trapezoidal rule,

when $t = 0$, $u_C\big|_{t=0} = 1.5$ m/s;

when $t = 5$ s, $u_C\big|_{t=5} = 6$ m/s.

Applying above numbers into Equation (3-10) gives

$$a_C\bigg|_{t=0} = \frac{\Delta u}{\Delta t}\bigg|_{C,t=0} + u_C\frac{\Delta u}{\Delta l}\bigg|_{C,t=0}$$

$$= \frac{6-1.5}{5} + 1.5 \times \frac{2-1}{2} = 0.9 + 0.75 = 1.65 \text{ m/s}^2$$

$$a_C\bigg|_{t=5} = \frac{\Delta u}{\Delta t}\bigg|_{C,t=5} + u_C\frac{\Delta u}{\Delta l}\bigg|_{C,t=5}$$

$$= \frac{6-1.5}{5} + 6 \times \frac{8-4}{2} = 0.9 + 12 = 12.9 \text{ m/s}^2$$

In this example, the velocity changes with both time and distance, so all the terms in Equation (3-10) need to be calculated.

【Example 3-2】 In a circular region in which the origin of the coordinate is at the center and the radius $r = 2$ m, the velocity field of fluid can be expressed as $u_x = x^2$ (m/s), $u_y = y^2$ (m/s), $u_z = z^2$ (m/s). Calculate the acceleration component and its modulus, as well as the acceleration at point $x = 1$ m, $y = 1$ m and $z = 1$ m.

Solution: This is a three-dimensional steady flow. There are the velocity components in x, y and z directions. Since it is steady flow, the local acceleration $\dfrac{\partial u}{\partial t} = 0$. From Equation (3-9),

$$a_x = \frac{du_x}{dt} = \frac{\partial u_x}{\partial t} + u_x \frac{\partial u_x}{\partial x} + u_y \frac{\partial u_x}{\partial y} + u_z \frac{\partial u_x}{\partial z}$$

$$= 0 + x^2 \frac{\partial x^2}{\partial x} + y^2 \frac{\partial y^2}{\partial y} + z^2 \frac{\partial z^2}{\partial z}$$

$$= 0 + 2x^3 + 0 + 0 = 2x^3$$

$$a_y = \frac{du_y}{dt} = \frac{\partial u_y}{\partial t} + u_x \frac{\partial u_y}{\partial x} + u_y \frac{\partial u_y}{\partial y} + u_z \frac{\partial u_y}{\partial z}$$

$$= 0 + x^2 \frac{\partial x^2}{\partial x} + y^2 \frac{\partial y^2}{\partial y} + z^2 \frac{\partial z^2}{\partial z}$$

$$= 0 + 0 + 2y^3 + 0 = 2y^3$$

$$a_z = \frac{du_z}{dt} = \frac{\partial u_z}{\partial t} + u_x \frac{\partial u_z}{\partial x} + u_y \frac{\partial u_z}{\partial y} + u_z \frac{\partial u_z}{\partial z}$$

$$= 0 + x^2 \frac{\partial x^2}{\partial x} + y^2 \frac{\partial y^2}{\partial y} + z^2 \frac{\partial z^2}{\partial z}$$

$$= 0 + 0 + 0 + 2z^3 = 2z^3$$

The modulus of the acceleration is

$$a = \sqrt{a_x^2 + a_y^2 + a_z^2} = \sqrt{4(x^6 + y^6 + z^6)}$$

$$= 2\sqrt{x^6 + y^6 + z^6}$$

$$a\big|_{x=1, y=1, z=1} = 2\sqrt{1^6 + 1^6 + 1^6} = 2\sqrt{3} = 3.464 \text{ m/s}^2$$

3.1.3 Some basic concepts of fluid movement

For the convenience of study and analysis on the motion of fluid, in fluid mechanics the movement of fluid is divided into different kinds of motion according to the property and characteristic of the flow. Some relevant concepts are required to help understand the principle of fluid movement.

3.1.3.1 Steady flow and unsteady flow

In the Eulerian method to describe the fluid movement, different movement elements are expressed as continuous functions of space coordinates and time.

If all the movement elements at any space point in a flow field do not change with time, then this kind of fluid motion is called steady flow. It means that if a fluid motion is steady flow, then at any given space point, all the movement elements will not change with time no matter what fluid particles pass. The movement elements are only the continuous functions of space coordinates and have nothing to do with time. For the velocity of the fluid:

$$\left.\begin{array}{l} u_x = u_x(x,y,z) \\ u_y = u_y(x,y,z) \\ u_z = u_z(x,y,z) \end{array}\right\} \tag{3-11}$$

Therefore, the partial derivative of all movement elements with time is zero:

$$\left.\begin{array}{l} \dfrac{\partial u_x}{\partial t} = \dfrac{\partial u_y}{\partial t} = \dfrac{\partial u_z}{\partial t} = 0 \\ \dfrac{\partial p}{\partial t} = 0 \\ \cdots\cdots \end{array}\right\} \tag{3-12}$$

If at least one movement element at any space point in a flow field changes with time, then this kind of fluid motion is called unsteady flow, i. e. $\dfrac{\partial u}{\partial t} \neq 0$ (or $\dfrac{\partial u_x}{\partial t} \neq 0$, $\dfrac{\partial u_y}{\partial t} \neq 0$, $\dfrac{\partial u_z}{\partial t} \neq 0$) or $\dfrac{\partial p}{\partial t} \neq 0$ ...

Some examples of unsteady flow are: flood flow in natural river, tide movement, the flow in a pipeline during the regulating of a valve, and the flow in the downstream discharge pipe when the water level in a reservoir is lowering down. Some examples of steady flow are: the pipe flow with a given supply flow rate from a constant head water tower, and the water flow over a spillway when a reservoir water level and the spillway gate opening height are unchanged. During the normal period and low-level-water period of a river, the change of flow velocity and discharge with time are small, and then such a flow in the river can be assumed to be approximately steady flow.

In steady flow, the analysis of fluid flow is simple, but in unsteady flow it is more complex than the steady flow. In engineering, an unsteady flow is usually analyzed by splitting it into several intervals (a period when the flow parameters change with time slowly), in which the flow is treated as approximately steady flow.

This chapter focuses on the analysis of steady flow.

3.1.3.2 Pathlines and streamlines

(1) Definition

In a flow field, a pathline traces the motion of a fluid particle, whilst a streamline is an instantaneous curve, which the velocity vectors of particles are tangent to.

The trace is a Lagrangian concept. Equation (3-1) is the trace parameter function of fluid particle. When t is eliminated, for given values of initial point (a, b, c), the trace of a liquid particle (a, b, c) can be expressed by x, y and z.

(2) Drawing a streamline

Assuming there is a point A_1 at time t_1 in a fluid field, the velocity vector of this point is u_1, and along the vector the point B_1 is made in a distance of Δl_1 from A_1 (see Fig. 3-3). At the same time, the velocity vector at point B_1 is u_2. Similarly, the point C_1 is made in a distance of Δl_2 from B_1 in the vector. If the velocity vector at point C_1 is u_3, then the point D_1 is made Δl_3 away from C_1 in the vector, and so on... By repeating this step, the polygonal line $A_1 B_1 C_1 D_1 \ldots$ will be generated. If the segment length Δl is taken to be close to zero, then the polygonal line will become a curved line, which is the streamline passing through Point A_1 at time t_1. In the same way, other streamlines can be drawn in the velocity field of flow, and then the series of streamlines can be drawn so a clear visualization of flow motion will be obtained. Therefore, the streamline is an important concept in analysing flow motion.

(3) Basic characteristics of a streamline

① If a flow is steady, the streamline does not change with time. When the flow is unsteady, the streamline will change with time.

② In a steady flow, the pathline coincides with the streamline. As shown in Fig. 3 - 3, assuming the curve $A_1 B_1 C_1 D_1 \ldots$ is a streamline, a particle starts moving from A_1 with the speed $\boldsymbol{u}_1$ at time t_1 and arrives at B_1 after Δt_1. Because the shape and position of a steady streamline does not change with time, at time $t_1 + \Delta t_1$, the flow velocity at B_1 is still the same as that at time t_1 and the direction of the particle is at the direction of $\boldsymbol{u}_1$. When the particle arrives at point B_1, the particle will move along the direction of $\boldsymbol{u}_2$. After Δt_2, the particle arrives at C_1, and it then moves along $\boldsymbol{u}_3$ and so on, so the trace along which the particle moves is the same as the streamline.

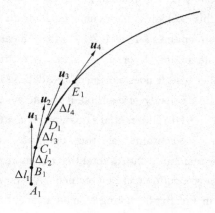

Fig. 3 - 3

In an unsteady flow, the pathline does not coincide with the streamline. The pathline only coincides with certain part (dl) of the streamline at different time. The pathline is an envelope line of the streamline.

As shown in Fig. 3 - 4, the streamline is drawn by the points. At time t_1, a particle moves from A_1 along the direction of $\boldsymbol{u}_1$, after Δt_1 the moving distance is $\Delta l_1 = \boldsymbol{u}_1 \Delta t_1$, and the particle arrives at Point B_1. Because the flow is unsteady, the streamline changes with time. At time $t_2 (= t_1 + \Delta t_1)$, the streamline is shown as the dashed line, and the velocity at Point

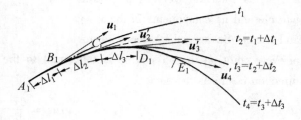

Fig. 3 - 4

B_1 becomes $\boldsymbol{u'}_2$. According to the definition of streamline, the particle must move along the direction of $\boldsymbol{u'}_2$ at time t_2. After Δt_2, the moving distance is $\Delta l_2 = \boldsymbol{u'}_1 \Delta t_2$ and the particle arrives at point C_1 at the time $t_3 (= t_2 + \Delta t_2)$, when the streamline is shown as the fine line, and the velocity at point C_1 becomes $\boldsymbol{u'}_3$. Thus the particle arriving at point C_1 moves along the direction of $\boldsymbol{u'}_3$ at time t_3, and after Δt_3, the particle arrives at D_1 after the distance Δl_3, so that the streamline has changed and the particle moves at a new direction, and so on. Therefore, at time t_1, the pathline coincides with the streamline over Δl_1; at time t_2, the coincided part is Δl_2; and at time t_3, the coincided part is $\Delta l_3 \ldots$ The pathline's trace is an envelope line of the thick line.

③ In a flow field, apart from the singular point (with infinite flow velocity), the stagnation point (with zero velocity) and the tangent points of streamlines, the streamlines generally do not intersect so that they are smooth curved lines without having turning points.

This characteristic of streamline can be proved by the method of apagoge as shown in Fig. 3 – 5: assuming the streamlines 1 and 2 intersect at point O, the corresponding flow velocity at point O is u_1 and u_2. The resultant velocity is $u = u_1 + u_2$, and this velocity u is not tangent to any of the streamlines (1 and 2), which means that either streamline 1 or 2 is actually a true streamline

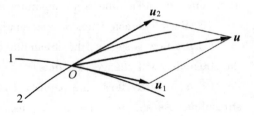

Fig. 3 – 5

because it does not meet the definition of the streamline, which is an imaginary line where the flow velocity of each point should be tangent to the line.

(4) Differential equations of a streamline

According to the definition of streamline, the differential equation of a streamline can be obtained. As shown in Fig. 3 – 6, taking a small segment dl (approximately as a straight line) along a streamline at Point M, the projections of dl in the x, y and z axis are dx, dy and dz, respectively. Let u and u_x, u_y, u_z as the flow velocity of liquid particle and its components in the x, y and z axis. Because the velocity vector

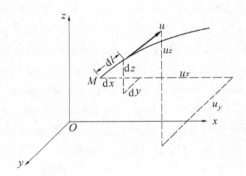

Fig. 3 – 6

of every point on the streamline is tangential to the streamline, i.e. u overlaps dl, and their direction cosine is the same:

$$\cos\alpha = \frac{dx}{dl} = \frac{u_x}{u}$$

$$\cos\beta = \frac{dy}{dl} = \frac{u_y}{u}$$

$$\cos\gamma = \frac{dz}{dl} = \frac{u_z}{u}$$

It can also be written as

$$\frac{dl}{u} = \frac{dx}{u_x}, \quad \frac{dl}{u} = \frac{dy}{u_y}, \quad \frac{dl}{u} = \frac{dz}{u_z}$$

That is

$$\frac{dx}{u_x} = \frac{dy}{u_y} = \frac{dz}{u_z} = \frac{dl}{u} \qquad (3-13)$$

Equation (3 – 13) is the differential equation of a streamline. The u_x, u_y and u_z is the function of x, y, z and t. The streamlines will be different at different times.

【Example 3 – 3】 There is a flat plate flow with the flow field: $u_x = 4y$, $u_y = -4x$. Find the form of the streamline.

Solution: From Equation (3 – 13), we have

$$u_y dx = u_x dy$$

Subsitituting the given values,

$$-4x dx = 4y dy$$

Then by integral:

$$\int y dy + \int x dx = 0$$

Thus we can obtain

$$x^2 + y^2 = C$$

It means that the streamlines are the concentric circles centered at the origin.

3.1.3.3 Streamtube, streamtube flow, total flow and cross-section, flow rate and cross-sectional velocity

(1) The streamtube is a small closed tube made up of numerous streamlines in a flow field, as shown in Fig. 3 – 7. Because the streamtube is surrounded by streamlines, the liquid cannot move into or out of the surface of the streamtube.

(2) The liquid flow in a streamtube is called a streamtube flow or flow beam, flow

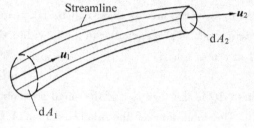

Fig. 3 – 7

filament. The limit of a streamtube is a streamline. In steady flow, the streamtube and the corresponding streamtube flow do not change with time. In unsteady flow, the streamtube and the corresponding streamtube flow will change with time.

The cross-section that is perpendicular to a streamtube flow is called the cross-section of flow and its area is represented by dA, which is very small. The movement variables at each point within the cross-section of a streamtube flow are considered to be uniform. That is, the flow velocity u at each point over the cross-section of a streamtube flow is the same, so is the pressure at each point.

The flow of fluid that is comprised of numerous small streamtubes is known as the total flow, such as pipe flow, which is commonly seen in engineering and daily life, and river flow. In the total flow, the cross-section that is perpendicular to the total flow is called the cross-section of total flow, denoted by C.S. In the case of parallel streamlines, the cross-section of flow is a flat plane; in a flow with non-parallel streamlines, the cross-section of flow is a curved surface. Fig. 3 – 8 shows the cross-sections of total flows with parallel, divergent and convergent streamlines.

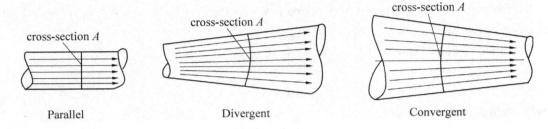

Fig. 3 - 8

In a unit time, the volume of a fluid passing through a cross-section of flow is called discharge (i. e. a volume flow rate), represented by the symbol Q. Its common unit is m³/s or L/s. In engineering, the concept of weight flow and mass flow is commonly used. Weight flow is the weight of fluid passing through a cross-section of flow per unit time, denoted by γQ, which has common unit in kN/h or in tons/hour(t/h). Mass flow is the mass of fluid passing through a cross-section of flow per unit time, denoted by ρQ, and its commonly used unit is kg/h.

In a total flow, considering a small streamtube with the cross-sectional area of dA, the flow velocity over the cross-section can be taken to be the same, which is assumed as u. Since the cross-section is perpendicular to the flow direction, the volume of fluid passing through the section dA at a unit time is

$$udA = dQ$$

where dQ is the flow rate of the small streamtube.

The total flow of the whole section of flow A is equal to the sum of innumerable small streamtube flows:

$$Q = \int_Q dQ = \int_A u dA \qquad (3-14)$$

The average velocity of the total flow over the cross-section is an imaginary flow velocity. If the velocity of each point in the cross-section is the same and equal to v, this will result in the same flow rate as the actual flow rate (usually not uniform). This flow velocity v is called the average velocity of the cross-section.

In fact, the flow velocity (u) of the every small streamtube is usually different from each other. Taking the pipe flow as an example, the velocity at different radii is different, depending on the size of the concentric circle. Let the velocity of every small circular streamtube as the same v, then

$$Q = \int_A u dA = \int_A v dA = v \int_A dA = vA$$

or
$$v = \frac{Q}{A} \qquad (3-15)$$

Equation (3 - 15) shows that the discharge of a total flow over a cross-section is equal to the product of the average velocity over the section and the area of the cross-section.

By introducing the concept of average flow velocity, a real three-dimensional flow can be reduced to a one-dimensional flow. In practical engineering, it is often required to know only the change of average flow velocity with space and time, not the precise velocity distribution of flow in a cross-section.

3.1.3.4 One-, two-, and three-dimensional flow

The flow variables at any point in a steady flow are related to only one independent variable of space (usually x coordinate as the flow direction), which is called one-dimensional flow. A streamtube flow is one-dimensional flow. The total flow with an average cross-sectional velocity v, which is used to represent the velocity distribution of a real flow, can be regarded as one-dimensional flow, one elemental flow or one-degree flow.

When flow variables are a function of the two space coordinates (x, y), the flow is called two-dimensional flow (plane flow, two-elemental flow, or two-degree flow). For example, in a shallow straight rectangular open channel flow, the velocity distribution changes with both distance (x axis) and depth (z axis), so the velocity is a function of x and z. Except on the vertical near the side wall of channel, the vertical velocity distribution (called velocity profile) along the width of channel (y axis) is the same. Therefore, this type of flow is plane flow, i.e. the vertical velocity distribution at any position away from the wall can represent the velocity distribution at any other position.

When flow variables are a function of the three space coordinates (x, y, z), the flow is called three-dimensional flow (space flow, three-elemental flow, or three-degree flow).

Strictly speaking, the flow of any real fluid is three-dimensional flow. The analysis of three-dimensional flow involves the three-dimensional space coordinate and the mathematical equations are very complex, in which there are also some difficulties to deal with mathematically. Only in some topics, such as the aeration cavitation of high-speed flow, turbulence fluctuations, and sediment transport, the three-dimensional method is required to study. For general engineering problems, their flows can be regarded as one-dimensional flow by introducing the concept of cross-sectional velocity. As a result, the error due to the simplification of the actual distribution of velocity can be corrected by a correction factor.

3.2 Continuity equation of steady total flow

Fluid movement must follow the universal law of mass conservation, and the continuity equation of steady total flow is a special form of mass conservation law in physics. The form of continuity equation obtained by the three-dimensional analysis method is different from the one derived by the one-dimensional analysis method. The derivation of the continuity equation for a steady, incompressible fluid flow is shown as follows.

In a steady flow, take a streamtube as shown in Fig. 3-9. There is no mass of fluid passing through the tube boundaries, but the fluid can only flow into or out of the two end cross-sections. The area of cross-section 1-1 is dA_1 while the area of cross-section 2-2 is dA_2. The flow velocities at the two ends are $\boldsymbol{u}_1$ and $\boldsymbol{u}_2$ respectively. The mass flow rate through the cross-section in a unit time is ρdQ, and the mass flow over a period time of dt is $\rho dQdt$. The mass of flow through the $1-1$ cross-section is $\rho_1 dQ_1 dt = \rho_1 u_1 dA_1 dt$. The mass of flow through the $2-2$ cross-section is $\rho_2 dQ_2 dt = \rho_2 u_2 dA_2 dt$. Because the fluid is incompressible, the density of ρ is constant, that is $\rho_1 = \rho_2 = \rho$. For a steady flow, the mass of flow in the streamtube does not change with time, i.e. the mass of fluid in the streamtube is neither increased nor decreased over a time period

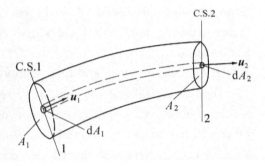

Fig. 3-9

of dt. Therefore, at the period dt, the mass of flow into the $1-1$ section should be equal to the mass of flow out of the $2-2$ section:

$$\rho u_1 dA_1 dt = \rho u_2 dA_2 dt$$

Eliminating the ρdt,

$$u_1 dA_1 = u_2 dA_2$$

where udA is the volume rate of flow dQ. Thus, this equation can be written as

$$dQ = u_1 dA_1 = u_2 dA_2 = \text{constant} \tag{3-16}$$

Equation (3-16) is the continuity equation for steady flow in a streamtube for an incompressible fluid.

The total flow is the sum of flow in countless streamtubes. Therefore, by integration of the streamtube flow over the whole cross-section, the total flow rate through the cross-section can be obtained:

$$Q = \int_Q dQ = \int_{A_1} u_1 dA_1 = \int_{A_2} u_2 dA_2$$

Using the average cross-sectional velocity v, then

$$Q = v_1 A_1 = v_2 A_2 \tag{3-17}$$

Equation (3-17) is the continuity equation of steady total flow for incompressible fluid, and it is one of the three basic equations of steady flow. Because it does not involve any force, it is a type of kinematic equation, also called the equation of motion.

Another form of Equation (3-17) is

$$\frac{v_1}{v_2} = \frac{A_2}{A_1} \tag{3-18}$$

Equation (3-18) shows how the average flow velocity of a cross-section changes with the cross-sectional area in a steady flow. The average velocity of cross-section is inversely proportional to the area of the cross-section. That is, when the flow rate is constant, the smaller the cross-sectional area, the greater the average flow velocity of the cross-section, and vice versa.

Equation (3-17) is suitable to the situation of single reach for a river without any branches between (i.e. there is no flow in or out between two ends of the river reach). Otherwise, for the case shown in Fig. 3-10a, the continuity equation should be

$$Q_2 = Q_1 + Q_3$$

For the case in Fig. 3 – 10b, the continuity equation is

$$Q_2 = Q_1 - Q_3$$

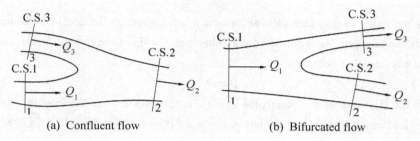

(a) Confluent flow (b) Bifurcated flow

Fig. 3 – 10

In practical engineering, the total discharge Q is commonly used, but for rectangular cross-section, the discharge per unit width is preferred to use, for example, in the calculation of anti-scour ability of spillway structure. The discharge per unit width is the volume rate of flow per unit width, denoted by the symbol q. The common unit of q is in $m^3/(s \cdot m)$ or $L/(s \cdot m)$ other than m^2/s. The calculation formula of discharge per unit width q is

$$q = \frac{Q}{b} = \frac{vA}{b} = \frac{vbh}{b} = vh \tag{3-19}$$

where b is the width of the rectangular cross-section and h is the depth of flow.

Although the continuity equation for steady total flow is simple, it is one of most important equations in solving the problems in fluid mechanics, especially for the problems of flow that require the average flow velocity of section for the given area of cross-section.

【Example 3 – 4】 A steel pipe is used to supply water with a flow rate of 0.032 m³/s. From the perspective of engineering economy, the average velocity of water flow is 1.1 m/s. Determine the diameter of the steel pipe. If another branch pipe with a diameter of 150 mm is connected to the existing water pipe, as shown in Fig. 3 – 11, half of the total flow is diverted in order to meet the water demand for other users. Under the same total flow, determine the average flow velocity in each branch.

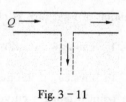

Fig. 3 – 11

Solution: According to the flow rate Q in Equation (3 – 17) and the relationship between the average velocity v and the area of the cross-section, the cross-sectional area of the pipe is

$$A = \frac{Q}{v} = \frac{0.032}{1.1} = 0.029 \text{ m}^2$$

Then the diameter of the pipe is

$$d = \sqrt{\frac{4A}{\pi}} = \sqrt{\frac{4 \times 0.029}{3.14}} = 0.192 \text{ m}$$

In general, an industrial pipe has a fixed size, which can be considered to be the inner diameter of the pipe. According to the specification of steel pipe, the steel pipe with a nominal

Fluid Mechanics in Civil Engineering

diameter of 200 mm now can be chosen. Therefore, the actual average flow velocity is

$$v_1 = \frac{Q}{A_1} = \frac{0.032}{\frac{3.14 \times 0.2^2}{4}} = 1.02 \text{ m/s}$$

where A_1 is the cross-section area of pipe, which is calculated by the nominal diameter.

When the water pipe is connected by another branch pipe to divert the water, the continuity equation in this situation becomes

$$Q = Q_1 + Q_2$$

where Q_1 is the flow rate of the main pipe and Q_2 is the flow rate of the branch pipe. Since the flow rate of the branch pipe is half of that in the main pipe and the diameter of the branch pipe is 150 mm, then the average flow velocity in the branch pipe is

$$v_2 = \frac{Q_2}{A_2} = \frac{0.032 \times 0.5}{\frac{3.14 \times 0.15^2}{4}} = 0.906 \text{ m/s}$$

where A_2 is the cross-sectional area of the branch pipe.

3.3 Energy equation of steady total flow

The energy equation of steady flow is a dynamic equation, which follows Newton's second law: $\sum F = ma$. This involves force, work done by the force, and mechanical energy. It follows the energy conversion and conservation law of fluid movement.

The energy equation of steady total flow can be derived in two steps: The first is to obtain the energy equation of a small streamtube, i.e. "break up the whole into parts", and then the energy equation of total flow is obtained by the integral method, i.e. "integrate parts into the whole".

3.3.1 Energy equation of steady streamtube flow of ideal fluid

In physics, if the mass of an object is m, under the action of force F, it is moved from point a to point b along a curve, and then the work done by F on the object is

$$W_{ab} = \int_a^b F \cos\alpha \, dl = \int_a^b F_\tau \, dl \qquad (3-20)$$

where F_τ is the tangent force along the curve.

According to Newton's second law

$$F_\tau = ma_\tau = m\frac{dv}{dt}$$

where a_τ and v are the tangent accelaration and tangent velocity along the curve, respectively.

Inserting it into Equation (3-20), then

$$W_{ab} = \int_a^b m\frac{dv}{dt} dl \qquad (3-21)$$

Since $\frac{dl}{dt} = v$ and the velocities at points a and b are v_a and v_b respectively, Equation (3-21) can

Chapter 3 Basic equations of steady total flow

be written as
$$W_{ab} = \int_{v_a}^{v_b} mv\,dv$$

By integrating, the following can be obtained:
$$W_{ab} = \frac{1}{2}mv_b^2 - \frac{1}{2}mv_a^2 \qquad (3-22)$$

where W_{ab} is the total work done by the external force, v_b is the final velocity of the object, and v_a is the initial velocity.

Equation (3-22) is the kinetic energy theorem: the increment of the kinetic energy of an object in a certain period of time is equal to the power of all the forces acting on the object.

The kinetic energy theorem, derived from Newton's second law, is also applicable to the flow of fluid. However, due to the special nature of the fluid flow, the kinetic energy equation has its special form.

As shown in Fig. 3-12, the flow sections of 2-2 and 1-1 in a steady streamtube are analyzed. The areas of the two end sections are dA_1 and dA_2 respectively, the flow velocities are u_1 and u_2, and the pressures are p_1 and p_2. Note that the u and p are the same over the cross-section. The centers of the two sections are z_1 and z_2 above the reference datum ($O-O$), respectively. After the dt period, the fluid at sections 1-1

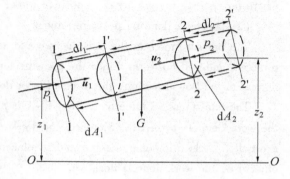

Fig. 3-12

and 2-2 has moved to sections 1'-1' and 2'-2', respectively. Next we will analyze the change of kinetic energy and the total work done by the external force.

3.3.1.1 Kinetic energy increment of flow segment

Consider the two flow segments: the initial segment $1-1 \sim 2-2$, and the final segment $1'-1' \sim 2'-2'$. Since it is a steady flow, the mass and velocity of the common fluid between $1'-1'$ and $2-2$ are the same, where there is no kinetic energy change. But there is a change of the kinetic energy between $2-2'$ and $1-1'$. Because the fluid is incompressible, the volumes of the flow in segments $1-1'$ and $2-2'$ are the same, which is dV. The flow process can simply be seen as the fluid element of dV is moved from position $1-1'$ to $2-2'$ (actually the flow as a whole moves forward). The weight of the fluid element is γdV and the mass (dm) is $\rho dV = \frac{\gamma}{g}dV$. Because the volume of dV is very small, the velocity of each point in $1-1'$ is approximately considered as u_1, and the velocity of each point in $2-2'$ is u_2, and then the momentum increment is

$$\frac{1}{2}dmu_2^2 - \frac{1}{2}dmu_1^2 = \frac{\gamma dV}{2g}(u_2^2 - u_1^2) \qquad (3-23)$$

83

3.3.1.2 Algebraic sum of work done by external force on the flow

The external forces acting on the flow segment $1-2$ include the body force: gravity, surface force: pressure force of fluid on the boundary, and the friction force of flow.

(1) The work done by the gravity W_1

In the time period of dt, when the fluid in the $1-2$ segment moves to the new position $1'-2'$, although the fluid particles between $1'-2$ move and change, the mass and position of the fluid have no change because of the condition that the fluid is uncompressible and steady. For the fluid between $1'-2$, there is no work done by the gravity. The gravity has just done work on the fluid from $1-1'$ to $2-2'$. Since the dV is small, and the center of gravity is considered to be at the centers of cross-sections $1-1$ and $2-2$, which have the height z_1 and z_2 above the reference datum $(O-O)$ respectively. From physics, the work done by gravity is

$$W_1 = G \cdot \Delta z = mg(z_1 - z_2) = \rho dV g(z_1 - z_2) = \gamma dV(z_1 - z_2) \quad (3-24)$$

where for $z_1 > z_2$, the gravity does positive work; for $z_1 < z_2$, the gravity does negative work.

(2) The work done by pressure force W_2

The pressure force of fluid on the two-end sections ($1-1$ and $2-2$) is parallel to the flow direction and will do work. Since the pressure force of fluid on the boundary of the streamtube is normal to the flow direction, the force will not do work.

The total pressure force on section $1-1$ is $p_1 dA_1$, and the moving distance $1-1'$ is dl_1, so the work done is $p_1 dA_1 dl_1$. Similarly the work done by the pressure force on section $2-2$ is $p_2 dA_2 dl_2$. If the direction of force and displacement are the same, the work done is positive; otherwise, the work done is negative. So

$$W_2 = p_1 dA_1 dl_1 - p_2 dA_2 dl_2$$
$$dA_1 dl_1 = dA_2 dl_2 = dV$$
$$W_2 = (p_1 - p_2) dV \quad (3-25)$$

(3) The work done by the friction force

Assuming the liquid is an ideal fluid (there is no viscosity), the external friction force is zero and the work done is zero. Applying Equations $(3-23)$, $(3-24)$ and $(3-25)$ to Equation $(3-22)$, then

$$\gamma dV(z_1 - z_2) + (p_1 - p_2) dV = \frac{\gamma dV}{2g}(u_2^2 - u_1^2)$$

Dividing by γdV and the resulting equation can then be written as

$$z_1 + \frac{p_1}{\gamma} + \frac{u_1^2}{2g} = z_2 + \frac{p_2}{\gamma} + \frac{u_2^2}{2g} \quad (3-26)$$

Equation $(3-26)$ is the energy equation of steady flow of incompressible ideal fluid, also called Bernoulli's equation. It was derived by the Swiss scientist Bernoulli in 1738 through the three-dimensional integral equation. Equation $(3-26)$ is derived from the streamtube, and a streamline is a special type of streamtube, so Equation $(3-26)$ is only suitable for a streamline. Two points on different streamlines cannot be used in Equation $(3-26)$. Meanwhile, it should be noted that in the derivation all the terms are divided by γdV, so Equation $(3-26)$ is the energy equation for a fluid body of unit weight, and the dimension of each term in the equation is [L] rather than [FL]. Therefore, the unit of each physical quantity is meter rather than the joule.

The first term z in the equation indicates the elevation of a certain point above the selected reference plane, which represents the position potential energy of the fluid per unit weight. The second term $\frac{p}{\gamma}$ is the pressure potential energy of the fluid per unit weight. The third term $\frac{u^2}{2g}$ represents the kinetic energy of the fluid per unit weight. This can be explained as follows: if the mass of a liquid particle is m and its flow velocity is u, the particle has the kinetic energy $\frac{1}{2}mu^2$. By dividing V, the kinetic energy per unit weight can be obtained:

$$\frac{\frac{1}{2}mu^2}{\gamma V} = \frac{\frac{1}{2}\rho V u^2}{\rho g V} = \frac{u^2}{2g} \tag{3-27}$$

where $m = \rho V$, $\gamma = \rho g$, and V is the volume of fluid. According to physics, the sum of kinetic energy and potential energy is mechanical energy. Therefore, Equation (3-26) indicates that for a steady flow of incompressible ideal fluid, in different cross-sections of the streamtube, the mechanical energy of fluid per unit weight remains the same. The components of mechanical energy such as position potential energy, pressure potential energy and kinetic energy can be converted to each other.

3.3.2 Energy equation of steady streamtube flow of real fluid

Equation (3-26) is the relationship between energy conservation and conversion of ideal fluid without viscosity for a steady streamtube flow. Because there is no viscosity, that is no relative motion, no friction and no energy consumption, the total mechanical energy in different sections maintains constant. But in a real fluid flow, the fluid has always viscosity and internal friction. The internal friction will consume the mechanical energy and will then be converted into heat energy and dissipated. Mechanical energy always decreases along the flow process, resulting in an energy loss. So in a real fluid,

$$z_1 + \frac{p_1}{\gamma} + \frac{u_1^2}{2g} > z_2 + \frac{p_2}{\gamma} + \frac{u_2^2}{2g}$$

Since the inherent law of flow friction is rather complex, which will be discussed in the next chapter, herein we introduce a general method of analysis: let the symbol h'_{w1-2} as the energy loss per unit weight of the fluid between cross-sections 1-1 and 2-2, then the energy equation should be written as

$$z_1 + \frac{p_1}{\gamma} + \frac{u_1^2}{2g} = z_2 + \frac{p_2}{\gamma} + \frac{u_2^2}{2g} + h'_{w1-2} \tag{3-28}$$

Equation (3-28) is the general energy equation for a steady flow of an incompressible real fluid.

This equation can be applied to the cases when both the velocity u and pressure p inside the fluid or on the boundary are required to be known.

3.3.3 Energy equation of steady total flow of real fluid

In practical engineering, the flow is total flow. The energy equation of streamtube flow can also be applied for the total flow. Thus the equation can be used to obtain the relationship among the average velocity v, the average pressure p and the elevation z of the center of a cross-section in a total flow. There are various types of total flows in practical engineering, but not all the flow movement has an integral solution, so only some special types of fluid movement do. Therefore, before the energy equation of a streamtube flow is integral to the total flow, we introduce some characteristics of total flow.

3.3.3.1 Uniform and non-uniform, gradually varied and rapidly varied flows

(1) Uniform and non-uniform flows

① Uniform flow (UF)

Uniform flow is a flow, in which the flow parameters do not change with the coordinate position. Mathematically, it means that $\frac{\partial u}{\partial l} = 0$, and $\frac{\partial p}{\partial l} = 0$... For a uniform flow, there is no convective acceleration, i.e. $u\frac{\partial u}{\partial l} = 0$ for one-dimensional flow, and the fluid is in uniform linear motion. The water flow in a pipe of the same diameter is a uniform flow.

Uniform flow has following characteristics:

a. The streamlines are straight lines parallel to each other.

b. The flow cross-section is a flat plane and its shape and size do not change along the flow.

c. The flow velocity u is the same at every point in a streamline. The velocity profile (i.e. the variation of flow velocity with the normal line to a streamline) at each cross-section is the same. Therefore, the average velocity v of the cross-section and flow depth h are not changed.

d. The distribution of dynamic pressure over a cross-section complies with the distribution of hydrostatic pressure. It means that the piezometric head $\left(z + \frac{p}{\gamma}\right)$ at any point in the same cross-section is constant.

In order to demonstrate this characteristic, take a small cylindrical element in a uniform flow, with its radial-axis n at an angle of α with the vertical line, as shown in Fig. 3 – 13. The heights of the two ends of the cylindrical element are z and $z + dz$ above the datum plane $O - O$, and their hydrodynamic pressures are p and $p + dp$ respectively. The projection of the hydrodynamic force acting on the two ends of the cylindrical element on the n axis is pdA and $(p + dp)dA$, where dA is the surface area. The projection of the weight of the element along the n direction is

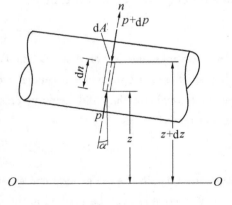

Fig. 3 – 13

$dG\cos\alpha = \gamma dAdn\cos\alpha$, where dn is the height of the element. Because $\cos\alpha = \dfrac{dz}{dn}$, $dG\cos\alpha = \gamma dAdz$. The hydrodynamic force on the side of the cylindrical element and the internal friction force of flow are orthogonal to the n axis, so their projections along the n direction are zero.

In the uniform flow, there is no acceleration along the flow direction, so the algebraic sum of the components of all the forces in the n direction is zero, i. e. forces in the equilibrium: $\sum F = 0$.

or
$$pdA - (p + dp)dA - \gamma dAdz = 0$$
where the force along the positive direction of n axis is positive; otherwise it is negative. Then,
$$d\left(\frac{p}{\gamma}\right) + dz = 0$$

By integration, it becomes
$$z + \frac{p}{\gamma} = \text{const} \tag{3-29}$$

Equation (3-29) indicates that the distribution of hydrodynamic pressure over a cross-section in a uniform flow satisfies the distribution of hydrostatic pressure. Therefore, the hydrodynamic pressure of any point within a cross-section or the total hydrodynamic force can be calculated by the equations and methods of hydrostatic pressure and total hydrostatic force.

The fourth characteristic of the uniform flow can be expressed by the geometric graph, as shown in Fig. 3-14. In the uniform flow, if two piezometer tubes are connected to any two sections 1-1 and 2-2 (note that the connection of tube must be normal to the pipe), the fluid level in the piezometer tube at any point within the cross-section must maintain the same, that is $z + \dfrac{p}{\gamma} = c$. However, the fluid levels in piezometer tubes on different sections are different. For example, at

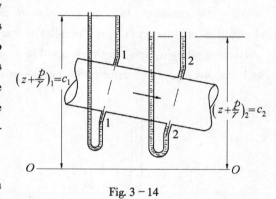

Fig. 3-14

section 1-1, $z_1 + \dfrac{p_1}{\gamma} = c_1$, at section 2-2, $z_2 + \dfrac{p_2}{\gamma} = c_2$, and $c_1 > c_2$.

② Non-uniform flow

Non-uniformflow is a flow, in which the flow parameters of the flow change with position, i. e. mathematically $\dfrac{\partial u}{\partial l} \neq 0$, $\dfrac{\partial p}{\partial l} \neq 0$...

(2) Gradually varied flow and rapidly varied flows

Based on the curvature of streamline, non-uniform flow can be further classified into gradually varied flow and rapidly varied flow.

① Gradually varied flow (GVF)

The gradually varied flow has the following characteristics:

a. The streamlines are almost parallel, when the angle between the streamlines is very small.

b. The streamlines are almost straight, when the curvature radii of streamlines are very large, i. e. their curvatures are very small.

Therefore, the streamlines of gradually varied flow are almost parallel lines. Uniform flow is a special case of gradually varied flow. Because the streamlines are almost parallel, the cross-section of the gradually varied flow is a plane. The dynamic pressure distribution in a cross-section is a linear distribution of hydrostatic pressure, that is $z + \dfrac{p}{\gamma} = $ const. But in the flows without any solid boundary, such as water jet, for which the water are fully in contact with the air and the pressure on a cross-section is the atmospheric pressure approximately, the dynamic pressure over the cross-section does not comply with the hydrostatic pressure distribution.

As to what degree the angle of the streamline or the size of curvature radius, a flow can be considered as gradually varied flow, there is no quantitative criterion. This depends on the accuracy requirement of a real project.

② Rapidly varied flow (RVF)

Rapidly varied flow is a flow that has very small radius of curvature of streamline or large angles between streamlines. Fig. 3 - 15 shows three types of rapidly varied flow. Case (a): the streamlines are parallel curved lines, for example, the pipe flow in a bend where the average flow velocity v is constant, but the direction of flow is changing along the pipe. Case (b): the streamlines are straight but not parallel to each other, for example, the flow in a diverging or converging straight pipe, in which the flow cross-section is no longer a plane but a curved surface and the average flow velocity of cross-section is changing along the pipe. Case (c): the

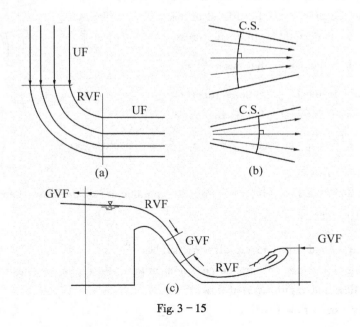

Fig. 3 - 15

streamlines are neither parallel nor straight lines; for example, the flow in a spillway of dam, in which the flow section is curved and its average velocity changes with distance.

In rapidly varied flows, the hydrodynamic pressure distribution in the cross-section does not follow the hydrostatic pressure distribution, so it means that the piezometric head $(z + \dfrac{p}{\gamma})$ on the same section is not a constant.

a. Curved streamline in the vertical direction

The main feature of rapidly varied flow is that the streamline of flow has large curvature, so the fluid particle in curvilinear motion is affected by both gravity and the centrifugal force of inertia. Fig. 3 – 16a shows a rapidly varied flow over a convex surface. For the simplicity of analysis, the streamlines are assumed as a cluster of concentric circular arc that are parallel to each other. When fluid particles move along the convex curves (pathlines and streamlines are the same for steady flow), the acceleration of fluid particle normal to the streamline is pointing to the center of curvature, with the centrifugal inertia force pointing outward normal from the center, which is opposite to the direction of gravity. Since the additional pressure generated by the centrifugal inertia force is negative, the hydrodynamic pressure force per unit weight is $p = \rho\left(g - \dfrac{u^2}{r}\right)h$, which is the resultant force of the gravity and centrifugal inertia force. This means that the hydrodynamic pressure force in a cross-section is smaller than the corresponding hydrostatic force at the fluid depth h. In the above expression, g is the gravitational acceleration caused by gravity, $\dfrac{u^2}{r}$ is the normal acceleration of a fluid particle in circular motion, u is the velocity of the particle, and r is the curvature radius of the circular arc. Fig. 3 – 16a shows the vertical distribution of both hydrodynamic pressure (thick solid line) and hydrostatic pressure (thin line) in the curved flow over a convex surface. If a fluid particle moves along an irregular curve, it is difficult to establish a general formula of hydrodynamic pressure, which can usually be obtained by measurement along the depth.

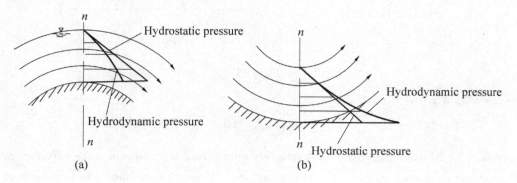

Fig. 3 – 16

If a fluid particle is moving along a concave streamline as shown in Fig. 3 – 16b, the direction of the normal acceleration of the particle is pointing upward away from the curvature center, the direction of the centrifugal inertia force is downward, like that of the gravity, i. e. the additional pressure generated by the centrifugal inertia force is the same as the direction of gravity, so the hydrodynamic force per unit weight is $p = \rho \left(g + \dfrac{u^2}{r} \right) h$. In this case, the hydrodynamic pressure in the section is larger than the corresponding hydrostatic pressure at the same depth h. The vertical distribution of hydrodynamic pressure (thick line) of flow on a concave boundary is shown in Fig. 3 – 16b.

b. Rapidly varied flow in a bend on horizontal plane

Flows in river bends or horizontally placed pipe bends, are horizontal rapidly varied flow in bend.

As shown in Fig. 3 – 17a, take a small circular element along the curved streamline with a radius of r, which moves with the speed v. The width of the element in the radial direction is dr. The areas of the inside and outside surfaces of the element are both dA. Over the increment dr, the hydrodynamic pressure on the outer surface dA is $p + dp$, and the mass of the element is $m = \rho dAdr$. Normal acceleration is $\dfrac{v^2}{r}$. Note that if the direction of the force is the same as the direction of the normal acceleration, the force is positive; otherwise it is negative. According to Newton's second law in the radial direction, we have

$$(p + dp)dA - pdA = \rho dAdr \dfrac{v^2}{r}$$

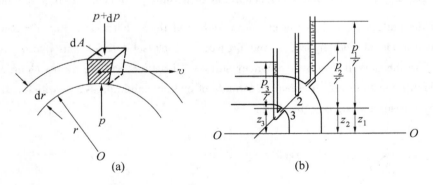

Fig. 3 – 17

Re-arranging it gives

$$dp = \rho \dfrac{v^2}{r} dr \qquad (3-30)$$

Equation (3 – 30) indicates that: dp is inversely proportional to r, and when the curvature radius of streamline r becomes infinite (the streamline tends to be straight line) the increment of the pressure is close to zero. Uniform flow and gradually varied flow are such types of this flow. Because ρ, v^2 and r cannot be negative, dp and dr have the same sign. So the hydrodynamic pressure on the outer streamline is larger than that on the inner streamline. Therefore in a river bend, the hydrodynamic pressure on the inner bank is larger than that on the outer bank; in a horizontally placed pipe bend, although the elevation of the two lateral walls are the same, the hydrodynamic pressure is not equal. The piezometric head $(z + \frac{p}{\gamma})$ at the outside of pipe is greater than that at the inside of pipe, as shown in Fig. 3-17b.

Based on the above analysis, we can conclude that the hydrodynamic pressure distribution over the section of a gradually varied flow follows the hydrostatic pressure distribution, which is $z + \frac{p}{\gamma}$ = constant, but this is not true for rapidly varied flow on either vertical or horizontal plane.

3.3.3.2 Energy equation of steady total flow of real fluid

(1) Derivation of energy equation

For a steady streamtube flow of incompressible real fluid, the energy equation, Equation (3-28), is written as

$$z_1 + \frac{p_1}{\gamma} + \frac{u_1^2}{2g} = z_2 + \frac{p_2}{\gamma} + \frac{u_2^2}{2g} + h'_{w1-2}$$

The total flow is comprised of innumerable streamtube flows. By integrating Equation (3-28) over the whole cross-section of total flow, this equation can be extended to be the energy equation of the total flow. If the flow rate of a streamtube flow is dQ, the corresponding weight flow rate is γdQ. Each term of Equation (3-28) is multiplied by γdQ and the result is integrated over the two end cross-sections at A_1 and A_2:

$$\int_Q \left(z_1 + \frac{p_1}{\gamma}\right)\gamma dQ + \int_Q \frac{u_1^2}{2g}\gamma dQ = \int_Q \left(z_2 + \frac{p_2}{\gamma}\right)\gamma dQ + \int_Q \frac{u_2^2}{2g}\gamma dQ + \int_Q h'_{w1-2}\gamma dQ \quad (3-31)$$

Equation (3-31) includes three kinds of integral:

① The first kind of integral: $\int_Q \left(z_1 + \frac{p_1}{\gamma}\right)\gamma dQ$

If the flow is gradually varied flow, the $z + \frac{p}{\gamma}$ is a constant and then it can be removed out of the sign of the integral.

$$\int_Q \left(z + \frac{p}{\gamma}\right)\gamma dQ = \left(z + \frac{p}{\gamma}\right)\gamma \int_Q dQ = \left(z + \frac{p}{\gamma}\right)\gamma Q \quad (3-32)$$

② The second kind of integral: $\int_Q \frac{u^2}{2g}\gamma dQ$

Because
$$dQ = udA$$

then
$$\int_Q \frac{u^2}{2g}\gamma dQ = \frac{\gamma}{2g}\int_A u^3 dA$$

For a total flow, the average flow velocity v is used to replace the point velocity u of streamline, so this is a one-dimensional flow. The relationship between the point flow velocity u and the average cross-sectional velocity v can be expressed as

$$u = v \pm \Delta u$$

where Δu is the difference between the point flow velocity u and the average cross-sectional flow velocity v.

Therefore, the second kind of integral can be written as

$$\frac{\gamma}{2g}\int_A (v \pm \Delta u)^3 dA = \frac{\gamma}{2g}\int_A (v^3 \pm 3v^2\Delta u + 3v\Delta u^2 \pm \Delta u^3) dA$$

$$= \frac{\gamma}{2g}\left[\int_A v^3 dA + 3v^2\int_A (\pm \Delta u) dA + 3v\int_A \Delta u^2 dA + \int_A (\pm \Delta u^3) dA\right] \quad (3-33)$$

Ignoring the high-order term of $\int_A (\pm \Delta u^3) dA$, according to the following equation:

$$Q = \int_A u dA = \int_A (v \pm \Delta u) dA = \int_A v dA + \int_A (\pm \Delta u) dA$$

$$= vA + \int_A (\pm \Delta u) dA = Q + \int_A (\pm \Delta u) dA$$

Then it follows

$$\int_A (\pm \Delta u) dA = 0$$

This means that the second and fourth item on the right side of Equation (3-33) can be removed. Thus,

$$\frac{\gamma}{2g}\int_A u^3 dA = \frac{\gamma}{2g}\left(v^3 A + 3v\int_A \Delta u^2 dA\right)$$

$$= \frac{\gamma}{2g}v^3 A\left(1 + 3\frac{\int_A \Delta u^2 dA}{v^2 A}\right)$$

Let

$$\eta = \frac{\int_A \Delta u^2 dA}{v^2 A}$$

Because both the numerator and the denominator are positive, η is always larger than zero. Then let $\alpha = 1 + 3\eta$ and α will always be larger than 1. Thus, we have

$$\int_A \frac{u^2}{2g}\gamma dQ = \alpha \frac{\gamma}{2g}v^3 A = \alpha \frac{v^2}{2g}\gamma Q \quad (3-33')$$

From Equation (3-33'), we can have

$$\alpha = \frac{\int_Q u^2 dQ}{v^2 Q} = \frac{\int_A u^3 dA}{v^3 A} \quad (3-34)$$

where α is called kinetic energy correction factor. It shows that the kinetic energy of a cross-section in a non-uniform flow can be obtained by the kinetic energy calculated by the average cross-sectional velocity (assumed uniform velocity in the cross-section) with a correct factor (α). In a real flow, the less the uniformity of flow velocity distribution, the larger the kinetic energy factor α is; the more the uniformity of flow velocity distribution, the closer the α tends to 1. For most gradually varied flows, $\alpha = 1.05 \sim 1.10$. Therefore, except where the flow velocity distribution is extremely non-uniform, Equation (3-34) can be used to calculate the kinetic energy correction factor α. For convenience, $\alpha = 1.0$ is often used, such as in the problems of this chapter.

③ The third kind of integral:

$$\int_Q h'_{w_{1-2}} \gamma \mathrm{d}Q$$

This kind of integral is different from the first and second kind of integral. It is not the value of integral over a cross-section but the value of integral along the flow path between two sections. $h'_{w_{1-2}}$ is the mechanical energy loss per unit weight due to the friction of fluid between cross-sections $1-1$ and $2-2$. Since different streamtube flows in the total flow have different flow paths, the $h'_{w_{1-2}}$ may not be the same. The expression of $\int_Q h'_{w_{1-2}} \gamma \mathrm{d}Q$ represents the total mechanical energy loss for the weight rate $\gamma \mathrm{d}Q$ of each streamtube flow moving along its own flow path to resist the friction. In other words, it is the mechanical energy loss when the total flow moves from cross-section $1-1$ to $2-2$. If $h_{w_{1-2}}$ is used to represent the average energy loss per unit weight of all the streamtube flows in a total flow, i.e. the energy loss per unit weight of each streamtube flow is the same, $h_{w_{1-2}}$ is constant. It can be removed out of the integral sign, i.e.

$$\int_Q h'_{w_{1-2}} \gamma \mathrm{d}Q = \gamma \int_Q h_{w_{1-2}} \mathrm{d}Q = h_{w_{1-2}} \gamma Q \qquad (3-35)$$

Applying Equations $(3-32)$, $(3-33')$ and $(3-35)$ to Equation $(3-31)$, then

$$\gamma Q \left(z_1 + \frac{p_1}{\gamma} \right) + \gamma Q \frac{\alpha_1 v_1^2}{2g} = \gamma Q \left(z_2 + \frac{p_2}{\gamma} \right) + \gamma Q \frac{\alpha_2 v_2^2}{2g} + \gamma Q h_{w_{1-2}} \qquad (3-36)$$

where subscripts 1 and 2 represent the values of physical quantities at cross-sections $1-1$ and $2-2$, respectively.

Equation $(3-36)$ is the energy equation of total flow in the weight rate of flow (γQ). In practical engineering, the energy per unit weight is commonly used. Therefore, Equation $(3-36)$ can be rewritten, divided by γQ, as

$$z_1 + \frac{p_1}{\gamma} + \frac{\alpha_1 v_1^2}{2g} = z_2 + \frac{p_2}{\gamma} + \frac{\alpha_2 v_2^2}{2g} + h_{w_{1-2}} \qquad (3-37)$$

Equation $(3-37)$ is the energy equation for a steady flow of incompressible real fluid, also called Bernoulli's equation. This equation is the basic and most important equation in fluid mechanics. It describes the relationship among the flow parameters z, p and v. Together with continuity equation, it can be used to solve many problems in hydraulics.

(2) Physical meanings of the energy equation

① Physical meaning of each term in the energy equation of steady real fluid flow

The dimension of each term in the energy equation is $[L]$. In hydraulics, "Head" is commonly used to represent the height. Therefore, z denotes the average position potential energy per unit weight of total flow on the cross-section, usually expressed by the elevation of the section center point or the surface position, also called the elevation head or "position head". $\frac{p}{\gamma}$ means the average pressure potential energy per unit weight of total flow over the cross-section. It represents the average pressure height at each point on the cross-section, called the pressure head. $z + \dfrac{p}{\gamma}$ represents the potential energy per unit weight, called the piezometric head. $\dfrac{\alpha v^2}{2g}$ means the average kinetic energy per unit weight of total flow over the cross-section, called the dynamic head or "velocity head". The total of the three heads described above is denoted by H, i. e. $H = z + \dfrac{p}{\gamma} + \dfrac{\alpha v^2}{2g}$, which represents the total mechanical energy per unit weight, called the total head. $h_{w_{1-2}}$ represents the average energy loss per unit weight of flow due to resisting the friction between two sections. This is called the head loss. The head loss is part of the effective mechanical energy, which is converted into heat energy and then dissipated. It cannot be converted into effective mechanical energy.

② Energy equation is actually a form of the conservation and conversion of energy. It expresses the relationship between the mechanical energy and the energy loss in a flow. In the flow, part of the mechanical energy is converted to the head loss for overcoming the resistance of flow. Potential energy and kinetic energy can be converted to each other. The potential energy will increase as the kinetic energy decreases; in other words, the kinetic energy will increase when the potential energy decreases. However, such energy conversion is not equal but equivalent to the head loss h_w. If the kinetic energy is constant, for example, in a pipe flow of the same diameter, the potential energy and the pressure energy can be converted into each other: the potential energy will increase with the decreasing pressure energy, or the pressure energy will increase as the potential energy decreases; however, the amount in the conversion is not equal but equivalent to the head loss. In high-rise buildings, the potential energy of the pipe flow is large in the higher floors, where the pressure may become smaller, which could result in no water being available because there are large head losses of flow in the supply pipe. Thus, in high buildings other water supply facilities have to be installed, such as the installation of pressure pumps, a water tank on the roof, or other means.

(3) Geometric diagram of energy equation—the total head line and the piezometric head line

In the energy equation, z, $\dfrac{p}{\gamma}$, $\dfrac{\alpha v^2}{2g}$ and h_w have the dimension of length. In a certain scale, they can be plotted in a diagram to show their variations with distance.

Fig. 3 – 18 shows a pipe flow with the following sections in order: uniform, divergent, uniform, convergent and uniform section. First select a reference datum, then determine the initial and the final end of each pipe section as its representative cross-section, and connect

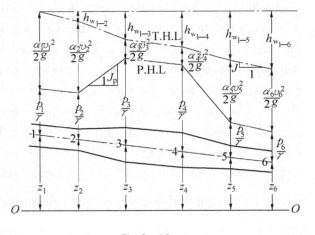

Fig. 3 – 18

the centroids of the selected sections, which will form the so-called central line of flow. The central line of flow represents the elevation z of the centroid of cross-section.

From the centroid of the cross-section, plus the vertical height that is equivalent to the corresponding average cross-sectional pressure head $\dfrac{p}{\gamma}$, we can get the piezometric head $z + \dfrac{p}{\gamma}$. The fluid levels in all the piezometers along the pipe are drawn together to form a line, called the piezometric head line abbreviated as P. H. L. The vertical distance between the piezometric head line and the flow centerline reflects the change of average pressure head with the flow. If the head line is above the centerline, then the pressure is positive, i.e. $p > p_{at}$. If the piezometric head line is below the flow centerline, then the pressure is negative, i.e. $p < p_{at}$, in which the pressure difference is a degree of vacuum. The piezometric head line can be increased or decreased along the flow, or even remain unchanged in certain conditions. For example, the piezometric head line in the divergent section is rising along the pipe, but the piezometric head line in the convergent section is declining along the pipe. In the open channel flow, the pressure on the free surface is the atmospheric pressure, which is $\dfrac{p}{\gamma} = 0$, so the water surface line is the piezometric head line.

When the vertical height of the velocity head $\dfrac{\alpha v^2}{2g}$ is added to the piezometric head, the total head of the cross-section $\left(H = z + \dfrac{p}{\gamma} + \dfrac{\alpha v^2}{2g}\right)$ is obtained. All the total head along the pipe is drawn to form a line, which is called the total head line or energy line, as abbreviated by T. H. L. Since the head loss always exists in a real fluid flow, the total head line always decreases along the flow (unless there is a source of external energy). The decreased height of the total head line between two cross-sections is the head loss per unit weight h_w.

The slope of the total head line relative to the pipe axis is called a hydraulic gradient or energy slope, which is represented by J. It indicates the head loss per unit distance along the flow direction. If the total head line is a straight line, then

$$J = \frac{h_w}{l} = -\frac{H_2 - H_1}{l} \tag{3-38}$$

where l is the flow length and h_w is the head loss of flow over the length.

If the total head line is a curve, then the hydraulic gradient is not a constant. For a water flow in a section, the hydraulic gradient is

$$J = \frac{dh_w}{dl} = -\frac{dH}{dl} \tag{3-39}$$

where the minus sign is used because the total mechanical energy H_2 downstream is always smaller than the total energy H_1 upstream in a case without external energy, i.e. $dH = H_2 - H_1 < 0$. However, in hydraulics, it is assumed that the hydraulic slope, which is declining downstream, is positive. In order to keep the same sign in both sides of Equations (3-38) and (3-39), a minus sign is added on the right hand in the equations.

The change of piezometric head (potential energy) per unit distance is called the slope or gradient of piezometric head line, denoted by J_P.

$$J_P = -\frac{d(z + \frac{p}{\gamma})}{dl} \qquad (3-40)$$

The piezometric head slope is positive if it declines along the flow direction, which is assumed in the same way as the hydraulic slope. Note that, J_P can be positive, negative or zero, but J is always positive.

The vertical distance between the piezometric head line and the total head line is the velocity head $\frac{\alpha v^2}{2g}$. In uniform flow, the average flow velocity does not change along the flow, so the piezometric head line is parallel to the total head line.

The diagram of the energy line can clearly illustrate how the mechanical energy per unit weight of flow changes along the flow. In the design of water supply pipeline over a long distance, the diagram of energy line can help to find out whether the pipeline layout is appropriate, and whether there is any section of the pipe in which the pressure exceeds the allowable negative pressure. If such a section exists, the design has to be modified. It also helps to check whether the flow velocity in the pipe is within the allowable range, in order to avoid the phenomenon of aeration, cavitation and so on.

(4) Energy equation with energy input and output

Equation (3 – 37) is the energy equation of flow between two cross-sections, from which there is no energy input or output. If a pump is installed in a water supply pipeline system, its mechanical power does the work and inputs energy to the flow. In a hydropower station, a turbine is installed in the pressure pipeline system and the water flow pushes the turbine blades to provide output energy. For the above cases, the energy equation should be rewritten as follows:

$$z_1 + \frac{p_1}{\gamma} + \frac{\alpha_1 v_1^2}{2g} \pm H_m = z_2 + \frac{p_2}{\gamma} + \frac{\alpha_2 v_2^2}{2g} + h_{w_{1-2}} \qquad (3-41)$$

where H_m is the gained or lost mechanical energy per unit weight from an external device in the flow system. If there is an energy input to the flow system, take $+ H_m$; if there is an energy extracted from the flow system, take $- H_m$. H_m can be calculated by the following equation:

$$H_m = \frac{\eta N_m}{\gamma Q} \qquad (3-42)$$

where η is the total efficiency of mechanical and power equipment; N_m is the power of the mechanical equipment, and its common unit is watts (W). Note that $1W = 1N \cdot m/s$.

(5) Application of the energy equation of steady flow of a real fluid

The conservation and conversion of energy is a universal law of nature, but the energy equation (3 – 37) is derived under certain assumptions. Therefore, there are a number of limitations in the application of Equation (3 – 37).

① Application conditions of the energy equation

a. The fluid is incompressible and must be steady flow.

b. The body force acting on the fluid is limited to gravity, i.e. there is no energy input or output between two cross-sections along the flow.

c. The flow in the two cross-sections must be gradually varied flow, so that it will satisfy the condition that the piezometric head on the cross-section is constant. However, the flow between the two cross-sections is not necessarily required to be gradually varied flow; in other words, it can be rapidly varied flow. If the flows over the cross-sections do not meet gradually varied flow conditions, in the application of energy equation specific boundary conditions are needed. For example, we need to know the average potential energy of the cross-section of non-gradually varied flow or the distribution of dynamic pressure in the section, and so on.

d. In the derivation of Equation (3 – 37), it is assumed that the flow rate does not change along the flow. For the flow that is diverted or confluent, the actual flow rate should be taken into consideration. For example, for the confluent flow in Fig. 3 – 10a, the equation should be written as

$$\gamma Q_1 \left(z_1 + \frac{p_1}{\gamma} + \frac{\alpha_1 v_1^2}{2g} \right) + \gamma Q_3 \left(z_3 + \frac{p_3}{\gamma} + \frac{\alpha_3 v_3^2}{2g} \right)$$
$$= \gamma Q_2 \left(z_2 + \frac{p_2}{\gamma} + \frac{\alpha_2 v_2^2}{2g} \right) + \gamma Q_1 h_{w1-2} + \gamma Q_3 h_{w2-3}$$

where Q_1 and Q_3 are the flow rates of two branches before the confluent junction respectively, Q_2 is the total flow rate after the junction, and $Q_2 = Q_1 + Q_3$ (continuity equation).

For the divergent case as shown in Fig. 3 – 10b, the energy equations can be written as

$$\gamma Q_1 \left(z_1 + \frac{p_1}{\gamma} + \frac{\alpha_1 v_1^2}{2g} \right) = \gamma Q_2 \left(z_2 + \frac{p_2}{\gamma} + \frac{\alpha_2 v_2^2}{2g} \right) + \gamma Q_3 \left(z_3 + \frac{p_3}{\gamma} + \frac{\alpha_3 v_3^2}{2g} \right) + \gamma Q_2 h_{w1-2} + \gamma Q_3 h_{w1-3}$$

where Q_1 is the flow before the junction, Q_2 and Q_3 are the flows after the junction, and $Q_1 = Q_2 + Q_3$.

In the application of the energy equation, the key point is the calculation of head loss, h_w, which is dependent on the physical characteristics of flow.

Among the above four conditions, (a) and (c) are the most important conditions. In short, the flow at the end section must be steady gradually varied flow.

② Application tips of the energy equation.

Equation (3 – 37) has a wide range of application in fluid mechanics. In order to use this equation correctly and intelligently, there are also some tips that can be known as "three selections and comprehensiveness". The so-called "three selections" means correct selection of the following three aspects: reference plane (datum), two end cross-sections with gradually varied flow, and calculation point on the selected cross-sections. "Comprehensiveness" means that the total head loss between two cross-sections of gradually varied flow should be calculated after a thorough analysis.

The detailed descriptions are as follows:

a. The reference level can be chosen randomly. However, once the reference level is selected, the position height, z, of representative point on two flow sections should be based on the same chosen reference level. The upstream and downstream flow sections cannot have different reference levels. Generally, the reference level is chosen at the lower elevation so that $z \geqslant 0$. For example, the reference level is usually chosen at the downstream flow surface, which is the horizontal plane at the center of a pipe outlet.

b. Any flow section with gradually varied flow is appropriate for the energy equation. However, to make it relatively easy to solve the equation, choose flow sections on which there are more known variables but fewer unknown variables. Usually a representative section is chosen at the initial cross-section of flow or a flow section in which unknown variable needs to be solved.

c. Although the potential energy of gradually varied flow over a section $z + \frac{p}{\gamma}$ is a constant, the difficulty in solving the equation remains when either the location height z or the pressure head $\frac{p}{\gamma}$ at a point on the representative section is difficult to determine. Therefore, a representative point should be chosen in order to simplify the calculation. For instance, the central point of pipe in a pipe flow (the center of pipe section) can be selected as the calculating point. If the reference level is also chosen at the level of the pipe axis, the location height z of the calculating point will be zero. If the pipe flow is discharging into the air, the relative pressure at the center point of pipe outlet equals zero. For open channel flow, the calculating point is chosen on the free surface, where the relative pressure equals zero. Thus, these calculating points are easy to identify for solving particular problems.

d. The dynamic pressure head $\frac{p}{\gamma}$ in the energy equation is usually used by relative pressure. However, if there exists a negative pressure, the absolute pressure can also be used so that the value of $\frac{p}{\gamma}$ is positive. In addition, the pressure head in the energy equation must have the dimension of length, [L], expressed by $\frac{p}{\gamma}$.

e. Strictly speaking, the kinetic energy correction factors (α_1 and α_2) on different fluid cross-sections are neither the same nor equal to 1. In practical use, for most gradually varied flow sections, they can, however, be approximately taken as $\alpha_1 = \alpha_2 = 1$.

As for the "comprehensiveness" tips, this will be dealt with in the next chapter on the head loss. For the problems given at the end of this chapter, the h_w is either negligible or is given. Usually the head loss is $h_w = \zeta \frac{v^2}{2g}$, which is expressed by a certain multiple of the velocity head.

3.3.3.3 Application examples of energy equation

(1) Application in measuring the flow rate

① Pitot tube for the measurement of flow velocity

The pitot tube was first introduced by a Frenchman Henri Pitot in 1730. After application and development of two centuries, there are several types of Pitot tubes. Among them, the Prandtl Pitot tube has shown relatively better performance. A Pitot tube does not need electricity, but can

Chapter 3 Basic equations of steady total flow

obtain relatively accurate measuring results.

The principle of the application of the Pitot tube is the energy conversion in the energy equation of the flow streamline (special case of streamtube). When a water flow, which is flowing forwards from a distance with the pressure energy $\dfrac{p_0}{\gamma}$ and the velocity u_0, meets an obstacle (as shown in Fig. 3 - 19), the velocity of the streamline that is facing the object will decrease from u_0 to zero. The point A is called the stagnation point. The reduction of the kinetic energy is converted to the pressure energy, and the pressure energy $\dfrac{p}{\gamma}$ at the stagnation point is equal to the sum of the pressure energy $\dfrac{p}{\gamma}$ and kinetic energy $\dfrac{u_0^2}{2g}$ of the same point at the original undisturbed state. The structure of Prandtl - Pitot tube is shown in Fig. 3 - 20. There is a small hole at the round head of the probe, which is opposite right to the flow direction, so the measured pressure at the stagnation point A is the total dynamic pressure that $\dfrac{p}{\gamma} = \dfrac{p_0}{\gamma} + \dfrac{u_0^2}{2g}$. This pressure is connected, by a small tube, to one end of a differential pressure gauge. A hole for measuring the hydrostatic pressure on a boundary of the probe is arranged at certain distance from the head, and the hole is normal to the flow direction. Thus, it measures the initial dynamic pressure energy $\dfrac{p_0}{\gamma}$ but does not include the flow head $\dfrac{u_0^2}{2g}$, and this pressure is connected to the other end of the differential pressure gauge through another tube. The differential pressure gauge shows the pressure difference between the two tubes $\Delta h = \dfrac{p_0}{\gamma} + \dfrac{u_0^2}{2g} - \dfrac{p_0}{\gamma} = \dfrac{u_0^2}{2g}$, which is $u_0^2 = 2g\Delta h$ or $u_0 = \sqrt{2g\Delta h}$. So u_0 can be obtained from measured Δh. Considering the difference of positions of two holes and the impact of the probe on the flow, a correction coefficient μ is introduced to take account of the impact, so the equation is modified as $u_0 = \mu\sqrt{2g\Delta h}$. The value of μ is determined by experiments and is typically between 0.98 and 1.0. The Pitot tube can usually be used for the flow with a velocity in the range between 0.15 and 2.0 m/s. In the pressurized pipe flow, the

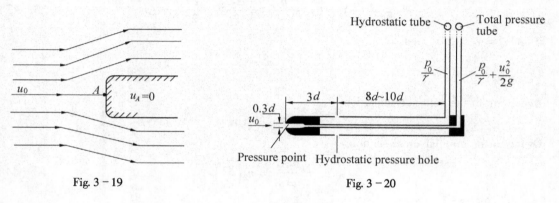

Fig. 3 - 19   Fig. 3 - 20

cylindrical Pitot tube can be used to measure the velocity, with a maximum flow velocity up to 6 m/s.

② Venturi meter

Venturi meter is a commonly used device for measuring the pipe flow. For example, in the water supply project from the Shenzhen reservoir to Hong Kong, the Venturi meter is used to measure the flow rate for calculating the total amount of water supplied. A Venturi meter consists of a contraction section, a throat and an expansion section (Fig. 3 - 21). At the uniform inlet section 1 - 1 and the throat section

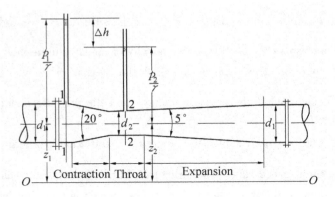

Fig. 3 - 21

2 - 2, there is a piezometric tube that is installed vertically in the wall of the pipe and can be connected to a differential pressure gauge. By measuring the piezometric head difference Δh (or the height difference of differential pressure gauge), the flow rate Q in the pipeline can be calculated.

The area A_2 of section 2 - 2 at throat is smaller than the area A_1 of the inlet section 1 - 1. According to the principle of continuity, the average flow velocity v_2 of cross-section 2 - 2 is larger than the average flow velocity v_1 of cross-section 1 - 1. Therefore, at section 2 - 2 the kinetic energy is increased but the potential energy is reduced, that is $z_2 + \frac{p_2}{\gamma} < z_1 + \frac{p_1}{\gamma}$. The difference is expressed by Δh. At the same time, because the two sections 1 - 1 and 2 - 2 are so close that the head loss of flow may be neglected, thus the energy equation for the two sections is

$$z_1 + \frac{p_1}{\gamma} + \frac{\alpha_1 v_1^2}{2g} = z_2 + \frac{p_2}{\gamma} + \frac{\alpha_2 v_2^2}{2g}$$

Since

$$(z_1 + \frac{p_1}{\gamma}) - (z_2 + \frac{p_2}{\gamma}) = \frac{\alpha_2 v_2^2}{2g} - \frac{\alpha_1 v_1^2}{2g} = \Delta h \qquad (3-43)$$

if let $\alpha_1 = \alpha_2 = 1.0$, then

$$\Delta h = \frac{v_2^2 - v_1^2}{2g}$$

According to the continuty equation

$$Q = A_1 v_1 = A_2 v_2$$

Owing to the circular cross-section

$$v_1 = \frac{v_2 A_2}{A_1} = v_2 \left(\frac{d_2}{d_1}\right)^2 \qquad (3-44)$$

where $A_1 = \dfrac{\pi d_1^2}{4}$, $A_2 = \dfrac{\pi d_2^2}{4}$.

Inserting Equation(3-44) into Equation(3-43),

$$v_2^2 - \frac{v_2^2 d_2^4}{d_1^4} = 2g\Delta h$$

$$v_2^2\left(1 - \frac{d_2^4}{d_1^4}\right) = 2g\Delta h$$

$$v_2 = \sqrt{\frac{2g\Delta h}{1 - \dfrac{d_2^4}{d_1^4}}}$$

$$Q' = v_2 A_2 = \frac{1}{\sqrt{1 - \left(\dfrac{d_2}{d_1}\right)^4}} \sqrt{2g} \, \frac{1}{4}\pi d_2^2 \sqrt{\Delta h}$$

Letting $\dfrac{1}{\sqrt{1 - \left(\dfrac{d_2}{d_1}\right)^4}} \sqrt{2g}\,\dfrac{1}{4}\pi d_2^2 = k$, when the values of d_1, d_2 are known, k is constant. Then

$$Q' = k\sqrt{\Delta h} \qquad (3-45)$$

In fact, there is energy loss between the two sections 1-1 and 2-2, so the flow rate will be smaller than the result calculated by Equation (3-45). Usually, a correction coefficient μ of less than 1 is multiplied in the equation. Thus the actual flow rate is

$$Q = \mu k\sqrt{\Delta h} \qquad (3-46)$$

where μ is called the discharge coefficient of Venturi meter, and its value is often between 0.95 and 0.98. The value of μ depends on the flow conditions and the geometrical shape of the contraction pipe, and should be determined and validated before the Venturi meter is used. In practical application, the tests are carried out to determine the relationship between Q and Δh, which is often plotted as a graph for ease of use.

In the continuity equation, if v_1 is used (expressed by v_2) for $Q' = v_1 A_1$, then another equation can be obtained:

$$Q = \mu \frac{\pi d_1^2}{4} \sqrt{\frac{2g}{\left(\dfrac{d_1}{d_2}\right)^4 - 1}} \sqrt{\Delta h} \qquad (3-46')$$

If a differential pressure gauge of mercury is connected to the Venturi meter, as shown in Fig. 3-22, we have

$$\left(z_1 + \frac{p_1}{\gamma}\right) - \left(z_2 + \frac{p_2}{\gamma}\right) = \frac{\gamma_m - \gamma}{\gamma}\Delta h = 12.6\Delta h$$

where Δh is the height difference in the mercury pressure gauge.

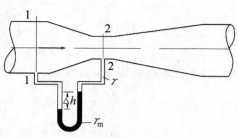

Fig. 3-22

Thus, the flow rate of the Venturi meter is

$$Q = \mu k \sqrt{12.6\Delta h} \qquad (3-47)$$

(2) Application examples

【Example 3-5】 Water in a large water tank, which has a constant depth of $h = 3$ m, is discharged by a vertical pipe of a diameter with $d_1 = 10$ cm from the bottom of the tank. At the outlet of the pipe, there is a nozzle. The length of the nozzle is 0.3 m and its diameter is $d_2 = 5$ cm. If the flow is open to the atmosphere and the water head loss is ignored, calculate the pressure at points A, B and C as shown in Fig. 3-23.

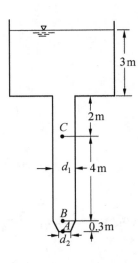

Fig. 3-23

Solution: First, the flow in this question is of kinematic motion, so the pressure is not hydrostatic pressure, i.e. the formula $p = \gamma h$ is not applicable in this case. Second, the large tank implies: when compared with the flow velocity of water in the pipe, the velocity of water in the tank is close to zero. Because the water surface of the tank is constant, the flow at both B and C in the pipe is uniform flow, and at the cross-section A the streamlines of flow are nearly parallel, so section A is approximately a steady gradually varied flow section.

Choose the nozzle exit plane as the reference datum, and then apply the energy equation:

(1) Establish the energy equation between C. S. at A at the nozzle outlet and the water surface of tank. Because the flow is discharged to the atmosphere, then $p_A = 0$, $\dfrac{p_A}{\gamma} = 0$,

$$(0.3 + 4 + 2 + 3) + 0 + 0 = 0 + 0 + \frac{v_A^2}{2g}$$

$$v_A = \sqrt{19.6 \times 9.3} = 13.5 \text{ m/s}$$

(2) Establish the energy equation between C. S. at B and the water surface of tank,

$$9.3 = 0.3 + \frac{p_B}{\gamma} + \frac{v_B^2}{2g}$$

From the continuity equation

$$v_B = \frac{v_A A_A}{A_B} = \frac{v_A \dfrac{\pi d_2^2}{4}}{\dfrac{\pi d_1^2}{4}} = v_A \left(\frac{d_2}{d_1}\right)^2 = 13.5 \times \left(\frac{0.05}{0.1}\right)^2$$

$$= 3.375 \text{ m/s}$$

Applying it to the above equation,

$$\frac{p_B}{\gamma} = 9.3 - 0.3 - \frac{3.375^2}{19.6} = 8.42 \text{ m (water column)}$$

$$p_B = 9.8 \times 8.42 = 82.52 \text{ kN/m}^2$$

(3) Establish the energy equation between C. S. at C and the water surface of tank,

$$9.3 = 4.3 + \frac{p_C}{\gamma} + \frac{v_C^2}{2g}$$

Since the diameters of C. S. at B and C are the same, $v_C = v_B = 3.375$ m/s, then

$$\frac{p_C}{\gamma} = 9.3 - 4.3 \frac{3.375^2}{19.6} = 4.42 \text{ m (water colomn)}$$

$$p_C = 9.8 \times 4.42 = 43.3 \text{ kN/m}^2$$

From this example, it can be seen that both the energy equation and continuity equation are combined to calculate the flow parameters p and v. Note that the energy equation has been used several times for the interesting sections (i. e. calculating sections) in which the unknown parameters are required.

【Example 3 - 6】 As shown in Fig. 3 - 24, a water pump is used to extract water in $Q = 5.56$ L/s from the water pool. The height of the water pump is $H_s = 5$ m. The diameter of the suction pipe is $d = 100$ mm. If the water head loss of the suction pipe is $h_w = 0.25$ m, then calculate the vacuum degree at the inlet section 2 - 2 of the water pump.

Solution: Near the inlet of the suction pipe, there is a significant change of cross-section, so the streamline is bent sharply, which is rapidly varied flow. Therefore, the inlet section of suction pipe cannot be taken as a calculation section for the energy equation. The flow velocity in the pool is small and the water head loss is small enough that it can be neglected. Hence, by choosing the water surface of the pool as the reference datum (i. e. the plane $O - O$), the energy equation between cross-sections 1 - 1 and 2 - 2 is

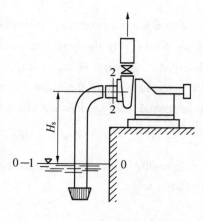

Fig. 3 - 24

$$0 + 0 + 0 = H_s + \frac{p_2}{\gamma} + \frac{\alpha_2 v_2^2}{2g} + h_w$$

Let $\alpha_2 \approx 1$ and $v_2 = \frac{4Q}{\pi d^2} = 0.71$ m/s. Applying them into the equation gives

$$\frac{p_2}{\gamma} = -5.28 \text{ m}$$

Therefore, the vacuum degree at cross-section 2 - 2 is 5.28 m in water column.

【Example 3 - 7】 Fig. 3 - 25 shows the water jet out of a hose. The velocity of water at the nozzle exit is $v_1 = 25$ m/s, at an angle of $60°$ to the horizontal plane. Ignoring the influence of the air friction, calculate the height H of the jet.

Solution: Because this is a water jet into the air, the pressure around the flow is atmospheric, and the air friction can be neglected. So this is a question about the kinetic energy being transferred into the potential energy.

Select the horizontal plane through the centroid of exit cross-section as the datum plane, and then write the energy equation between the exit cross-section 1−1 and the highest cross-section 2−2 of the jet. Thus, $z_1 = 0$, $p_1 = 0$, $z_1 = H$, $p_1 = 0$, $h_{w_{1-2}} = 0$, then

$$\frac{\alpha_1 v_1^2}{2g} = H + \frac{\alpha_2 v_2^2}{2g}$$

Let $\alpha_1 = \alpha_2 = 1.0$, then

$$H = \frac{v_1^2}{2g} - \frac{v_2^2}{2g}$$

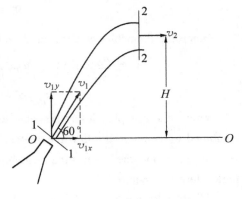

Fig. 3−25

The average velocity (v_1) at the exit cross-section can be divided into the horizontal velocity $v_{1x} = v_1 \cos 60° = 25 \times 0.5 = 12.5$ m/s and the vertical velocity v_{1y}. Because of the gravity, the vertical velocity of the jet becomes smaller and smaller until it reaches zero in the highest cross-section. If the influence of air resistance can be ignored, the horizontal velocity does not change, so $v_2 = v_{2x} = v_{1x} = 12.5$ m/s. Then

$$H = \frac{25^2}{2 \times 9.8} - \frac{12.5^2}{2 \times 9.8} = 23.9 \text{ m}$$

In reality, since air resistance always exists, the actual height of the jet is smaller than this calculated value.

3.4 Momentum equation of steady total flow

In addition to the continuity and energy equations, the momentum equation is another fundamental equation that is used to solve fluid mechanics problems. It is the application of the momentum conservation law for the fluid movement, which describes the relationship between the change of fluid momentum and the force. The momentum, energy and continuity equations are called the three basic equations of fluid mechanics.

The energy equation and the continuity equation are very useful for analyzing the problems in fluid mechanics. However, they do not reflect the relationship between the fluid flow and the force on the boundary of an object and thus cannot be applied for calculating the total hydrodynamic force of the fluid on the boundary. In addition, the application of the energy equation requires the head loss due to the flow, so its application is limited when the head loss is difficult to evaluate accurately. Therefore, the momentum equation is needed here. It is the equation that can be used to calculate the total dynamic force for the flow as a whole when the parameters of flow between two gradually varied flow sections are not required.

3.4.1 Derivation of the momentum equation

The derivation of the momentum equation for steady flow is still based on Newton's second law, i.e. $\sum F = ma = m\dfrac{\mathrm{d}v}{\mathrm{d}t}$. For steady flow of an incompressible fluid (the density $\rho =$ constant), the mass of flow $m = \rho V$ is also constant and can be moved inside the differential sign, so $\sum F = \mathrm{d}\dfrac{(mv)}{\mathrm{d}t}$, which can also be expressed in a finite difference form as

$$\sum F = \frac{\Delta(mv)}{\Delta t} = \frac{mv_2 - mv_1}{\Delta t} \qquad (3-48)$$

where $\sum F$ is the sum of the external forces on the object and mv is the momentum, i.e. the mass multiplied by the velocity of the object. The mv_2 is the momentum at the end of the time period Δt and the mv_1 is the momentum at the beginning of the time period Δt. Since the velocity is a vector, the momentum is also a vector with the same direction as the velocity. The common unit is kg·m/s, denoted by the letter M(Momentum).

Equation (3-48) signifies that the momentum change per unit time of an object equals the sum of external forces on the object. Actually, this is another statement about the momentum law in physics. According to Equation (3-48), if the rate of change of fluid momentum during the time Δt is known, the resultant external forces on the object can be obtained. Among the resultant forces, if only the force R' of the flow acting on the solid boundary needs to be calculated, while the other external forces are known, then the unknown force R' can be obtained. According to the relation between a force and its reacting force, i.e. Newton's third law, the total dynamic force R of the fluid on the solid boundary is the same as R' in magnitude but in the opposite direction. The force and reacting force always co-exist.

The following is the derivation of the momentum equation for a steady total flow:

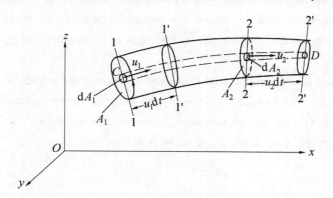

Fig. 3-26

In a steady total flow, consider a control element of flow between two cross-sections $1-1$ and $2-2$ for study, as shown in Fig. 3-26. After the time $\mathrm{d}t$, the control element moves from the position $1-1 \sim 2-2$ to a new position $1'-1' \sim 2'-2'$. If at the original place, the momentum of the control element is $M_{1\text{-}2}$ and at the new position, the momentum is $M_{1'\text{-}2'}$, then the change of momentum is

$$\Delta M = M_{1'\text{-}2'} - M_{1\text{-}2}$$

However, M_{1-2} is the sum of the momentum of fluid in $1-1'$ and $1'-2$, which is

$$M_{1-2} = M_{1-1'} + M_{1'-2}$$

and similarly,

$$M_{1'-2'} = M_{1'-2} + M_{2-2'}$$

Under the condition of steady flow, the geometric shape and flow rate of the fluid in section $1'-2$ do not change with time. Therefore, $M_{1'-2}$ does not change with time. Thus,

$$\Delta M = M_{1'-2} + M_{2-2'} - M_{1-1'} - M_{1'-2}$$
$$= M_{2-2'} - M_{1-1'}$$

Now calculate $M_{2-2'}$ and $M_{1-1'}$. Take an arbitrary streamtube flow CD in the total flow. At section $1-1$, the cross-section area of the streamtube is dA_1, and its flow velocity and flow rate are u_1 and dQ respectively; after the time dt, the flowing length is $u_1 dt$, the volume is $u_1 dt dA_1$, the mass is $\rho u_1 dt dA_1$ and the momentum is $\rho u_1 dt dA_1 u_1 (= \rho dQ dt u_1)$. Then by integrating over cross-section A_1, the momentum of the fluid in section $1-1'$ is

$$M_{1-1'} = \int_{Q_1} \rho u_1 dQ dt$$

Similarly,
$$M_{2-2'} = \int_{Q_2} \rho u_2 dQ dt$$

Since the velocity distribution and the point velocity u on a cross-section are unknown, their integrals actually cannot be obtained in practice. In engineering, the momentum $\rho Q v dt$, where v is expressed by the average velocity v, is used to replace the term $\int_Q \rho u dQ dt$. This requires that the direction of velocity at each point on the cross-section is either the same as or related to that of the average velocity on the cross-section. This condition can only be met when the flow is uniform or gradually varied in which case the streamlines tend to be parallel. Therefore, another limitation in applying the momentum equation is that on two ends of a control volume, the flow should be gradually varied flow.

Owing to the fact that the momentum expressed by the average velocity v of the cross-section (v is the same over the cross-section) is not really the same as the momentum integral obtained by the actual point velocity over the cross-section, a momentum correction coefficient β should be introduced. Based on this statement, the β value is defined as

$$\beta \rho Q dt v = \int_Q \rho u dQ dt$$

Therefore,
$$\beta = \frac{\int_Q u u dA}{v Q}$$

In the cross-section of gradually varied flow, the angle between the point velocity u and the momentum projection axis is the same as that between the velocity u and v of the cross-section. Assume the angle is θ, then $u = u\cos\theta$, $v = v\cos\theta$. Therefore,

$$\beta = \frac{\int_A u^2 dA}{v^2 A} \tag{3-49}$$

The momentum correction coefficient β is the ratio of the actual momentum through the cross-section at unit time to the momentum of flow with the average velocity passing through the cross-section at unit time. Similar to the kinetic energy correction coefficient α, β reflects the non-uniformity of velocity distribution in a cross-section. The more uniform the velocity distribution, the smaller the value of β is, in which case β tends to 1. The more non-uniform the velocity distribution, the larger the value of β is. β is usually larger than 1 but less than the kinetic energy correction coefficient α for the same velocity distribution in a cross-section. On a cross-section of gradually varied flow, the value of β is usually between 1.02 and 1.05. In order to simplify calculation, β is usually taken as 1.0.

Notes:

According to Equation (3-49)

$$\beta = \frac{\int_A u^2 dA}{v^2 A}$$

Consider the relationship between the point velocity u and the average velocity v of the cross-section

$$u = v \pm \Delta u$$

Then
$$\int_A u^2 dA = \int_A (v \pm \Delta u)^2 dA = \int_A (v^2 \pm 2v\Delta u + \Delta u^2) dA$$

Because
$$Q = \int_A u dA = \int_A (v \pm \Delta u) dA = vA \pm \int_A \Delta u dA$$

According to the continuity equation $Q = vA$, then $\int_A \Delta u dA = 0$,

$$\int_A u^2 dA = v^2 A + \int_A \Delta u^2 dA$$

Applying this to Equation (3-49), then

$$\beta = 1 + \int_A \frac{\Delta u^2}{v^2 A} dA = 1 + \eta$$

where

$$\eta = \int_A \frac{\Delta u^2}{v^2 A} dA$$

which η is always larger than zero. Since the kinetic correction coefficient $\alpha = 1 + 3\eta$, α is larger than the momentum correction coefficient $\beta(\beta = 1 + \eta)$ on a cross-section of the same velocity distribution.

Through the analysis above, then

$$\boldsymbol{M}_{1-1'} = \int_{Q_1} \rho \boldsymbol{u}_1 dQ dt = \beta_1 \rho Q_1 \boldsymbol{v}_1 dt$$

$$\boldsymbol{M}_{2-2'} = \int_{Q_2} \rho \boldsymbol{u}_2 dQ dt = \beta_2 \rho Q_2 \boldsymbol{v}_2 dt$$

According to the continuity equation

$$Q_1 = Q_2 = Q$$

Therefore,
$$\Delta M = \rho Q(\beta_2 v_2 - \beta_1 v_1)\Delta t$$
According to the momentum equation (3-48)
$$\sum F = \frac{\Delta(mv)}{\Delta t}$$
then
$$\sum F = \rho Q(\beta_2 v_2 - \beta_1 v_1) \qquad (3-50)$$

The left hand side of Equation (3-50) represents the sum of all external forces on the control volume. These forces include the total dynamic pressure force P_1 on the upstream cross-section 1-1, the total dynamic pressure force P_2 on the downstream cross-section 2-2, the gravity, and the total force R' of the solid boundary on the control volume. The right hand side of the equation represents the change between the momentum passing through the downstream cross-section and the upstream cross-section.

Equation (3-50) is the momentum equation for the steady total flow of an incompressible real fluid. However in practical application, the following component equations in the Cartesian coordinate system are used:

$$\left.\begin{array}{l}\sum F_x = \rho Q(\beta_2 v_{2x} - \beta_1 v_{1x}) \\ \sum F_y = \rho Q(\beta_2 v_{2y} - \beta_1 v_{1y}) \\ \sum F_z = \rho Q(\beta_2 v_{2z} - \beta_1 v_{1z})\end{array}\right\} \qquad (3-51)$$

where v_{1x}, v_{1y} and v_{1z} are the three components of the average velocity v_1 in the x, y and z direction respectively; v_{2x}, v_{2y} and v_{2z} are the corresponding components of the average velocity v_2. $\sum F_x$, $\sum F_y$ and $\sum F_z$ are the algebraic sum of three projections of all external forces along the x, y, and z axis on the control volume.

In Equation (3-51), there is no difference between β_1 and β_2 in x, y and z directions. Assuming $\beta_1 = \beta_2 = 1$, the difference can be neglected.

In the above derivation, the momentum equation is obtained for a control volume element with only two sections: one upstream and one downstream cross-section. In reality, the momentum equation can be extended to apply to any closed control volume in a flow field. In this condition, the right hand side of Equation (3-51) can be interpreted as the difference in the total momentum after a change. The left hand side of the equation can be taken as the sum of all external forces on the whole close control volume. Take a Y-branch pipe as example for analysis, as shown in Fig. 3-27.

Through the Y-branch pipe, a flow from the cross-section 1-1 is split into two parts: one leaving from cross-section 2-2 and the other leaving from cross-section 3-3. Consider the closed control volume, which consists of above three cross-sections, on which the flows are gradually varied flow. Owing to the changes on the boundary, the force R' on the fluid causes the change of momentum in the fluid. The sum of momenta over the two cross-sections 2-2 and 3-3 is the momentum after the change. The momentum over cross-section 1-1 is the momentum before the change. Therefore, the increment of momentum ΔM should be that the sum of momenta over

Chapter 3 Basic equations of steady total flow

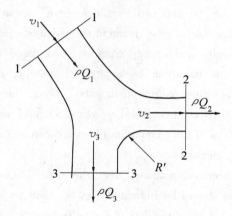

Fig. 3 – 27

cross-sections 2 – 2 and 3 – 3 minus the momentum over cross-section 1 – 1. Equation (3 – 50) should be written as

$$\sum F = \rho Q_2 \beta_2 v_2 + \rho Q_3 \beta_3 v_3 - \rho Q_1 \beta_1 v_1$$

Equation (3 – 51) can also be written into

$$\sum F_x = \rho Q_2 \beta_2 v_{2x} + \rho Q_3 \beta_3 v_{3x} - \rho Q_1 \beta_1 v_{1x}$$
$$\sum F_y = \rho Q_2 \beta_2 v_{2y} + \rho Q_3 \beta_3 v_{3y} - \rho Q_1 \beta_1 v_{1y}$$
$$\sum F_z = \rho Q_2 \beta_2 v_{2z} + \rho Q_3 \beta_3 v_{3z} - \rho Q_1 \beta_1 v_{1z}$$

3.4.2 Conditions and tips in the application of the momentum equation

(1) Conditions of application

① The fluid is incompressible and must be steady flow.

② The two end cross-sections in the control volume should be the cross-sections of gradually varied flow. The point velocity u over the cross-section should have the same direction as the average velocity v of the cross-section, but the flow between the two end cross-sections may be rapidly varied flow.

③ In the normal condition, the tangent friction of the external boundary on the flow body is usually ignored. Therefore, the segment of rapidly varied flow in the control volume should not be too long and it is preferable for the two end cross-sections to be chosen as close as possible to the rapidly varied flow.

(2) Tips of application

The momentum equations are usually used along with the continuity equation and energy equations. Generally, the average velocity v of the cross-section needs to be calculated first by the continuity equation and then the dynamic pressure p can be obtained by the energy equation. Thus the total dynamic pressure force P on the cross-sections in the control volume can be calculated, and finally, the remaining unknown variable is the force R' by the solid body on the flow boundary in the control volume, which can be calculated by the momentum equations. The total dynamic force R then can be obtained by the force R' according to Newtown's third law $R = R'$. Because the application of the momentum equations is based on the successful implementation of the continuity and energy equations, we need to note that the momentum equations are vector equations in which each physical quantity should be projected on the coordinate axis with the value either positive or negative. Thus, the momentum equations are the most difficult to apply among three fundamental equations in fluid mechanics.

The tips for the application of the momentum equations can be summarized as "two selections and one summation", which means:

① In applying the momentum equations, first, the projection axis should be chosen and the positive directions of the axis should be indicated clearly; then project the velocity and external forces on the projection axis. If the directions of the velocity and external forces are the same as the direction of projection axis, their values are positive; otherwise the values are negative. The projection axis can be chosen arbitrarily but it should be convenient for projection and calculation. For example, a horizontal axis is selected to be x axis with a positive direction to the right; the vertical axis is y axis with the positive direction vertically upward. Also the axis of a pipe can be chosen as the x axis with the positive direction following the flow, and then the external normal to the pipe axis will be y axis with its positive direction pointing outside the pipe.

② Second, the control volume of flow should be chosen carefully. Usually, the boundary of the whole total flow is used as the boundary of the control volume and the two end cross-sections should be of gradually varied flow. The segment length of rapidly varied flow within the control volume should be chosen to be as short as possible so that this will not result in too much errors by ignoring the tangent friction on the boundary.

③ "One summation" is meant by analyzing all the external forces on the control volume as a whole. The external forces include the following:

    a. The total hydrodynamic pressure forces P_1 and P_2 on the two end cross-sections. If the average pressure p_c on the end cross-section is known, then $P = p_c A$ where A is the cross-sectional area.

    b. The tangent frictional stress on the solid boundary T, which distributes over the boundary surface of the control volume. When T is smaller than any other external forces, it can be ignored and otherwise it will be included into the total force R' of the solid boundary on the control volume.

    c. The gravity G on the control volume. If the control volume is placed horizontally, then the projection of the gravity on the horizontal axis is zero.

    d. The total force R' of the solid boundary on the control volume. This is the force that makes the momentum of the flow changing over the control volume. It is also the force that needs to be calculated by the momentum equations. The corresponding reacting force to this force is the total hydrodynamic force R of the flow in the control volume acting on the solid boundary. Therefore,

$$\sum F = P_1 + P_2 + R' + G$$

The projection equations of force on the axis, for example in x axis,

$$\sum F_x = P_{1x} + P_{2x} + R'_x + G_x$$

Similarly,

$$\sum F_y = P_{1y} + P_{2y} + R'_y + G_y$$

$$\sum F_z = P_{1z} + P_{2z} + R'_z + G_z$$

Since the direction of R' is unknown beforehand, its direction can first be assumed as one direction. If the calculated R' is positive based on the momentum equations, it means the assumed direction is right; otherwise if the value is negative, the assumed direction is wrong, i.e. its actual direction is opposite to the assumed direction.

If the flow is on a plane, only two projection equations are needed. For example, when the projection equations on the x and y axis are needed to calculate R'_x and R'_y, and the resultant force R' can be calculated from the following.

$$R' = \sqrt{R'^2_x + R'^2_y}$$

$$R = -R'$$

where R is the total hydrodynamic force of rapidly varied flow acting on the control volume.

④ On the right hand sides of Equations (3-50) and (3-51), the momentum change that must be the momentum over the outlet minus that over the inlet. Note that this order cannot be reversed.

⑤ The unit of each physical quantity in the momentum equations should be consistent, usually in metric system (SI), i.e. m, kg, N(kN) and s.

3.4.3 Application examples of the momentum equation

3.4.3.1 The force of flow on an elbow

【Example 3-8】 A vertically placed elbow is shown in Fig. 3-28, where at the inlet the diameter of cross-section is $d_1 = 0.3$ m, and the pressure $p_1 = 68.6$ kN/m^2; the diameter of cross-section at the outlet is $d_2 = 0.2$ m. The flow rate of steady flow in the elbow is $Q = 0.3$ m^3/s and the volume of the water in the elbow is $V = 0.085$ m^3. Ignoring the water head loss and the friction force, calculate the total dynamic force R of the flow acting on the elbow.

Solution:

The total dynamic force R has the same magnitude as the reacting force of R' (i.e. the force of the elbow acting on the flow) but in the opposite direction. By calculating R' in the momentum equation, the average velocities (v_1 and v_2) and the average dynamic pressures p_1 (known) and p_2 at the two end cross-sections will be used. Therefore, the momentum

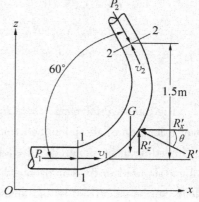

Fig. 3-28

equations should be used together with the continuity equation and energy equation. By using the continuity equation,

$$v_1 = \frac{4Q}{\pi d_1^2} = \frac{4 \times 0.3}{\pi \times 0.3^2} = 4.24 \text{ m/s}, \quad \frac{v_1^2}{2g} = \frac{4.24^2}{19.6} = 0.92 \text{ m}$$

$$v_2 = \frac{4Q}{\pi d_2^2} = \frac{4 \times 0.3}{\pi \times 0.2^2} = 9.54 \text{ m/s}, \quad \frac{v_2^2}{2g} = \frac{9.54^2}{19.6} = 4.63 \text{ m}$$

By taking the horizontal plane along the centroid of the inlet cross-section as the datum plane, the energy equation between C. S. 1 and C. S. 2 is

$$0 + \frac{68.6}{9.8} + \frac{4.24^2}{19.6} = 1.5 + \frac{p_2}{9.8} + \frac{9.54^2}{19.6}$$

Then
$$p_2 = 17.54 \text{ kN/m}^2$$

and
$$P_2 = p_2 A_2 = 17.54 \times \frac{\pi \times 0.2^2}{4} = 0.55 \text{ kN}$$

According to the known conditions

$$P_1 = p_1 A_1 = 68.6 \times \frac{\pi \times 0.3^2}{4} = 4.85 \text{ kN}$$

$$G = \gamma A = 9.8 \times 0.085 = 0.833 \text{ kN}$$

Now we can apply the momentum equations to calculate the external force R'. Choose the closed part formed by C. S. 1 and C. S. 2 and the boundary of the elbow pipe as the control volume. The horizontal axis is x axis with the positive direction to the right. The vertical axis is z axis with the positive direction upwards. All the forces and velocities are shown in Fig. 3 – 28. The direction of R' is assumed as also shown in the figure.

The momentum equation in the x axis:

$$P_1 - R'_x + P_2 \cos 60° = \rho Q(-v_2 \cos 60° - v_1)$$
$$R'_x = 4.85 + 0.55 \times 0.5 + 1 \times 0.3 \times (9.54 \times 0.5 + 4.24) = 7.83 \text{ kN}$$

The momentum equation in the z axis:

$$R'_z - P_2 \sin 60° - G = \rho Q(v_2 \sin 60° - 0)$$
$$R'_z = 0.55 \times 0.866 + 0.833 + 1 \times 0.3 \times (9.54 \times 0.866) = 3.79 \text{ kN}$$

$$R' = \sqrt{R'^2_x + R'^2_z} = \sqrt{7.83^2 + 3.79^2} = 8.7 \text{ kN (to the upper left)}$$

$$R = -R' = -8.7 \text{ kN (to the downward right)}$$

$$\theta = \arctan \frac{R'_z}{R'_x} = \arctan \frac{3.79}{7.83} = 25.8°$$

The total hydrodynamic force R of flow acting on the elbow may cause the elbow to vibrate. Over a long time, the joints between the elbow and pipe may be broken or displaced in position, thus needing to be repaired. In engineering, usually an anchor block (a concrete block of large volume) is used to fix a horizontal large elbow. The volume of the concrete used for the anchor block can be calculated by the following equation:

$$V \geq \frac{R}{f\gamma}$$

where R is the total hydrodynamic force of flow on the elbow, f is the friction coefficient between the anchor block and ground, and γ is the bulk density of the concrete.

3.4.3.2 Application of momentum equations on a Y-branch.

【Example 3-9】 In a diversion power station, the main pressurized pipe is divided into two branched pipes to draw water to a hydro-turbine. As shown in Fig. 3-29, all the center lines of the pipe are on the same plane. The diameter of the main pipe is $d_1 = 3$ m and the diameters of the branched pipes are $d_2 = d_3 = 2$ m. The bend angle $\alpha = 60°$ and the total flow rate is $Q = 35$ m³/s. The pressure head on cross-section 1-1 is $\dfrac{p_1}{\gamma} = 30$ m (water column). If the water loss head is ignored, calculate the total dynamic force of flow on the bend.

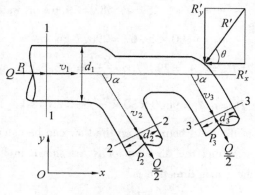

Fig. 3-29

Solution: Firstly, choose the coordinate axis and let the horizontal axis as x axis with the positive direction to the right, the vertical axis as y axis with the positive direction upwards. Secondly, choose the control volume, which is comprised of C.S. 1-1, C.S. 2-2, C.S. 3-3 and the boundary of the bend.

According to the continuity equation, the following can be obtained.

$$v_1 = \frac{4Q}{\pi d^2} = \frac{4 \times 35}{\pi \times 3^2} = 4.95 \text{ m/s}$$

$$\frac{v_1^2}{2g} = \frac{4.95^2}{2 \times 9.8} = 1.25 \text{ m}$$

$$v_2 = v_3 = \frac{4 \times \dfrac{Q}{2}}{\pi d_2^2} = \frac{4 \times \dfrac{35}{2}}{\pi \times 2^2} = 5.57 \text{ m/s}$$

$$\frac{v_2^2}{2g} = \frac{v_3^2}{2g} = \frac{5.57^2}{2 \times 9.8} = 1.583 \text{ m}$$

Since the pipe is placed horizontally, it means that the centroid height z of each cross-section is the same; thus they can be cancelled out in the energy equation. Furthermore, this is a problem of a Y-branch pipe, so different weight flow of cross-section should be considered in the application of energy equation. Assuming $\alpha_1 = \alpha_2 = \alpha_3 = 1$ and taking the horizontal plane on each cross-section's centroid as the datum plane, the energy equation for C.S. 1-1, C.S. 2-2 and C. S. 3-3 is

$$\left(\frac{p_1}{\gamma} + \frac{v_1^2}{2g}\right)\gamma Q = \left(\frac{p_2}{\gamma} + \frac{v_2^2}{2g}\right)\gamma \frac{Q}{2} + \left(\frac{p_3}{\gamma} + \frac{v_3^2}{2g}\right)\gamma \frac{Q}{2}$$

Because the water head loss is ignored and $\frac{v_2^2}{2g} = \frac{v_3^2}{2g}$, thus $\frac{p_2}{\gamma} = \frac{p_3}{\gamma}$. Therefore, the equation above can be written as

$$\frac{p_1}{\gamma} + \frac{v_1^2}{2g} = \frac{p_2}{\gamma} + \frac{v_2^2}{2g}$$

Substituting $\frac{v_1^2}{2g} = 1.25$ m, $\frac{v_2^2}{2g} = 1.583$ m and $\frac{p_1}{\gamma} = 30$ m(water column), then

$$\frac{p_2}{\gamma} = \frac{p_1}{\gamma} + \frac{v_1^2}{2g} - \frac{v_2^2}{2g} = 30 + 1.25 - 1.583 = 29.67 \text{ m(water column)}$$

That is
$$p_2 = 9800 \times 29.67 = 290.7 \text{ kN/m}^2$$

$$P_2 = p_2 A_2 = 290.7 \times \frac{\pi \times 2^2}{4} = 913.5 \text{ kN}$$

$$P_1 = p_1 A_1 = 9800 \times 30 \times \frac{\pi \times 3^2}{4} = 2\,078 \text{ kN}$$

Finally, the force R' of the solid boundary on the flow can be calculated using the projection equations of momentum. Assuming the direction of R' as shown in Fig. 3-28, the projection equation of momentum in the x axis direction is

$$P_1 - 2P_2\cos\alpha - R'_x = \rho\frac{Q}{2}(v_2\cos\alpha + v_3\cos\alpha) - \rho Q v_1$$

$$= \rho Q(v_2\cos\alpha - v_1)$$

$$R'_x = P_1 - 2P_2\cos\alpha - \rho Q(v_2\cos\alpha - v_1)$$

$$= 2\,078 - 2 \times 913.5 \times \frac{1}{2} - 1 \times 35 \times \left(5.57 \times \frac{1}{2} - 4.95\right)$$

$$= 1\,240.3 \text{ kN}$$

The projection equation of momentum in the y axis direction is

$$-R'_y + 2P_2\sin\alpha = \rho\frac{Q}{2} \cdot (-v_2\sin\alpha) + \rho\frac{Q}{2} \cdot (-v_3\sin\alpha)$$

$$= -\rho Q v_2 \sin\alpha$$

$$R'_y = 2P_2\sin\alpha + \rho Q v_2 \sin\alpha$$

$$= 2 \times 913.5 \times 0.866 + 1 \times 35 \times 5.57 \times 0.866$$

$$= 1\,751 \text{ kN}$$

$$R' = \sqrt{R'^2_x + R'^2_y} = \sqrt{1\,240.3^2 + 1\,751^2} = 2\,145.8 \text{ kN}$$

$$\theta = \arctan\frac{R'_y}{R'_x} = \arctan\frac{1\,751}{1\,240.3} = 54.69°$$

$$R = -R' = -2\,145.8 \text{ kN (dynamic total force)}$$

The direction of the force is to the right, upwards.

3.4.3.3 Total horizontal force of flow on overflow dam

【Example 3-10】 Fig. 3-30 shows an overflow dam. When the water flows over the spillway of the dam, the streamlines are bent sharply, which is rapidly varied flow. The horizontal force R_x of water on the dam and the force R'_x of the dam acting on the flowing water is a pair of force and the reacting force, which has the same value but in opposite direction. This force can be obtained by the momentum equations.

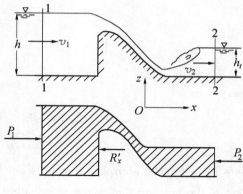

Fig. 3-30

Solution:

Firstly, choose the x axis as the horizontal axis with the positive direction to the right. Then choose the control volume, which is comprised of cross-sections $1-1$ and $2-2$ of gradually varied flow, the dam surface and the free surface of flow, shown in Fig. 3-30.

Assume the width of the overflow dam is b and the upstream depth at section $1-1$ is h. The average velocity at cross-section $1-1$ is v_1 and the downstream depth of the dam is h_1. The average velocity at cross-section $2-2$ is v_2. The external forces on the control volume in the x axis direction include the dynamic pressure force P_1 on cross-section $1-1$, the pressure force P_2 on cross-section $2-2$, and the force R'_x of water flow on the overflow dam surface. R'_x includes the x-axis projection of the flow frictional force on the dam surface. There is no projection of gravity G on the x axis.

The dynamic pressure forces P_1 and P_2 can be calculated by the hydrostatic force distribution based on $P = p_c A = \gamma h_c b h$, where p_c is the average pressure at the centroid of cross-section. The depth of centroid on cross-section $1-1$ is $\dfrac{h}{2}$, and the depth of centroid on cross-section $2-2$ is $\dfrac{h_1}{2}$. Therefore,

$$P_1 = \frac{1}{2}\gamma b h^2, \quad P_2 = \frac{1}{2}\gamma b h_1^2$$

If the hydrostatic pressure distribution diagram is used for analysis, the same result can be obtained.

The momentum equation along the x axis is

$$P_1 - P_2 - R'_x = \rho Q(\beta_2 v_{2x} - \beta_1 v_{1x})$$

Then

$$\frac{1}{2}\gamma b h^2 - \frac{1}{2}\gamma b h_1^2 - R'_x = \rho Q(\beta_2 v_{2x} - \beta_1 v_{1x})$$

where,

$$v_{1x} = v_1 = \frac{Q}{A_1} = \frac{Q}{bh}$$

$$v_{2x} = v_2 = \frac{Q}{A_2} = \frac{Q}{bh_1}$$

letting

$$\beta_1 = \beta_2 = \beta$$

gives

$$R'_x = \frac{1}{2}b\left[\gamma h^2 - \gamma h_1^2 - \frac{2\beta\rho Q^2}{b^2}\left(\frac{1}{h_1} - \frac{1}{h}\right)\right]$$

$$R_x = -R'_x$$

where R_x is the total horizontal force of water flow acting on the overflow dam, and R'_x is the total forces of the overflow dam surface (including the upstream and downstream dam surfaces) acting on the water flow.

3.4.4 Similarities and differences between the momentum equation and energy equation

(1) Their derivations are both based on steady flow, and they require that the cross-section of flow should be gradually varied flow.

(2) They both are the dynamic equations based on Newton's second law on the fluid flow: $\sum F = ma$. The energy equation is related to work and energy and is often used to calculate the kinematic parameters (e.g. pressure in common); the momentum equation is related to the external forces acting on the fluid and is often used to calculate the total dynamic force on a session with rapidly varied flow.

(3) The energy equation is a scalar equation, which does not need projection. The momentum equation is a vector equation, whose projection on the positive coordinate axis needs to be considered. In practical application, it is more convenient to apply the momentum equation along the direction of axis.

(4) By applying the energy equation, the length of the flow between two cross-sections of gradually varied flow is of no limit and can be long or short, and depends on requirements. When applying the momentum equation, the length of any rapidly varied flow between two cross-sections of gradually varied flow should be as short as possible, so that the tangent friction of flow against the solid body can be neglected.

(5) When applying the energy equation, all the head loss of flow should be taken into account as a whole. When applying the momentum equation, all the external forces on the control volume should be taken into account as a whole.

Chapter summary

This chapter introduced the basic concepts and approaches for describing the motion of a fluid, as well as the basic equations of governing the actual motion of the fluid. It is one of the most important chapters in this book on fluid mechanics.

Chapter 3 Basic equations of steady total flow

1. Two approaches have been used to describe the motion of a fluid: Lagrangian and Eulerian. The Eulerian approach, which has been widely used in the study of fluid motion, takes the whole space point of the flow field as a subject and gathers the movement of fluid particles at every space point at any moment, in order to describe the entire motion of fluid.

2. Using the Eulerian approach, the flow is divided into steady and unsteady flow; one-dimensional, two-dimensional and three-dimensional flow; uniform and non-uniform flow. In the Eulerian approach, a streamline is a vector line which directly reflects the instant flow direction in a flow field. On the basis of streamline, streamtube, flow cross-section, streamtube flow, total flow, volume rate of flow (discharge), average velocity of cross-section, and other concepts are introduced.

3. Three basic equations of total flow are established: continuity equation, energy equation, and momentum equations.

Continuity equation: $v_1 A_1 = v_2 A_2$

Energy equation: $z_1 + \dfrac{p_1}{\gamma} + \dfrac{\alpha_1 v_1^2}{2g} = z_2 + \dfrac{p_2}{\gamma} + \dfrac{\alpha_2 v_2^2}{2g} + h_w$

Momentum equations:

$$\left.\begin{array}{l} \sum F_x = \rho Q(\beta_2 v_{2x} - \beta_1 v_{1x}) \\ \sum F_y = \rho Q(\beta_2 v_{2y} - \beta_1 v_{1y}) \\ \sum F_z = \rho Q(\beta_2 v_{2z} - \beta_1 v_{1z}) \end{array}\right\}$$

The above three basic equations actually respresent the principles of conservation of mass, conservation of energy, and conservation of momentum of the total flow, which are also the most important principles in fluid mechanics.

Multiple-choice questions (one option)

3-1 The acceleration of fluid particle $\boldsymbol{a}$ in Eulerian approach equals _____.

(A) $\dfrac{\partial \boldsymbol{u}}{\partial t}$ (B) $(\boldsymbol{u} \cdot \nabla)\boldsymbol{u}$

(C) $\dfrac{\partial \boldsymbol{u}}{\partial t} + (\boldsymbol{u} \cdot \nabla)\boldsymbol{u}$ (D) $\dfrac{\partial \boldsymbol{u}}{\partial t} - (\boldsymbol{u} \cdot \nabla)\boldsymbol{u}$

3-2 Steady flow is the flow in a flow field, _____.

(A) which has the same distribution of flow velocity in every section

(B) in which the streamlines are parallel straight lines

(C) where the flow parameters don't change with time

(D) in which the flow changes with time in a certain law

3-3 One-dimensional flow refers to _____.

(A) the flow in which the flow parameters are a function of spatial coordinates and time

(B) the flow with the linear distribution of velocity

(C) the uniform linear flow

(D) the flow in which the flow prameters change with time

3−4　The _____ acceleration of uniform flow is zero.
　　(A) local　　　　　　　　　　(B) convective
　　(C) centripetal　　　　　　　　(D) particle

3−5　In _____ flow, streamline coincides with pathline.
　　(A) steady　　　　　　　　　　(B) unsteady
　　(C) incompressible　　　　　　(D) one-dimensional

3−6　The continuity equation means that the movement of fluid follows the principle of the conservation of _____.
　　(A) energy　　　　　　　　　　(B) momentum
　　(C) mass　　　　　　　　　　　(D) flow rate (discharge)

3−7　In the _____ flow, the Bernoulli equation is false.
　　(A) steady　　　　　　　　　　(B) ideal liquid
　　(C) incompressible　　　　　　(D) compressible

3−8　In the Bernoulli equation of total flow, velocity v means the velocity of _____.
　　(A) a point　　　　　　　　　　(B) cross-sectional average
　　(C) the centroid of cross-section　(D) the largest at the cross-section

3−9　In the Bernoulli equation of total flow, $z + \dfrac{p}{\gamma}$ has to be the _____ in the cross-section of uniform flow or gradually varied flow.
　　(A) piezometric head of any point
　　(B) average piezometric head
　　(C) piezometric head at the centroid
　　(D) largest piezometric head

3−10　Venturi meter is used to measure _____.
　　(A) point velocity of flow　　　(B) pressure
　　(C) density　　　　　　　　　　(D) discharge

3−11　Pitot tube is used to measure _____.
　　(A) point velocity of flow　　　(B) pressure
　　(C) density　　　　　　　　　　(D) discharge

3−12　When applying the energy equation of total flow, between the two cross-sections _____.
　　(A) it must be a gradually varied flow
　　(B) it must be a rapidly varied flow
　　(C) the rapidly varied flow cannot occur
　　(D) the rapidly varied flow may occur

3−13　When the momentum equation of total flow is applied for the resultant force of fluid on an object, the inlet and outlet pressure should be _____.
　　(A) absolute pressure　　　　　(B) relative pressure

Chapter 3 Basic equations of steady total flow

(C) atmospheric pressure (D) vacuum degree

3-14 In the energy equation, $z + \dfrac{p}{\gamma} + \dfrac{\alpha v^2}{2g}$ means _____.
(A) the mechanical energy per unit mass of liquid
(B) the mechanical energy per unit weight of liquid
(C) the mechanical energy per unit volume of liquid
(D) the total mechanical energy of cross-section

3-15 The total head of the steady total flow of viscous liquid is to _____.
(A) decrease along the flow (B) increase along the flow
(C) remain the same (D) be unknown

3-16 The piezometric head line of the steady total flow of viscous liquid is to _____.
(A) decrease along the flow (B) increase along the flow
(C) remain the same (D) be unknown

3-17 The direction of the steady total flow should be _____.
(A) from higher position to lower position
(B) from the position with higher pressure to that with lower pressure
(C) from the place with higher velocity to one with lower velocity
(D) from the position with higher mechanical energy per unit weight to that with lower mechanical energy per unit weight

Problems

3-1 Determine the accelerations in the velocity field, where $u_x = x + t$, $u_y = -y + t$, and $u_z = 0$.

3-2 If the moving path of a fluid particle is:
$$x = 2 + 0.001 \sqrt{t^5}$$
$$y = 2 + 0.001 \sqrt{t^5}$$
$$z = 2$$
find the acceleration of the liquid particle at $x = 8$ m

3-3 The component of velocity at a point in a fluid flow: $u_x = u$, $u_y = 0$, $u_z = 0$, in which u is constant. Find the streamline equation of the fluid flow.

3-4 The component of velocity at a point in a fluid flow: $u_x = -\dfrac{Ky}{x^2 + y^2}$, $u_y = -\dfrac{Kx}{x^2 + y^2}$, $u_z = 0$, where K is constant. Find the streamline and pathline of the fluid flow.

3-5 As shown in Fig. 3-31, identify whether the flow in the AB section of each pipe is steady flow, unsteady flow, uniform flow, or non-uniform flow.

3-6 Draw the flow pattern (streamline) in the various configurations of pipe as shown in Fig. 3-32.

3-7 The flow from a water tank moves along three pipes in series (with pipe diameters being $d_1 = 10$ cm, $d_2 = 5$ cm, $d_3 = 2.5$ cm respectively) and ends up to the atmosphere, as shown

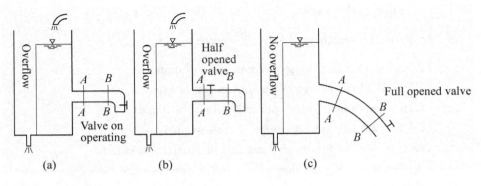

Fig. 3-31

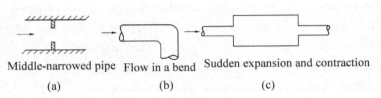

Fig. 3-32

in Fig. 3-33. The flow velocity at exit is 1 m/s. Find the discharge and average cross-sectional velocities in AB and BC.

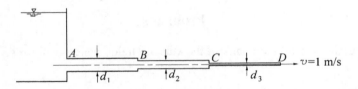

Fig. 3-33

3-8 A piping system is shown in Fig. 3-34, where $d_1 = 0.25$ cm, $d_2 = 0.15$ cm, $d_3 = 0.1$ cm, $v_3 = 1.0$ m/s, $q_1 = 50$ L/s, and $q_2 = 21.5$ L/s. Calculate the average velocity of flow in all the pipes.

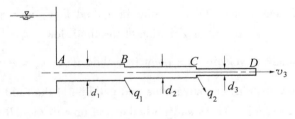

Fig. 3-34

3-9 As shown in Fig. 3-35, from a water tank water is distributed to the customers through a main pipeline of AB connected by a branched pipe of BC and BD. Given $d_1 = 0.4$ m, $v_1 = 2.15$ m/s, $d_2 = 0.3$ m, $v_3 = 5$ m/s, find the flow rate in the main pipeline.

3-10 The cross-section of a ventilation tunnel is 50×50 cm^2, as shown in Fig. 3-36,

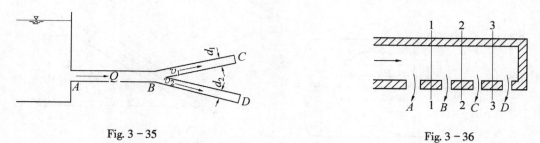

Fig. 3-35   Fig. 3-36

where the air is transported through four outlets at A, B, C and D with each having an area of 40×40 cm^2. When the velocity of 5 m/s is required at every air supply outlet, find the average cross-sectional velocites in sections $1-1$, $2-2$ and $3-3$.

3-11 A water supply pipe is designed to have a carrying capacity of 2.94×10^3 kN/h, and the flow velocity is limited within $0.9 \sim 1.4$ m/s. Determine the pipe diameter and the flow velocity for the selected pipe diameter (it has to be a multiple of 50 mm).

3-12 As shown in Fig. 3-37, the diameter of the front end of a steam main $d_0 = 50$ mm, the flow velocity $v_0 = 25$ m/s, and $\rho_0 = 2.62$ kg/m^3. The connected branch pipes are $d_2 = 40$ mm and $d_1 = 45$ mm, and $\rho_1 = 2.24$ kg/m^3, $\rho_2 = 2.30$ kg/m^3. If the mass per unit time in each branch pipe is the same, find the mean velocities (v_1 and v_2) of the two branch pipes.

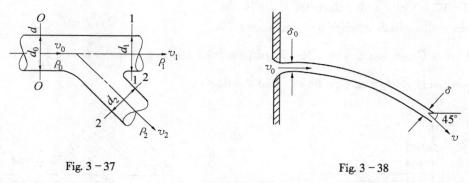

Fig. 3-37   Fig. 3-38

3-13 As shown in Fig. 3-38, a water jet was ejected horizontally from a narrow opening with a thickness of $\delta_0 = 0.03$ m and a mean velocity $v_0 = 8$ m/s. Assuming that the water jet is affected by gravity and the horizontal speed remains the same, calculate

(1) the average velocity v when the inclined angle is $\theta = 45°$;

(2) the thickness δ.

3-14 The flow velocity of a pipe is axis-symmetrical distributed as shown in Fig. 3-39. The velocity distribution function is $u = \dfrac{u_{max}}{r_0^2}(r_0^2 - r^2)$, where u is the flow velocity at the position with a distance of r to the axis. Given $r_0 = 3$ cm and $u_{max} = 0.15$ m/s, find the flow discharge Q and the average flow velocity v_0 of cross-section.

3-15 A pipe of varying diameter is shown in Fig. 3-40. Given the diameter $d_1 = 15$ cm in the cross-section $1-1$, its center pressure is $p_1 = 70$ kN/m^2, the diameter of cross-section $2-2$ is

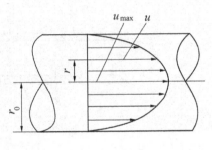
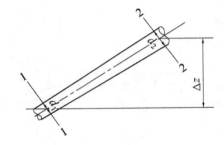

Fig. 3 – 39

Fig. 3 – 40

$d_2 = 30$ cm, its center pressure is $p_2 = 60$ kN/m², the mean velocity is $v_2 = 1.5$ m/s, and the elevation difference between the centers of two cross-sections is $\Delta z = 1$ m. Determine the water direction in the pipe and the head loss h_w between the two cross-sections.

3 – 16 A pipe is used to draw water from a large water tank ($v_0 \approx 0$), as shown in Fig. 3 – 41. Given $H = 4$ m, $d_1 = 2d_2$, $d_2 = 20$ cm, and the overall head loss (h_w) of the pipe triples the velocity head of the pipe d_2, find the flow velocity and discharge of the two pipes.

3 – 17 A pipe with a diameter of $d = 10$ cm is to draw water from a water tank, as shown in Fig. 3 – 42. Given that the head loss h_w of pipe from entrance to outlet is $0.8 \dfrac{v^2}{2g}$ (v is the mean cross-sectional velocity of pipe flow), find the flow Q.

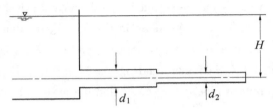

Fig. 3 – 41

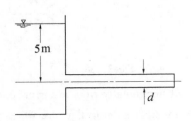
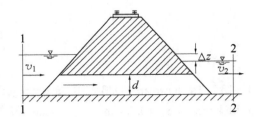

Fig. 3 – 42

Fig. 3 – 43

3 – 18 A division channel will pass through a railway, where a circular cross-section culvert will be built under the base of the railway, as shown in Fig. 3 – 43. Given the design flow capacity of the culvert is $Q = 1$ m³/s, the water level difference between the upstream and downstream of the culvert is $\Delta z = 0.3$ m, the head loss of the culvert is $h_w = 1.47 \dfrac{v^2}{2g}$ (v is the flow velocity in the culvert). If the flow velocity of the upstream and downstream channel of the culvert is approximately equal ($v_1 \approx v_2$), find the diameter (d) of the culvert.

3 – 19 As shown in Fig. 3 – 44, when the valve K is completely closed, the reading of the pressure gauge is 98 kPa; when the valve K is open, the reading of the pressure gauge reduces to

58.8 kPa. If the water level in the water tank remains the same, and the head loss of the pipe before the gauge is 0.5 m water column, find the average cross-sectional velocity and discharge of the pipe.

3-20 The flow passes through a vertically placed Venturi meter, as shown in Fig. 3-45. Given $d_1 = 40$ mm, $\dfrac{d_1}{d_2} = 2$, the reading of the mercury differential manometer is $\Delta h = 30$ mm, and the head loss between the cross-sections is $0.05 \dfrac{v_2^2}{2g}$, find the average cross-sectional velocity and discharge of flow at the throat.

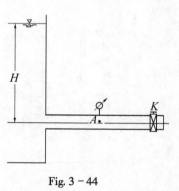

Fig. 3-44

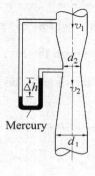

Fig. 3-45

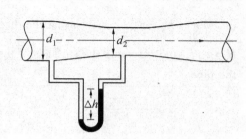

Fig. 3-46

3-21 In a horizontally placed Venturi meter, as shown in Fig. 3-46, the pressure difference measured by a mercury differential manometer is $\Delta h = 20$ cm. Given the diameter of the pipe is $d_1 = 15$ cm and the diameter at the throat is $d_2 = 10$ cm, if the head loss is neglected, find the flow rate of pipe Q.

3-22 As shown in Fig. 3-47, the height of piezometric tube measured at cross-section 1-1 in a pipe is $\dfrac{p_1}{\gamma} = 1.5$ m. The flow area (A_1) of cross-section 1-1 is 0.05 m^2, the area of cross-section 2-2 is 0.02 m^2, the head loss h_w between the two cross-sections is $0.5 \dfrac{v_1^2}{2g}$, and the flow rate (Q) of the pipe is 20 L/s. Find the level of piezometric tube $\dfrac{p_2}{\gamma}$ at the 2-2 cross-section. Given $z_1 = 2.6$ m, $z_2 = 2.5$ m.

3-23 The suction pipe of an oil pump is shown in Fig. 3-48. Given that the exit pressure is -4.08 m (the height of oil column), the diameter of the suction pipe at the bottom is $d_1 = 1$ m, the diameter on the top is $d_2 = 0.5$ m, the energy loss in the pipe is ignored, and the specific weight of oil is $\gamma_{oil} = 8330$ N/m^3, find the flow rate and the pressure at the bottom of the oil suction pipe.

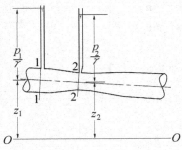
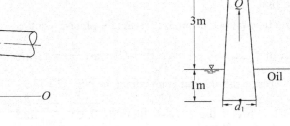

Fig. 3-47   Fig. 3-48

3-24 As shown in Fig. 3-49, a pipe with the diameter of $d = 0.5$ m is used to extract water from the river into the water-collecting well, and the overall head loss of flow from pipe to the well is $h_w = 5\dfrac{v^2}{2g}$. Given that the discrepancy of water level between the river and the well is $H = 2$ m, and assuming the surfaces of the river and the well are very large, find the flow rate in the pipe.

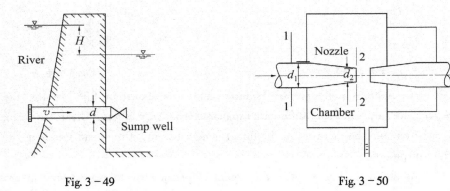

Fig. 3-49   Fig. 3-50

3-25 The pipe diameter of a water jet is $d_1 = 50$ mm, and the orifice diameter is $d_2 = 25$ mm, as shown in Fig. 3-50. The height of pressure in the cross-section 1-1 is 1180 mm of mercury column, and the flow rate Q is 10 L/s. If the head loss of the nozzle is neglected, find the degree of vacuum in the working chamber.

3-26 A pumping system is shown in Fig. 3-51, where the water in the container M can be extracted out by using the negative pressure caused by a water jet in the throat cross-section. Given H, b, h, and the negligible head loss, under what conditions between the cross-sectional area A_1 at the throat and the nozzle cross-sectional area A_2 will the pumping system start to work?

3-27 A transition section of the pressurized water pipe of a hydropower station is as shown in Fig. 3-52. The diameters $d_1 = 1.5$ m and $d_2 = 1$ m, the pressure p_1 at the starting section is 400 kN/m² (relative pressure), and the flow rate Q is 1.8 m³/s. If the head loss is neglected, find the axial thrust on the anchor block at the transition section.

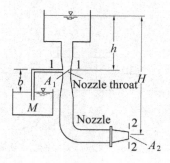
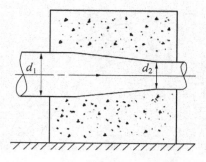

Fig. 3 – 51   Fig. 3 – 52

3 – 28 A horizontally placed 90° pressured pipe bend is shown in Fig. 3 – 53. Given the diameter $d = 15$ cm and flow velocity $v = 2.5$ m/s, the hydrodynamic pressure in cross-sections 1 – 1 and 2 – 2 is $\dfrac{p_1}{\gamma} = \dfrac{p_2}{\gamma} = 14$ m (water column). In order to prevent the movement of the bend due to the flow of water, the bend is fixed with a concrete anchor block. The friction coefficient of the anchor block against the ground is $f = 0.3$, and the specific weight of concrete is $\gamma = 23.62$ kN/m^3. Find how many cubic meters of concrete are needed to build the anchor block.

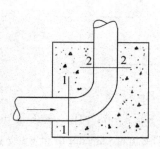
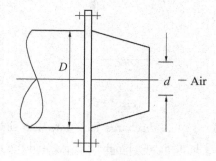

Fig. 3 – 53   Fig. 3 – 54

3 – 29 As shown in Fig. 3 – 54, the exit diameter of the nozzle is $d = 5$ cm, the diameter (D) of pipe is 15 cm. The nozzle is connected to the pipe with 4 bolts. Given the velocity in the pipe $v_1 = 3.0$ m/s and the head loss is ignored, find the flow rate Q of pipe and the tensile force of every bolt.

3 – 30 A horizontal pipe, as shown in Fig. 3 – 55, includes a bending section where the diameter gradually changes from $d_1 = 30$ cm to $d_2 = 20$ cm. Given that the pressure in the center of the cross-section 1 – 1 is $p_1 = 35$ kN/m^2, and the flow rate is $Q = 150$ L/s, neglecting the head loss, find the magnitude and the direction of horizontal force of flow acting on the curved section.

3 – 31 The 90° transition bending section of a diversion pipe is shown in Fig. 3 – 56. The center line of the pipe is on a horizontal plane. The diameter of the entry section 1 – 1 is $d_1 = 25$ cm, the relative pressure is $p_1 = 200$ kN/m^2, the diameter of exit cross-section 2 – 2 is $d_2 = 20$ cm, the flow rate Q is 150 L/s, and the head loss is neglected. Find the force needed to fix the bending section.

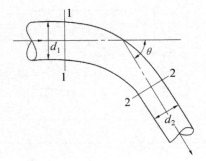

Fig. 3 – 55

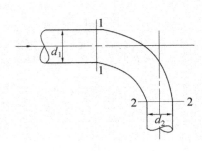

Fig. 3 – 56

3 – 32 The width of a flat floodgate is $b = 2$ m, the water depth in front of the gate is $H = 4$ m when the flow rate under the gate $Q = 8$ m³/s, and the water depth at the contraction section after the floodgate is $h_c = 0.5$ m, as shown in Fig. 3 – 57. Find the total water pressure force acting on the gate (the friction is not considered).

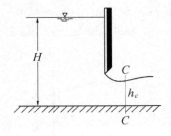

Fig. 3 – 57

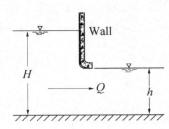

Fig. 3 – 58

3 – 33 A discharge lock gate with supporting wall is shown in Fig. 3 – 58. The width of the opening is 3 m, the height is 2 m, and the flow rate $Q = 45$ m³/s. The upstream depth of water is $H = 4.5$ m and the downstream depth of water is $h = 2$ m. The bottom of the lock is horizontal. Calculate the horizontal force on the supporting wall of the gate and compare this with the results if the hydrostatic pressure distribution is considered.

3 – 34 When the flow rate through an overflow dam is $Q = 40$ m³/s and the upstream depth of water is $H = 10$ m, the water depth at the contraction section of the dam is $h_c = 0.5$ m. The length of dam is $l = 7$ m. As shown in Fig. 3 – 59, calculate the horizontal total force of flow on the dam.

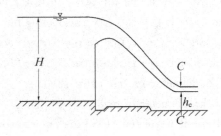

Fig. 3 – 59

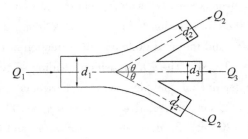

Fig. 3 – 60

3 - 35　A four-way fork tube is shown in Fig. 3 - 60 and all the axis lines are in the horizontal plane. The flow rate in two ends is $Q_1 = 0.2$ m³/s and $Q_3 = 0.1$ m³/s. The dynamic pressure at the corresponding cross-section is $p_1 = 20$ kN/m² and $p_3 = 15$ kN/m². The other two branch pipes are directly open to the atmosphere. The diameters of each pipe are $d_1 = 0.3$ m and $d_3 = d_2 = 0.2$ m, and $\theta = 30°$. Calculate the force of flow on the wall at the forks (the friction is ignored).

Chapter 4 Types of flow and head loss

In Chapter 3, the flow head loss (h_w) was introduced in the energy equation of total flow, but what factors are related to the head loss h_w and how to calculate it have not been discussed. Since the head loss is related to the type of flow, the different flow types follow different rules. Therefore, this chapter will focus on two kinds of flow and the calculation of corresponding head losses.

4.1 The classification of flow resistance and head loss

From the energy viewpoint, energy loss is the dissipative heat energy that is converted from effective mechanical energy. In fluid mechanics, the loss of mechanical energy is called the head loss. There are internal and external causes for the head loss. The internal cause is that a real fluid has viscosity and a transverse velocity gradient $\frac{du}{dy}$, which both generate a frictional shear stress τ. Such a frictional resistance is due to the relative motion between fluid particles attached on the solid boundary and their adjacent fluid particles of a different velocity, rather than between the flow of fluid and a solid boundary, because there is no relative sliding between the flow of fluid and the solid boundary. Therefore, this frictional resistance is called internal frictional resistance. Hence, this frictional resistance is called internal frictional resistance. Internal frictional resistance always exists as long as the fluid is in motion. To overcome the resistance, the fluid must consume part of its mechanical energy to do work. The external cause is due to either the local sudden change of a solid boundary in either shape or size, such as a sudden expansion or contraction, bends in pipes and open-channels, or the partial blockage, such as valves in a pipeline or piers in open-channels. In such cases, the internal structure of the fluid flow will produce separation and vortices, curved streamline, and the re-distribution of the flow velocity. These changes will increase the relative movement of fluid, which will result in pressure resistance, so part of the mechanical energy will be dissipated. The internal cause is the basis while the external cause is the condition. For an ideal fluid without viscosity, even though there are significant changes in the shape or size of boundary, it will only result in the transformation of mechanical energy mutually, but no head loss.

4.1.1 The classification of flow resistance

The flow resistance includes the resistance of friction and pressure. The frictional resistance is composed of shear stress, which is due to the viscosity of real fluid. The resistance of pressure (or form resistance) is generated by the pressure difference between the upstream and downstream faces of a solid boundary.

Now, consider flow around a cylinder as an example to explain how the resistance of pressure is generated.

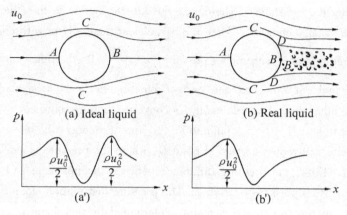

Fig. 4 − 1

Fig. 4 − 1a shows the flow of an ideal fluid around a cylinder. Now analyze the streamline that is facing the center of the cylinder. When the fluid particle arrives at the stagnation point A, with the velocity u_0 along the streamline decreased to zero as the flow is blocked, all the kinetic energy will be converted into the pressure energy, $\dfrac{u_0^2}{2g} = \dfrac{p}{\gamma}$, in other words, $p = \dfrac{1}{2}\rho u_0^2$. In the convex section AC outside the cylindrical boundary, the area of flow cross-section becomes smaller and then the streamlines become dense. Such an area is an accelerating region with lower pressure, in other words, $\dfrac{\partial p}{\partial x} < 0$, where x is the coordinate of flow path. When the particle reaches point C, the flow velocity increases to a maximum and the pressure decreases to a minimum. Because it is an ideal fluid, which has no viscosity and head loss, the reduced value of pressure energy is converted into the equivalent kinetic energy at point C. On the contrary, in the downstream section CB of the cylinder, the area of flow cross-section increases and then the streamlines become sparse. So such an area is a decelerating region with higher pressure, in other words, $\dfrac{\partial p}{\partial x} > 0$. When the particle reaches point B, the reduced value of kinetic energy will be converted into the equivalent pressure energy at point B. Because there is no head loss, after the two processes of conversion (pressure energy into kinetic energy, and then kinetic energy into pressure energy), the pressure at point B is equal to that at point A, and they both equal $\dfrac{1}{2}\rho u_0^2$, so there is no pressure difference. Its pressure distribution along the boundary is shown in Fig. 4 − 1a′.

Fig. 4 − 1b shows the flow of a real fluid around the cylinder. The main difference between the real and ideal fluid is that the real fluid has viscosity, internal friction and head loss, while the ideal fluid does not have. When the particles of fluid pass through the area of AC, which is an accelerating region with lower pressure, i. e. $\frac{\partial p}{\partial x} < 0$, the reduced value of pressure energy cannot be fully converted into the kinetic energy at point C. In this process, part of mechanical energy is used for overcoming the resistance. Therefore, the kinetic energy is less than that of the ideal fluid flow under the same condition. When the particles sequentially pass through the area of CB, which is a decelerating region with higher pressure, i. e. $\frac{\partial p}{\partial x} > 0$, the flow cannot remain the same kinetic energy of incident flow, so one part of kinetic energy is dissipated to overcome the resistance and the other part of kinetic energy is converted into pressure energy. Therefore, when the particle reaches point D (before point B), the kinetic energy will be dissipated completely and the flow velocity will reduce to zero at point D, where the pressure is smaller than that at the stagnation point A. Thus, the pressure difference between the upstream and downstream of the solid boundary leads to a pressure resistance. The pressure distribution along the path is shown in Fig. 4-1b'. The pressure resistance is mainly determined by the shape of object, so it is also called the form resistance. The pressure resistance can be expressed as the product of the kinetic energy $\frac{\rho u_0^2}{2}$ in unit volume and a representative area, multiplied by a drag coefficient (C_p):

$$F_p = C_p \frac{\rho u_0^2}{2} A_p \qquad (4-1)$$

where C_p is also called the coefficient of pressure resistance, which is a complicated parameter depending on the shape of object and the flow boundary conditions. The values of C_p are usually determined by experiments and presented as a diagram for use. ρ is the density of fluid and u_0 is the velocity of incoming flow. A_p is the projection area of object on the plane perpendicular to the direction of flow.

Point D is also called the separation point. At this point, the main flow breaks away from the cylindrical boundary whereas in the region downstream the boundary it will form a wake zone with a strong eddy due to the flow continuum. In the wake zone, the formation, exchange, break-up, and reformation of the vortex needs the mechanical energy from the flow so that the head loss will increase.

Another separation phenomenon, which is similar to the stagnation separation mentioned above, is called the inertial separation. Due to the existence of inertia, the flow cannot adapt itself to the rapid change of boundary or the sharp change at the corners of boundary. The fluid will follow the main flow due to its own inertia, so some separations exist between the main flow and the boundary, as shown in Fig. 4-2.

The common characteristics of the stagnation and inertial separations are given as follows:
- They both produce the separation phenomenon in the eddy area.
- The velocity distribution of flow varies rapidly. Fig. 4-2a shows a sudden expansion of

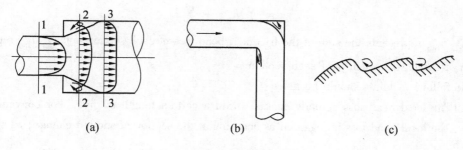

(a)   (b)   (c)

Fig. 4−2

pipe. The flow velocity distribution at cross-section 1 − 1 is of larger average velocity and gradient. At cross-section 2 − 2 in the separation area of inertia, there exist both positive and negative velocities. In addition, the velocities distribution at cross-section 3 − 3 is of lower average velocity and gradient. Obviously, such a rapid change in the velocity distribution will accelerate the relative movement of fluid particles so that the loss of energy will increase.

The frictional resistance of flow always exists in a real fluid. However, the pressure resistance (or form resistance) depends on the pressure distribution on the surface of an object.

4.1.2 The classification of head losses

In hydraulics, the head loss is divided into frictional head loss and local head loss.

(1) Frictional head loss

The frictional head loss is the head loss that is the dissipated energy per unit weight by the work of flow done to overcome friction at a boundary, denoted by h_f, which represents the frictional head loss per unit weight of fluid. It increases with distance along the flow path. The frictional head loss is the main loss in long pipelines and open channels.

(2) Local head loss

The local head loss occurs in the local area around a solid boundary, which has a sudden change in shape or size. This loss is also known as the local friction loss (minor loss), denoted by h_j, which represents the local head loss per unit weight of fluid. The local head loss is mainly used to maintain the eddy motion and relates strongly to the shape of boundary. It usually occurs in the regions of rapidly varied flow, for example, the entrances of pipelines or open channels, sudden expansions or contractions, bends and valves.

4.1.3 The superposition principle of head losses

In practical engineering, there usually exist both the frictional and local head losses. For simplicity of calculation in engineering, the effects of the frictional and local head losses are considered separately, i.e. their effects do not affect each other so the two losses can be applied by superposition. The sum of the frictional and local head losses in a flow system is called the total head loss. Therefore, the total head loss of real fluid in the energy equation of total flow can be expressed by the following formula:

$$h_{w_{1-2}} = \sum h_{f_{1-2}} + \sum h_{j_{1-2}} \qquad (4-2)$$

where $\sum h_{f_{1-2}}$ represents the sum of the frictional head loss of each section and $\sum h_{j_{1-2}}$ represents the sum of the local head loss of each section.

The following points should be noted:

(1) The local head loss actually occurs within a certain length of pipe. For convenience of analysis, the local head loss is regarded as occuring at the section of sudden change, as shown in Fig. 4 – 3.

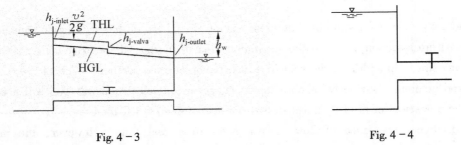

Fig. 4 – 3   Fig. 4 – 4

(2) The frictional head loss within the length of flow is considered as an independent head loss, which is not affected by the local head loss.

(3) If two local head losses are close to each other (e. g. the entry to a pipe followed by a valve in Fig. 4 –4), the local head losses calculated by the superposition method tend to be larger than the actual losses. In such cases the overall value of h_j should be determined by the experiments.

(4) The head loss is usually expressed as a multiple of velocity head, and the reasons are as follows:

• The different positions on a cross-section have different values of potential energy z and pressure energy $\frac{p}{\gamma}$, but the velocity head $\frac{v^2}{2g}$ (expressed by the average cross-sectional velocity) is the same for each point on the section.

• Observations in practical engineering show that the head loss is often not related to the potential energy or pressure energy directly, but is closely related to flow velocity. For example, the head loss increases with increasing flow velocity, and vice versa.

Therefore, it is appropriate to use a certain multiple of the velocity head to express the head loss.

4.2 Two regimes of real fluid flow

4.2.1 Reynolds' experiment

In 1880, the British scientist Osborne Reynolds carried out a well-known test. The schematic diagram of the test apparatus is shown in Fig. 4 – 5. The tank has an overflow device for keeping a constant water level so that the water flow is constant in the long glass tube with a diameter of d. The tube has a bellmouth inlet inside the sidewall of water tank (D), and the tube, selected from a large number of ordinary glass tubes, was almost uniform with the diameter difference being less than 0.794 mm. Reynolds selected three tubes with the diameter being 0.026 8 m, 0.015 27 m and 0.007 886 m, respectively. Near the end of glass tube is a control valve (K), which is far from the test section. The needle-typed tube (A) inside the bellmouth of glass tube can be used to inject a filament of coloured water and the coloured water should be chosen to be of similar density of the test liquid. The other end of the needle-typed tube (A) was connected to the base of tank C, which held the coloured water. The flow velocity of the coloured water was controlled by the small valve P.

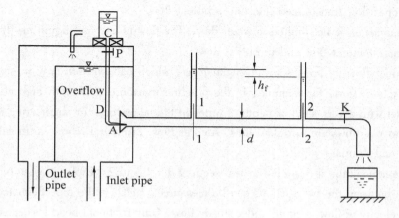

Fig. 4-5

In the test, Reynolds paid particular attention to two aspects:
- Try to keep the water in the tank static.
- Try to keep the temperature of water in the tank uniform.

At the beginning of test, the valve K was opened slightly so that the water in the tube moved slowly and at the same time the coloured water moved at a small velocity, and thus in the entire glass tube, the coloured water formed a single smooth filament. Afterwards, the valve K was gradually opened to increase the flow velocity. The coloured water continued to maintain a smooth filament until the valve K was opened to a certain degree; in other words, the flow velocity was increased to a certain degree, at which point the

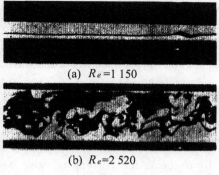

(a) $Re = 1\,150$

(b) $Re = 2\,520$

Fig. 4-6

filament of coloured water at a certain distance from the entry of tube suddenly changed from a linear to a wavy line, and the wavy line was bent and thickened gradually, as shown in Fig. 4-6a. The flow velocity was continued to increase until a certain value when the coloured water would diffuse to the entire tube with a vigorous eddy motion, as shown in Fig. 4-6b.

This experiment shows that the same fluid in a pipe can form two types of flow patterns with different characteristics, which depend on the flow velocity. For a small velocity, the fluid particles move linearly and do not influence each other. This type of flow pattern is called laminar flow. For a large velocity, the fluid particles are mixed with each other in the movement along the direction of tube axis. In terms of movement path of each particle, although it is disorganized and random, the flow generally moves forward along the tube axis, but with transverse fluctuations normal to the main direction of flow. This type of flow pattern is called turbulent flow.

When the experiment was carried out in the reverse process, in which the flow was reduced by controlling valve K, the observed phenomenon was repeated in the opposite manner. However, the flow velocity with which the turbulent flow transformed into laminar flow was smaller than that when the laminar flow transformed into the turbulent flow.

The phenomenon is not limited to water flow. The flow of any real liquid has these two kinds of flow patterns: laminar flow and turbulent flow.

The laminar flow is rarely seen in engineering, which can be seen only in small pipes with very low velocity of flow, for example, in the pipeline transporting oil with high viscosity or in a physical model with a very shallow depth of water in the laboratory. In engineering and daily life, the vast majority of flows are turbulent flow, such as flows in natural rivers, artificial channels, or water supply and drainage pipes.

If a piezometer tube is installed at two sections 1-1 and 2-2 of a glass tube respectively, the head loss between the two sections can be measured. Because the pipe diameter is the same, the average velocity of flow is equal under steady flow. Only frictional head loss exists. Therefore, the energy equation can be written as

$$z_1 + \frac{p_1}{\gamma} = z_2 + \frac{p_2}{\gamma} + h_{f_{1-2}}$$

so

$$h_{f_{1-2}} = \left(z_1 + \frac{p_1}{\gamma}\right) - \left(z_2 + \frac{p_2}{\gamma}\right)$$

It means that the water column difference in two piezometer tubes is the frictional head loss from 1-1 to 2-2.

For various openings of the valve K, we can obtain a series of velocity v (measure Q by using volume method and then calculate $v = \frac{Q}{A}$ from continuity equation) and the corresponding frictional head loss h_f. By changing the opening of valve monotonously either from large to small or from small to large, a series of corresponding values of v and h_f can be obtained.

Letting $\lg v$ be the abscissa and $\lg h_f$ the vertical coordinate, the measured data can be shown as in Fig. 4-7. The flow velocity that occurs at the

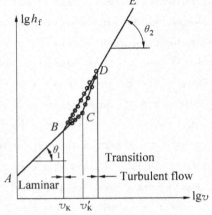

Fig. 4-7

transition of fluid movement is called the critical flow velocity. In the test, as the velocity increases, the laminar flow remains until at point C, when it changes to the turbulent flow. The flow velocity of point C is called the upper critical velocity. If the test is carried out in the reverse process, as the flow velocity decreases, then the turbulent flow will start to change into the laminar flow at point B. The flow velocity of point B is called the lower critical velocity. The fluid flow between B and D can be either turbulent flow or laminar flow, depending on the test procedure. The flow between B and D is called the transition region, which is unstable, as seen the scattered data in Fig. 3 – 7. The line AC and DE are both straight lines, which can be expressed by the following equation:

$$\lg h_f = \lg k + m \lg v$$

where $\lg k$ is the intercept of AC on the vertical coordinate and m the slope of the line.

The formula above can also be expressed as an exponential form:

$$h_f = kv^m \quad (4-3)$$

The experimental results show that the laminar flow is suitable for line AC, $\theta_1 = 45°$, so that $m = \tan\theta_1 = \tan 45° = 1$ and $h_f = kv^1$. Therefore, the frictional head loss of laminar flow is linearly proportional to the flow velocity. The turbulent flow is suitable for the line DE, $\theta_2 = 59° \sim 63.5°$, and the corresponding m equals $1.75 \sim 2$, so that $h_f = kv^{1.75 \sim 2}$. Thus, the frictional head loss of turbulent flow is proportional to the $1.75 \sim 2$ power of the flow velocity. It can be seen that the laminar flow and turbulent flow have different laws of the frictional head loss.

4.2.2 The identification of laminar and turbulent flows

Since the frictional head loss h_f complies with different laws in laminar flow and turbulent flow, the first step is to determine whether a flow is laminar flow or turbulent flow, and then calculate the frictional head loss accordingly. Thus the critical flow velocity could be used as a criterion to identify the flow regime, because it represents the transition between two types of fluid flow. However, further tests show that the critical velocity (v_K) will change with diameter (d) and kinematic viscosity coefficient (v) of liquid, so that it is not a feasible parameter. Furthermore, no matter how d and v change, the dimensionless number of $\dfrac{v_K d}{\nu}$ is roughly constant. $\dfrac{v_K d}{\nu}$ is called the critical Reynolds number, denoted by Re_K. The Reynolds number corresponding to the upper critical flow velocity v'_K is called the upper critical Reynolds number, i.e. $Re'_K = \dfrac{v'_K d}{\nu}$. The number corresponding to the lower critical flow velocity v_k is called the lower critical Reynolds number, i.e. $Re_K = \dfrac{v_K d}{\nu}$. For any actual flow velocity v, the corresponding Reynolds number is $Re = \dfrac{vd}{\nu}$. Through a large number of experiments, it has been proved that the lower critical Reynolds number Re_K of flow in a circular pipe is a relatively stable value.

$$Re_K \approx 2000 \quad (4-4)$$

However, the upper critical Reynolds number is an unstable value, i.e. $Re'_K = 12\,000 \sim$

Fluid Mechanics in Civil Engineering

20 000; in certain cases, it can also be as high as 40 000 ~ 50 000, which depends on the steadiness of flow and whether the flow has disturbance or not. In practical engineering, the disturbance in flow is ubiquitous so that the upper critical Reynolds number is not suitable as a criterion. Therefore, the lower critical Reynolds number is used as a criterion for determining the flow regime. When an actual Reynolds number Re is greater than the lower critical Reynolds number Re_K, the flow is called turbulent flow; otherwise, the flow is laminar flow when Re is less than Re_K.

Flow in open channels can either be laminar or turbulent flow. The lower critical Reynolds number of flow in open channels is

$$Re_K = \frac{v_K R}{\nu} \approx 500 \qquad (4-5)$$

where R is the hydraulic radius, $R = \frac{A}{\chi}$. A is the area of cross-section and χ is the perimeter length of the cross-section which is in contact with the fluid, also named the wetted perimeter.

For the flow in open channels, when $Re = \frac{vR}{\nu} < 500$, the flow is laminar flow; when $Re > 500$, the flow is turbulent flow.

【Example 4 – 1】 A circular pipe with a diameter of 2.0 cm has the average flow velocity of 1 m/s, and the temperature of water is 15 ℃. Determine the flow regime in the pipe. When oil passes the pipe under the same condition, is the flow laminar or turbulent? (note that $t = 15\,℃$, $\nu_{oil} = 0.6$ cm²/s, $\nu_{water} = 0.0114$ cm²/s)

Solution: For water flow in the pipe,

$$Re = \frac{vd}{\nu} = \frac{100 \times 2}{0.0114} = 17\,540 > 2000$$

It is turbulent flow.

For oil flow in the pipe,

$$Re = \frac{vd}{\nu} = \frac{100 \times 2}{0.6} = 333.3 < 2000$$

It is laminar flow.

4.2.3 The physical meaning of Reynolds number

(1) The Reynolds number represents the ratio of inertia force to viscous force of flow.

The inertia force (F_I) equals the product of mass (m) and acceleration (a). Since the mass is equal to the product of density (ρ) and volume (V), the dimension of mass can be expressed as $[m] = [\rho L^3]$.

The acceleration $a = \frac{du}{dt} = \frac{du}{dx} \cdot \frac{dx}{dt} = u \cdot \frac{du}{dx}$, then

$$[a] = \left[v \cdot \frac{v}{L}\right]$$

Therefore, the inertia force F_I has the following dimension:

$$[F_I] = \left[\rho L^3 v \frac{v}{L}\right] = [\rho L^2 v^2]$$

The viscous force F_τ = area (A) × shear stress (τ) = $A\mu \frac{du}{dy}$, so the viscous force has the following dimension:

$$[F_\tau] = \left[L^2 \mu \frac{v}{L}\right] = [\mu L v]$$

Thus, the dimension of the ratio of inertia force to viscous force can be expressed as

$$\frac{[F_I]}{[F_\tau]} = \frac{[\rho L^2 v^2]}{[\mu L v]} = \left[\frac{vL}{\nu}\right]$$

The dimension of the Reynolds' number is $\left[\frac{vL}{\nu}\right]$, where v is a characteristic velocity that can be a representative flow velocity, such as the mean cross-sectional velocity v, or incident velocity u_0; L is a characteristic length, which can be the pipe diameter, d, for pipe flow or hydraulic radius, R, for open-channel flow.

When the Reynolds number is large, it means that the inertia force is larger than the viscous force. The inertia force and the active force are action and reaction forces, which have the same magnitude but act in opposite direction. A large inertia force means a large active force, so the active force accelerates the movement of fluid. The viscous force is the force that resists the relative motion of fluid. If the viscous force is small, the resistance is small, i. e. the relative motion is accelerated, so the flow is turbulent flow. On the contrary, a small Reynolds number means that the inertia force is smaller than the viscous force, i. e. the resistance to the relative movement of fluid is larger, so the flow is laminar flow. When the Reynolds number is less than the critical Reynolds number Re_K, which indicates that the viscous force is the predominant force, the viscous force is sufficient to dampen the increased relative motion of fluid. That is the reason why the flow is laminar flow when the Reynolds number Re is less than the critical Reynolds number Re_K.

(2) The Reynolds number is a dimensionless number, which does not change with selected units. This can be proved as follows:

The dimension of flow velocity v is $[LT^{-1}]$, the dimension of pipe diameter is $[L]$, and the dimension of viscosity of fluid is $[L^2 T^{-1}]$. The dimension of the Reynolds number becomes

$$[Re] = \left[\frac{vd}{\nu}\right] = \frac{[LT^{-1}][L]}{[L^2 T^{-1}]} = [L^0 T^0] = [1]$$

, so it is the dimensionless number, in other words, cardinal number.

In practical calculations, it should be noticed that the units of flow velocity (v) and diameter of pipe (d) or hydraulic radius (R) should be consistent with the viscosity coefficient (ν). The criterion Re_K for determining laminar flow or turbulent flow does not change with the selection of units.

4.3 The relationship between frictional head loss and shear stress of uniform flow

The viscosity of fluid and the shear stress τ, which are caused by the transverse velocity gradient on the solid boundary, lead to the frictional head loss h_f. The remaining question is whether there exists any relationship between τ and h_f, which will be discussed in this section.

4.3.1 The relationship between frictional head loss and wall shear stress

Consider the segment from $1-1$ to $2-2$ of steady uniform flow in a pipe, as shown in Fig. 4 -8. The length of flow segment is l, the cross-sectional area is A, the wetted perimeter is χ, and the wall surface area is A_b. The vertical distances of the centroids of cross-sections $1-1$ and $2-2$ to the datum plane are z_1 and z_2 respectively. The dynamic pressures at the two centroids are p_1 and p_2 respectively. The mean cross-sectional velocity is v. The specific weight of fluid is γ and the average shear stress on the boundary surface is τ_0.

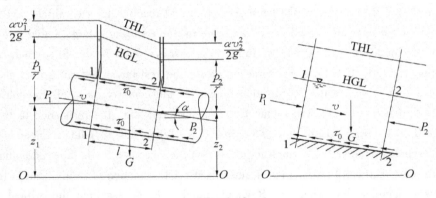

Fig. 4 – 8

The external forces acting on the entire flow segment are as follows.

(1) Hydrodynamic pressure force

The hydrodynamic pressure force acting on the cross-section $1-1$ is $P_1 = p_1 A$ and its direction is the same as that of flow. The hydrodynamic pressure force acting on the cross-section $2-2$ is $P_2 = p_2 A$ and its direction is opposite to the flow direction. Since the hydrodynamic pressure force on the wall is perpendicular to the direction of flow, its projection along the flow direction is zero.

(2) Gravity

$$G = \gamma V = \gamma A l$$

The component of gravity in the flow direction is $G_x = \gamma A l \sin\alpha$, where α is the inclined angle of the pipe axis to the horizontal line and $\sin\alpha = \dfrac{z_1 - z_2}{l}$.

(3) Frictional resistance

Since the internal frictional forces within the flow exist in a pair of action and reaction (the same in magnitude but opposite in direction), they can offset each other; but the internal friction T cannot be offset between the fluids on the wall and those close to the wall. Supposing that the average shear stress on the wall is τ_0, the total friction force can be expressed as: $T = \tau_0 A_b = \tau_0 \chi l$, which is opposite to the flow direction.

Because the flow is steady uniform flow, the flow velocity does not change with time and distance, in other words, there is no acceleration. The forces are in equilibrium, i.e. $\sum F = 0$. Taking the pipe centerline as the coordinate axis, the dynamic equation in equilibrium can be written as follows:

$$P_1 - P_2 + G\sin\alpha - T = 0$$

then

$$p_1 A - p_2 A + \gamma A l \frac{z_1 - z_2}{l} - \tau_0 \chi l = 0$$

Dividing by γA and rearranging,

$$\left(z_1 + \frac{p_1}{\gamma}\right) - \left(z_2 + \frac{p_2}{\gamma}\right) = \frac{\tau_0 \chi l}{A \gamma} \qquad (4-6)$$

Since the velocity heads at cross-sections $1-1$ and $2-2$ are the same, i.e. $\frac{v_1^2}{2g} = \frac{v_2^2}{2g}$, the left side of Equation (4-6) is the frictional head loss h_f. Meanwhile, the hydraulic radius $R = \frac{A}{\chi}$, then Equation (4-6) becomes

$$h_f = \frac{\tau_0 l}{R \gamma} \qquad (4-7)$$

Since the hydraulic slope (energy slope) $J = \frac{h_f}{l}$ is the frictional head loss in unit length, Equation (4-7) can be written as

$$\tau_0 = \gamma R J \qquad (4-8)$$

Equation (4-7) or (4-8) is the relationship between shear stress and the frictional head loss for steady uniform flow. It is called the basic equation of uniform flow.

Equation (4-7) shows that the frictional head loss h_f is inversely proportional to the hydraulic radius R. Hydraulic radius is $R = \frac{A}{\chi}$, so for given cross-sectional area A, the smaller the wetted perimeter χ, the larger R, then the smaller the frictional head loss is. Geometrically, for the same area A, the shape of the cross-section that has a minimum wetted perimeter χ is circular. In engineering, circular pipes and trapezoidal channels that are close to a circle in shape are widely used in order to minimize the frictional head loss.

4.3.2 The relationship between frictional head loss and shear stress

Internal frictional shear stress τ always exists on all flow layers of fluid flow. For many concentric cylindrical flows with different diameters that are all less than d, their shear stresses can be obtained as

$$\tau = \gamma R'J \tag{4-9}$$

where R' is the hydraulic radius of a concentric cylindrical flow and less than d; J is the hydraulic slope of uniform flow.

From Equations (4-8) and (4-9), we can obtain

$$\frac{\tau}{\tau_0} = \frac{R'}{R}$$

For the circular pipe, $R = \dfrac{A}{\chi} = \dfrac{\frac{\pi d^2}{4}}{\pi d} = \dfrac{d}{4} = \dfrac{2r_0}{4} = \dfrac{r_0}{2}$, where the r_0 is the radius of pipe. Thus

$$\frac{\tau}{\tau_0} = \frac{\frac{r}{2}}{\frac{r_0}{2}} = \frac{r}{r_0}$$

$$\tau = \frac{r}{r_0}\tau_0 = Cr \tag{4-10}$$

where r is the radius of the concentric cylinder with its centerline at the coordinate axis, and the value of $C = \dfrac{\tau_0}{r_0}$ is constant.

It can be seen that in uniform pipe flow the shear stress on a cross-section is linearly distributed. At the center of pipe ($r=0$) the shear stress is zero, and then it will linearly increase to reach the maximum ($\tau = \tau_0$) on the wall ($r = r_0$), as shown in Fig. 4-9a.

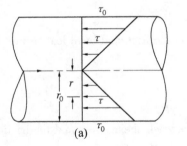
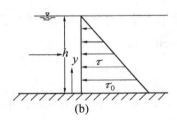

Fig. 4-9

In a similar way, the distribution of shear stress of uniform flow in open channels with a water depth h can be obtained by

$$\tau = (1 - \frac{y}{h})\tau_0 \tag{4-11}$$

where y is the vertical distance from the bottom of channel upwards, h is the depth of flow, τ is the shear stress at the point y, and τ_0 is the bed shear stress of channel.

Equation (4-11) shows that the shear stress on a cross-section is also linearly distributed for uniform flow of wide and shallow open channel. On the surface, $y = h$, $\tau = 0$; on the channel bed, $y = 0$, $\tau = \tau_0$, which is the maximum value. The distribution is shown in Fig. 4-9b.

4.3.3 The general calculation formula for frictional head loss

In Section 4.3.1, the derived Equation (4-1), i.e. $h_f = \dfrac{\tau_0 l}{\gamma R}$, shows the relationship between h_f and τ_0, in which τ_0 is unknown so far. Because of the complexity of turbulent flow, it is difficult to deduce a theoretical formula for the frictional head loss, and therefore it is usually obtained through experiments.

Experiments on laminar and turbulent flows in pipes and open channels show that the shear stress of wall (T) is proportional to the product of contact surface area A, velocity head $\dfrac{v^2}{2g}$ and specific weight of fluid γ. Namely,

$$T \propto A_b \gamma \dfrac{v^2}{2g}$$

It can be written as

$$T = \psi A_b \gamma \dfrac{v^2}{2g}$$

Thus,

$$\tau_0 = \dfrac{T}{A_b} = \psi \gamma \dfrac{v^2}{2g} \qquad (4-12)$$

where ψ is the dimensionless coefficient.

According to Equation (4-8), $\tau_0 = \gamma RJ$, subsitute it into Equation (4-12),

$$\gamma RJ = \psi \gamma \dfrac{v^2}{2g}$$

where $J = \dfrac{h_f}{l}$ and l is the length of the flow. Thus,

$$h_f = \psi \dfrac{l}{R} \dfrac{v^2}{2g} = 4\psi \dfrac{l}{4R} \dfrac{v^2}{2g} = \lambda \dfrac{l}{4R} \dfrac{v^2}{2g} \qquad (4-13)$$

where $\lambda = 4\psi$, which is an undetermined dimensionless number, called the friction coefficient, or drag coefficient.

Equation (4-13) is called the Darcy-Weisbach formula and it is suitable for both laminar flow and turbulent flow in pipes and open channels. Thus Equation (4-13) is also called the general formula of frictional head loss.

For pipe flow, the hydraulic radius $R = \dfrac{d}{4}$, namely $4R = d$. Thus Equation (4-13) becomes

$$h_f = \lambda \dfrac{l}{d} \dfrac{v^2}{2g} \qquad (4-14)$$

Obviously, Equation (4-14) is only suitable to pipe flow.

Since $\lambda = 4\psi$, then $\psi = \dfrac{\lambda}{4}$. Substituting it into Equation (4-12),

$$\tau_0 = \psi \gamma \dfrac{v^2}{2g} = \dfrac{\lambda}{4} \dfrac{\rho v^2}{2} = \dfrac{\lambda}{8} \rho v^2 \qquad (4-15)$$

or
$$\frac{\tau_0}{\rho} = \frac{\lambda}{8}v^2 \tag{4-15'}$$

where the squared root of $\frac{\tau_0}{\rho}$ is called the friction velocity v_*, which can be explained by the following analysis. From Equation (4-8), $\tau_0 = \gamma RJ$. Then,

$$\sqrt{\frac{\tau_0}{\rho}} = \sqrt{\frac{\gamma RJ}{\rho}} = \sqrt{gRJ} \tag{4-16}$$

The dimension on the left side of Equation (4-16) is $\left[\frac{\frac{F}{A}}{\frac{M}{V}}\right]^{\frac{1}{2}} = \left[\frac{MLT^{-2}}{\frac{L^2}{L^3}}\right]^{\frac{1}{2}} = [L^2T^{-2}]^{\frac{1}{2}} =$ $[L^1T^{-1}]$, in which $[L^1T^{-1}]$ is also the dimension of velocity. Because the energy slope (hydraulic gradient) J includes the term of frictional head loss, which is related to the internal friction resistance, this velocity is called the frictional velocity. No such velocity actually exists in the flow, so v_* is only a 'reference' parameter. Thus,

$$v_* = \sqrt{\frac{\tau_0}{\rho}} = \sqrt{gRJ} \tag{4-16'}$$

Subsituting Equation (4-16') into (4-15'),

$$\lambda = 8\frac{v_*^2}{v^2} \tag{4-17}$$

or

$$v_* = \sqrt{\frac{\lambda}{8}}\,v \tag{4-18}$$

【Example 4-2】 An experiment on a pipeline with 300 mm diameter shows the flow velocity of $v = 3$ m/s, and the frictional coefficient $\lambda = 0.015$. In the test, the water temperature is 15℃ and the corresponding density of water is $\rho = 999.1$ kg/m³. Determine the shear stress of pipe wall τ_0 and the shear stress at the location of $r = 0$, $r = 0.2r_0$, $r = 0.5r_0$, and $r = 0.75r_0$, which is from the center of the pipe. Note that r_0 is the diameter of pipe.

Solution: According to Equation (4-14), $h_f = \lambda \frac{l}{d}\frac{v^2}{2g}$, where

$$J = \frac{\lambda}{d}\frac{v^2}{2g} = \frac{0.015}{0.3} \times \frac{3^2}{2 \times 9.8} = 0.0229$$

$$\tau_0 = \gamma RJ = \rho g \times \frac{d}{4} \times J = 999.1 \times 9.8 \times \frac{0.3}{4} \times 0.0229 = 16.8 \text{ N/m}^2$$

$$\tau_{r=0.75r_0} = \gamma \frac{0.75r_0}{2}J = 224.2 \times 0.375 \times 0.15 = 12.6 \text{ N/m}^2$$

$$\tau_{r=0.5r_0} = \gamma \frac{0.5r_0}{2}J = 224.2 \times 0.25 \times 0.15 = 8.41 \text{ N/m}^2$$

$$\tau_{r=0.2r_0} = \gamma \frac{0.2r_0}{2}J = 224.2 \times 0.1 \times 0.15 = 3.36 \text{ N/m}^2$$

Chapter 4 Types of flow and head loss

$$\tau_{r=0} = 0$$

4.4 Laminar flow in circular pipes

The motion of uniform laminar flow in a circular pipe can be regarded as the movement of a set of concentric cylinders of infinite thickness, which is independent and in relatively simple motion. Therefore, the velocity distribution, average flow velocity, flow rate and frictional head loss in laminar pipe flow can all be deduced through theoretical analysis.

4.4.1 The velocity distribution of laminar flow

For laminar flow, the shear stress of flow can be expressed by the Newton inner friction law,

$$\tau = \mu \frac{du}{dy} = -\mu \frac{du}{dr} \qquad (4-19)$$

where y is the radial distance away from the wall and r is the distance from the axis (the centerline of pipe), see Fig. 4-10. If the radius of pipe is r_0, $y = r_0 - r$. The symbol u is the flow velocity at the point that is r away from the axis. If the flow velocity increases as increasing y, the value of $\frac{du}{dy}$ is positive; otherwise, if the flow velocity deceases as increasing y, the value of $\frac{du}{dr}$ is negative.

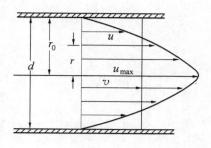

Fig. 4-10

From Equation (4-9) that $\tau = \gamma R'J$, where $R' = \frac{r}{2}$, the shear stress at the point with the radius r is

$$\tau = \gamma \cdot \frac{r}{2} J \qquad (4-20)$$

Combining Equations (4-19) and (4-20),

$$\frac{\gamma r J}{2} = -\mu \frac{du}{dr}$$

By separating the variables

$$du = -\frac{\gamma J}{2\mu} r dr$$

Integrating it,

$$u = -\frac{\gamma J}{2\mu} \int r dr = -\frac{J r^2}{4\mu} + C$$

where the integral constant C can be determined by $u = 0$ at $r = r_0$, so $C = \frac{\gamma J r_0^3}{4\mu}$. Thus,

$$u = \frac{\gamma J}{4\mu}(r_0^2 - r^2) \qquad (4-21)$$

143

Equation (4-21) shows that for uniform laminar flow in a pipe the velocity distribution is a parabola, as shown in Fig. 4-10. When $r=0$, the flow has the maximum flow velocity at the centerline of pipe:

$$u_{max} = \frac{\gamma J}{4\mu} r_0^2 = \frac{\gamma J}{16\mu} d^2 \qquad (4-22)$$

4.4.2 The mean flow velocity of laminar flow

The average flow velocity v of cross-section can be obtained from the following relationship between discharge Q and the area of cross-section A.

$$v = \frac{Q}{A} = \frac{\int_A u dA}{A} = \frac{\int_0^{r_0} u 2\pi r dr}{\pi r_0^2}$$

$$= \frac{\gamma J}{4\mu} \frac{\int_0^{r_0} (r_0^2 - r^2) 2\pi r dr}{\pi r_0^2}$$

$$= \frac{\gamma J}{2\mu r_0^2} \int_0^{r_0} (r_0^2 - r^2) r dr$$

$$= \frac{\gamma J}{2\mu r_0^2} \left(r_0^2 \cdot \frac{r^2}{2} \Big|_0^{r_0} - \frac{r^4}{4} \Big|_0^{r_0} \right)$$

$$= \frac{\gamma J r_0^2}{8\mu} = \frac{\gamma J}{32\mu} d^2$$

$$(4-23)$$

It can be seen that in the uniform laminar flow of pipe, the average cross-sectional flow velocity v is the half of the maximum velocity u_{max} at the center of pipe. Namely,

$$v = \frac{1}{2} u_{max} \qquad (4-24)$$

Equation (4-24) is the important basis for determining whether flow in a pipe is laminar flow.

4.4.3 The flow rate of laminar flow

$$Q = v \cdot A = \frac{\gamma J}{32\mu} d^2 \cdot \frac{\pi d^2}{4} = \frac{\gamma J \pi d^4}{128\mu} = \frac{\gamma h_f \pi d^4}{128\mu l} \qquad (4-25)$$

It shows that the flow rate of laminar flow is proportional to the diameter of pipe to the fourth power and that therefore the diameter will significantly affect the flow rate. Blood flow in the body is laminar flow, and when the flow section of blood vessel is reduced due to high cholesterol and so on, an insufficient flow rate of blood will inevitably occur.

4.4.4 The frictional head loss of laminar flow

According to Equation (4-23)

$$v = \frac{\gamma J}{32\mu} d^2$$

then,
$$J = \frac{32\mu v}{\gamma d^2}$$

since $J = \dfrac{h_f}{l}$,

$$h_f = \frac{32\mu v}{\gamma d^2} l \tag{4-26}$$

Equation (4-26) is the calculation formula for the frictional head loss of uniform laminar flow in a pipe. It shows that the frictional head loss is proportional to the average cross-sectional velocity v, which is consistent with the results of the Reynolds experiment.

Equation (4-26) can be restructured to the form of the Darcy-Weisbach formula:

$$h_f = \frac{32\mu v l}{\gamma d^2} = \frac{64}{\underbrace{vd}_{\nu}} \frac{l}{d} \frac{v^2}{2g} = \lambda \frac{l}{d} \frac{v^2}{2g}$$

where

$$\lambda = \frac{64}{Re} \tag{4-27}$$

It can be seen from Equation (4-27) that the frictional resistance coefficient of uniform laminar flow in a pipe is only related to the Reynolds number and is inversely proportional to the Reynolds number.

4.4.5 The kinetic correction coefficient of laminar flow

According to Equation (3-34)

$$\alpha = \frac{\int_A u^3 dA}{v^3 A}$$

where the velocity of uniform laminar flow in a pipe is

$$u = \frac{\gamma J}{4\mu}(r_0^2 - r^2) = \frac{\gamma J}{4\mu} r_0^2 \left[1 - \left(\frac{r}{r_0}\right)^2\right]$$

$$= u_{max}\left[1 - \left(\frac{r}{r_0}\right)^2\right]$$

and $\qquad dA = 2\pi r dr, \quad v = 0.5 u_{max}, \quad v^3 = 0.125 u_{max}^3, \quad A = \pi r_0^2$

Inserting these into Equation (3-34),

$$\alpha = \frac{\int_0^{r_0} u_{max}^3 \left[1 - \left(\frac{r}{r_0}\right)^2\right]^3 2\pi r dr}{0.125 u_{max}^3 \pi r_0^2}$$

$$= \frac{16}{r_0^2}\left\{\int_0^{r_0}\left[1 - 3\left(\frac{r}{r_0}\right)^2 + 3\left(\frac{r}{r_0}\right)^4 - \left(\frac{r}{r_0}\right)^6\right] r dr\right\}$$

$$= \frac{16}{r_0^2}\left(\frac{r_0^2}{2} - \frac{3}{r_0^2} \cdot \frac{r_0^4}{4} + \frac{3}{r_0^4} \cdot \frac{r_0^6}{6} - \frac{r_0^8}{8 r_0^6}\right)$$

$$= 16\left(\frac{1}{2} - \frac{3}{4} + \frac{3}{6} - \frac{1}{8}\right)$$

$$= 16\left(\frac{12 - 18 + 12 - 3}{24}\right) = 2$$

This shows that the parabolic velocity distribution of uniform flow in a pipe is not uniform and that α is larger than 1.

Similarly, the corresponding parameters of uniform laminar flow in an open channel are given as follows:

$$u = \frac{\gamma J}{2\mu} y(2h - y), \text{ i. e. a parabolic velocity distribution}$$

and

$$v = \frac{\gamma J}{3\mu} h^2$$

$$h_f = \lambda \frac{l}{4R} \frac{v^2}{2g}$$

$$\lambda = \frac{24}{Re}$$

where y is the vertical distance from the channel bed and h is the depth of flow. The meanings of other symbols are the same as those for pipe flow.

4.5 The basic concepts of turbulent flow

Turbulent flow is a type of fluid motion that exists widely in nature and engineering. It plays an important role in the study of fluid mechanics. The research of turbulence represents the development progress of hydraulics and fluid mechanics. It is closely related to energy loss, cavitation of high-speed flow, the laws governing sediment transport, erosion and deposition, temperature density currents of power plant, and many more practical topics in hydraulics.

Since the Reynolds experiment one hundred years ago, many important results of research on turbulent flow have been achieved. However, due to the complexity of turbulence, our understanding of turbulence still cannot meet all engneering design needs.

There are two research theories for turbulent flow: statistical and semi-empirical. The statistical theory of turbulence focuses on turbulence structure by using the strict method of probability and statistics. The semi-empirical theory focuses on the time averaged flow laws based on some assumptions of flow structure. This section will only introduce some fundamentals of the semi-empirical theory of turbulent shear flow.

4.5.1 Developing process of turbulent flow

From the viewpoint of internal structure of flow, the fundamental difference between laminar flow and turbulent flow is that liquid particles of each flow layer are immiscible in the laminar flow and there exist regular "instantaneous streamlines", while in the turbulent flow there are different sized eddies that fluctuate between flow layers and mix them.

It can be seen that the transition from laminar flow into turbulent flow has to meet two essential conditions:

(1) The formation of an eddy;
(2) The developed eddy moves out of its original flow layer into the adjacent layers and mixes with the new layers.

The formation of an eddy is based on two physical phenomena as follows:

The first is that the liquid has viscosity, which is the internal cause. The viscosity of fluid and the flow resistance of boundary cause the relative motion of fluid particles so that the motion will produce the internal friction shear stress between the adjacent flow layers. For a selected flow layer, the shear stress added by the layer with larger velocity follows the flow direction, whilst the stress added by the layer with lower velocity is against the flow direction, as shown in Fig. 4-11a. Such a shear stress has the tendency to produce a moment, which triggers the formation of eddy.

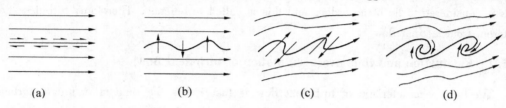

(a)   (b)   (c)   (d)

Fig. 4-11

The second physical phenomenon that causes the formation of eddy is the influences of disturbance (either external or internal) on the flow, which is the external reason. Supposing that a certain flow layer has a slight fluctuation due to disturbance, the fluctuation will lead to the adjustment of local flow velocity and pressure. In the convex side, the flow velocity of the upper section will increase due to the reduced area, so the pressure is reduced according to the Bernoulli equation; on the contrary, the pressure of the lower section is increased due to the increased cross-sectional area and flow velocity. Thus, there exists the higher pressure in the upper region but lower pressure in the lower region, so the transverse pressure force will be produced on the interface, as seen in Fig. 4-11b. This transverse pressure force will make the convex streamline more convex, the concave streamline more concave, so that the streamline becomes more wavy. When the amplitude of wavy streamline increases to a certain extent, the eddy will be formed due to the combination effect of both the transverse pressure and shear stress, as shown in Figs. 4-11c and 4-11d.

How is an eddy developed to start the mixing movement after it is formed?

After the formation of eddy, in the side where the flow direction is the same as the rotating direction of eddy, the flow velocity increases but the pressure decreases; on the opposite side, the situation is the reverse. The pressure difference on both sides of the eddy will form a lift (or sinking) force upwards, as shown in Fig. 4-12. This lift (or sinking) force has the tendency

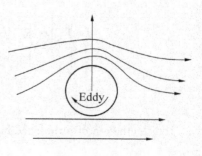

Fig. 4-12

Fluid Mechanics in Civil Engineering

to move the eddy from its original layer and to mix it with adjacent layers. However, in the process of the eddy entering into a new layer from its original layer, it has to overcome two kinds of resistance. One is the inertia force in the starting and accelerating process of eddy. The other one is the viscous resistance of fluid to eddy motion. Thus, after the formation of eddy, it is not necessary for the eddy to remove from the original flow layer, but the separation occurs only when the eddy rotation strength reaches a certain value. Namely, when the lateral pressure P is enough to overcome the resistance, then separation will take place and the eddy moves from its original layer into the new layer. According to the continuity of flow, something has to be replaced in the original place of eddy, i. e. some kind of exchange exists. Due to its complex nature, it is impossible to describe each eddy motion mathematically. Because of this, the inner structure of flow is considered to be of no order, and this is called turbulence. Therefore, turbulent flow is different from laminar flow.

4.5.2 Fluctuation and time averaged motion of turbulent flow

The basic characteristic of turbulent flow is that it has the random nature of turbulence structure and the random motion of an eddy, which cause the fluctuations of velocity, pressure and temperature. The turbulent flow is full with many differently sized eddies. In such flow, the magnitude and direction of velocity at a space point are constantly changing. If the flow velocity at any point is measured by an instantaneous velocity meter, the recording velocity of point may appear to be as shown in Fig. 4 – 13a. The variation of velocity is extremely random and it seems initially that there is no clear tendency to follow. The variation of velocity with time is called the fluctuation of velocity. The recording of instantaneous pressure shows a similar variation with that of the instantaneous flow velocity, as seen in Fig. 4 – 13b.

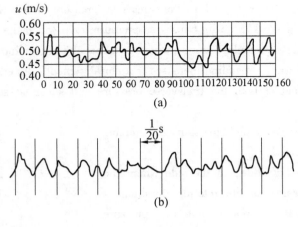

Fig. 4 – 13

The fluctuation of turbulence has influences on practical engineering in many aspects, which are:

(1) The energy loss will increase. In the laminar flow, the frictional head loss h_f is linearly

proportional to the average velocity v of cross-section, namely, $h_f \propto v^{1.0}$. While in the turbulent flow, the frictional head loss h_f is proportional to the average velocity v in the power of $1.75 \sim 2$, namely, $h_f \propto v^{1.75 \sim 2}$.

(2) The velocity distribution of turbulent flow tends to be much uniform. The fluctuation of velocity must cause the exchange of fluid between adjacent layers, resulting in the exchange and fluctuation in momentum, pressure, temperature and concentration. The velocity distribution complies with a logarithmic function. The ratio of maximum velocity to average cross-sectional flow velocity is usually between $1.05 \sim 1.3$.

(3) The fluctuating pressure will increase the instantaneous load on hydraulic structures, and the downstream erosion of structure may increase $30\% \sim 40\%$. It may also cause the vibration of the structure and cavitation.

(4) The fluctuation of flow parameters in turbulent flow is the driving force for the diffusion, pollution and suspension of solid phase (sediment, pollutant, dust, etc.) in two-phase flow.

Because both flow velocity and pressure are the fluctuating variables with random nature, they vary with both time and space, but their time-averaged quantities may show some interesting characteristics, for which certain relationships exist. Now we use the time averaging method to deal with the turbulent flow. This method is widely used for studying the time-averaged value of flow parameter over a sufficiently long time T (typically more than 100 times the interval of a wave). Consider instantaneous quantities (velocity and pressure) as the arithmetric sum of time-averaged and fluctuating values. Take the velocity as an example to write its mathematical expression as follows:

$$u_x = \bar{u}_x + u'_x \qquad (4-28)$$

where u_x is the instantaneous velocity of point, $\bar{u}_x$ is the time-averaged velocity of point, and u'_x is the fluctuating velocity of point.

The concept of time-averaged velocity can be explained through the graph shown in Fig. 4 – 14, which shows the variation of u_x with time. The time-averaged component velocity $\bar{u}_x$ is the area of integrating u_x with time divided by the period of time T, namely

$$\bar{u}_x = \frac{1}{T}\int_0^T u_x \mathrm{d}t \qquad (4-29)$$

It can be seen that $\bar{u}_x$ is the height of the rectangle, whose area is the same as the area of the curve above the horizontal axis within the T period.

Fig. 4 – 14 shows that in the period T, the sum of positive fluctuating values is the same as that of the negative values, i.e. the total area of horizontal-line pattern is equal to the total area of oblique-line pattern. It can be concluded that the time-averaged value of fluctuating velocity is zero. This can mathematically be proved.

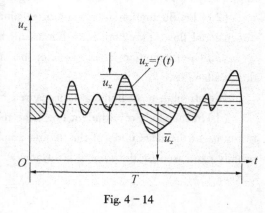

Fig. 4 – 14

$$\overline{u'}_x = \frac{1}{T}\int_0^T (u_x - \overline{u}_x)\,dt = \frac{1}{T}\int_0^T u_x\,dt - \frac{1}{T}\int_0^T \overline{u}_x\,dt = \overline{u}_x - \overline{u}_x = 0 \qquad (4-30)$$

Other variables of turbulent flow can be dealt with the same method. Take instantaneous pressure as an example.

$$p = \overline{p} + p'$$
$$\overline{p} = \frac{1}{T}\int_0^T p\,dt$$
$$\overline{p'} = \frac{1}{T}\int_0^T p'\,dt = 0$$

Turbulent flow is simplified as the time-averaged flow without fluctuations after the flow parameters have been taken time-averaged. The concepts of steady flow, unsteady flow, streamline, streamtube and total flow are all in terms of the time-averaged value. The steady flow means that the time-averaged flow property does not change with time, while the unsteady flow means that time-averaged flow property changes with time, as shown in Fig. 4 - 15. The streamline is the streamline of time-averaged velocity field, and the streamtube is the streamtube within the time-averaged velocity field.

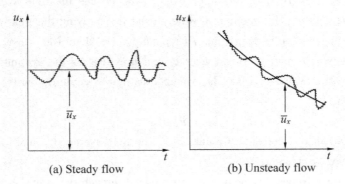

(a) Steady flow   (b) Unsteady flow

Fig. 4 - 15

For the fluctuating of flow, it is necessary to emphasize the following points:

(1) The fluctuating value cannot be treated as an infinitesimal quantity, because it may be large, maybe up to 1/3 of the time-averaged value.

(2) The fluctuation is always three-dimensional. Although the main flow is one- or two-dimensional flow, they both have fluctuating velocities in three directions.

(3) The fluctuation is the result of the mixing and exchanging of eddy but not the motion of fluid molecules.

The strength of turbulence is measured by turbulence intensity:

(1) In the study of turbulent flow, the root-mean-square (RMS) of fluctuating velocity is used to represent the magnitude of fluctuating amplitude. In the mathematical statistics, this value is usually expressed by the symbol σ. Namely,

$$\sigma = \sqrt{\overline{u'^2_x}} = \left(\frac{1}{T}\int_0^T u'^2_x\,dt\right)^{\frac{1}{2}} \qquad (4-31)$$

where the square of the u_x' value is taken in order to avoid that the positive and negative fluctuating velocity will offset to each other. The bigger σ is, the stronger fluctuation is; the smaller σ is, the weaker fluctuation is.

(2) Although σ represents the fluctuation amplitude, it is a dimensional number with the dimension of $[\text{LT}^{-1}]$. Dividing the σ by the time-averaged $\bar{u}_x$ of longitudinal velocity, a dimensionless value can be obtained, and this ratio is expressed by T_u, which is called the turbulence intensity.

$$T_u = \frac{\sigma}{\bar{u}_x} = \frac{\sqrt{\overline{u'^2_x}}}{\bar{u}_x} = \frac{\sqrt{\frac{1}{3}(\overline{u'^2_x} + \overline{u'^2_y} + \overline{u'^2_z})}}{\bar{u}_x} \qquad (4-32)$$

$$T_{u_x} = \frac{\sigma_x}{\bar{u}_x} = \frac{\sqrt{\overline{u'^2_x}}}{\bar{u}_x}$$

$$T_{u_y} = \frac{\sigma_y}{\bar{u}_x} = \frac{\sqrt{\overline{u'^2_y}}}{\bar{u}_x}$$

$$T_{u_z} = \frac{\sigma_z}{\bar{u}_x} = \frac{\sqrt{\overline{u'^2_z}}}{\bar{u}_x}$$

For homogeneous turbulent flow, $\overline{u'^2_x} = \overline{u'^2_y} = \overline{u'^2_z}$, so that

$$T_u = \frac{\sqrt{\overline{u'^2_x}}}{\bar{u}_x}$$

From the experimental data of open channel, the variation of turbulence intensity with the depth of flow is shown in Fig. 4–16, where H is the depth of flow and y is the distance from the bottom of channel. T_{u_x} is the turbulence intensity of velocity in the horizontal direction x, and T_{u_y} is the turbulence intensity of velocity in the vertical direction y. Fig. 4–16 shows that the turbulence intensity is large near the bed of the channel but small near the surface. The velocity gradient and shear stress is relatively large near the bed because of the influence of the rough wall, where vorticity originates and is generated.

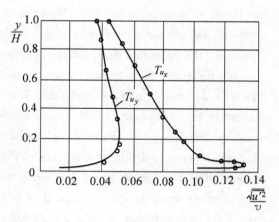

Fig. 4–16

However, in the region that is very close to the wall, the turbulence intensity will rapidly decrease from a maximum value and reduce to zero at $y = 0$. This can be explained as follows: when the value y is very small, the vortex is restricted by the wall, on the other hand, near the wall the velocity of the particle is low and the viscous effect predominates relative to inertia effect.

4.5.3 The shear stress and Prandtl's theory of turbulent flow

In laminar flow, the viscous shear stress that is caused by the relative motion of flow layers can be calculated by

$$\tau = \mu \frac{du}{dy}$$

However, the motion of fluid in turbulent flow is different. In addition to the relative motion between flow layers, there also exists the transverse exchange of eddies. Therefore, the calculation of turbulent shear stress should also adopt the concept of time-averaged value. The time-averaged shear stress $\bar{\tau}$ of turbulent flow should include two parts. One is the viscous shear stress $\overline{\tau_1}$, which is generated by the relative motion of time-averaged velocity in two adjacent layers; the other one is the additional shear stress $\overline{\tau_2}$ of turbulent flow, which results from the fluctuations. Thus, the total shear stress of turbulent flow is

$$\bar{\tau} = \overline{\tau_1} + \overline{\tau_2} \qquad (4-33)$$

where $\overline{\tau_1} = \mu \dfrac{du_x}{dy}$.

For the additional shear stress, it is often described by a semi-empirical theory, known as Prandtl's momentum transfer theory at present, in order to deal with turbulence problems through some simple assumptions about flow structure. Prandtl stipulated that the instantaneous velocity remains unchanged when an eddy moves transversely, so that the momentum remains the same. However when the eddy enters a new layer, its momentum will have a sudden change to gain or lose the same momentum as that of liquid particle at the new layer. According to the momentum principle, the change rate of momentum should equal the external force, which can establish a formula for the additional shear stress $\overline{\tau_2}$ of the turbulence.

To illustrate this theory, now consider two-dimensional, steady uniform turbulent flow of open channel in the xOy plane, which has relatively simple boundary conditions. Consider an elemental area of dA_y, which is perpendicular to the y axis in Fig. 4-17. Suppose that at a certain time, the liquid particle at point a moves upwards with the speed of u'_y, passes through dA_y and achieves at point a'. If the density of liquid is ρ, then the mass that transversely passes through dA_y in unit time is

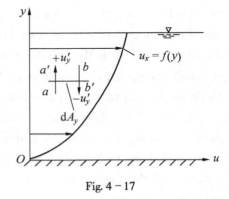

Fig. 4-17

$$m = \rho u'_y dA_y$$

If the time-averaged velocity in the x direction at point a is $\bar{u}_{x_a}$, the initial instantaneous momentum in the period of dt is $\rho u'_y dA_y \bar{u}_{x_a}$. Assuming that the momentum remains unchanged in the process of transverse motion, when the eddy's motion has completed and then mixes fully with the new local eddy to gain a new momentum at point a', which has the instantaneous momentum of $\rho u'_y \mathrm{d}A_y \, \overline{u}_{x_{a'}}$, the increment of momentum is $\rho u'_y \mathrm{d}A_y (\overline{u}_{x_{a'}} - \overline{u}_{x_a})$. The change of momentum is resulted from the completely mixing process of turbulence. From the principle of momentum, the increment of momentum in unit time should equal the external force, so the additional resistance of turbulence can be written as

$$F = \rho u'_y \mathrm{d}A_y (\overline{u}_{x_{a'}} - \overline{u}_{x_a}) = -\rho u'_y \mathrm{d}A_y (\overline{u}_{x_a} - \overline{u}_{x_{a'}})$$

where $\overline{u}_{x_a}$ has two meanings: at point a, it is the time-averaged velocity of that point while at point a', it is the instantaneous velocity for point a'. According to the relationship between the instantaneous and time-averaged velocity, the fluctuating velocity is

$$u'_x = u_x - \overline{u}_x$$

Replacing u_x by $\overline{u}_{x_a}$, and $\overline{u}_{x'}$, by $\overline{u}_{x_{a'}}$, it can be obtained that $u'_x = \overline{u}_{x_a} - \overline{u}_{x_{a'}}$.
Therefore,

$$\tau_2 = \frac{F}{\mathrm{d}A_y} = -\rho u'_x u'_y$$

where τ_2 is the additional shear stress of turbulent flow and it is also called Reynolds shear stress. It can be written into the formula with time-averaged value.

$$\overline{\tau_2} = -\rho \, \overline{u'_x u'_y} \tag{4-34}$$

The minus sign can also be obtained from the analysis of the direction of shear stress. It is assumed that the fluctuating velocity (u'_y) in the y axis is positive upwards and negative downwards, and that the fluctuating velocity (u'_x) in the x axis is positive to right and negative to left. When u'_y is positive, namely the eddy moves transversely from the lower to the upper layer, the time-averaged velocity at the lower layer is smaller than that at the upper layer, so the value of u'_x is negative and the product of $u'_x u'_y$ is negative. When u'_y is negative, which means the eddy moves transversely from the upper layer b to the lower layer b', the time-averaged velocity at the upper layer is bigger than that at the lower layer, so the product of $u'_x u'_y$ is negative as the value of u'_x is positive. The additional shear stress produced by the transverse movement of eddy has an effect on the flow: the liquids in the upper layer appear to be retarded by the liquids from the lower layer, and the liquids in the lower layer appear to be speeded up by the liquids from the upper layer. The additional shear stress has the same direction as the viscous shear stress. In order to make additional turbulent shear stress a positive value, a minus sign is needed on the right side of Equation (4-34).

In terms of time-averaged motion, because the viscous shear stress always exists, the total shear stress of turbulent flow is the sum of the viscous shear stress $\mu \dfrac{\mathrm{d}\overline{u}_x}{\mathrm{d}y}$ and additional turbulent shear stress $-\rho \, \overline{u'_x u'_y}$.

$$\overline{\tau} = \overline{\tau_1} + \overline{\tau_2} = \mu \frac{\mathrm{d}\overline{u}_x}{\mathrm{d}y} - \rho \, \overline{u'_x u'_y} \tag{4-35}$$

It should be noted that both shear stresses vary with the condition of flow. When the Reynolds number is small and the turbulence is weak, $\overline{\tau_1}$ predominates. As the Reynolds number and the intensity of turbulence increase, $\overline{\tau_2}$ will increase gradually. When the Reynolds number is so large that the flow is fully developed turbulent flow, $\overline{\tau_2}$ predominates and the influence of $\overline{\tau_1}$ can be ignored.

Fig. 4-18 shows the shear stress distribution, which was measured along the centerline in a rectangular wind tunnel. The tunnel is 1 m wide and the height H is 0.244 m. The maximum velocity is 1 m/s. In the figure, both $-\overline{u'_x u'_y}$ and $\dfrac{\tau}{\rho}$ are plotted. The difference between $\dfrac{\tau}{\rho}$

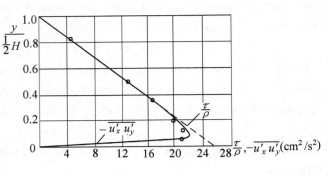

Fig. 4-18

and $-\overline{u'_x u'_y}$ is $\nu \dfrac{d\bar{u}_x}{dy}$. It can be seen that at the side wall, $y = 0$ and $-\overline{u'_x u'_y} = 0$. The shear stress on the wall is the viscous shear stress, and it reaches a maximum. When y reaches a certain value, the additional shear stress of turbulence predominates and the influence of viscosity tends to zero. When $y = \dfrac{1}{2}H$, the total shear stress tends to be zero, which means that there is no shear stress at the center of the cross-section.

Equation (4-34) is the expression of additional shear stress of turbulent flow in terms of fluctuating velocity. However, it is yet unknown about the relationship between the additional shear stress and the time-averaged velocity.

In the analysis of turbulence exchange, Prandtl introduced the concept of turbulent mixing length l, which is similar to the length of free movement of gaseous molecules.

There are a lot of eddies in turbulent flow. These eddies have longitudinal and lateral motion but keep their longitudinal velocities within a certain distance l', such that the eddy will mix with the surrounding fluid. Supposing that an eddy at $y-l'$ has the time-averaged velocity of $\bar{u}_{y-l'}$ where the subscript denotes the space positon, when the eddy moves upwards at a distance of l' with the fluctuating velocity $+u'_y$ and reaches the y positon, as shown in Fig. 4-19. In the process, the eddy maintains its original time-averaged velocity $\bar{u}_{y-l'}$ which is smaller than that at the new positon y. This difference can be written as

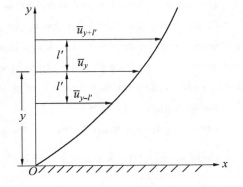

Fig. 4-19

$$\Delta u_1 = \bar{u}_y - \bar{u}_{y-l'} \approx l' \dfrac{d\bar{u}_x}{dy}$$

Similarly, supposing that the eddy at the point $y + l'$ has the time-averaged velocity of $\bar{u}_{y+l'}$, when the eddy moves downwards at a distance of l' with the fluctuating velocity of $+u'_y$ and reaches the y positon, the eddy maintains its original time-averaged velocity of $\bar{u}_{y+l'}$, which is larger than that at the new position y. This difference can be written as

$$\Delta u_2 = \bar{u}_{y+l'} - \bar{u}_y \approx l' \frac{d\bar{u}_x}{dy}$$

Prandtl considered the difference of time-averaged velocity as the transverse fluctuating velocity due to the turbulence exchange in the longitudinal direction. Therefore, no matter the l' is positive or negative; the longitudinal fluctuating velocity can be expressed as

$$u'_x = l' \frac{d\bar{u}_x}{dy} \tag{4-36}$$

where l' is the distance of motion, which is called the mixing length by Prandtl.

For the transverse component u'_y of fluctuating velocity, Prandtl described it as follows: In both the $y+l'$ layer and the $y-l'$ layer, an eddy has always the potential to reach the middle layer at y. If the eddy with larger time-averaged velocity is behind the one with smaller velocity, in other words, the slower eddy is in front of the faster one, the two eddies will collide with each other at the velocity of $2u'_x$, which will force the liquid between the two eddies to move upwards and downwards to produce transverse movement. If the eddy with larger time-averaged velocity is in front of the eddy with smaller velocity, in other words, the faster eddy is in front of the slower one, the two eddies will be separated from each other with a relative velocity of $2u'_x$ and then the surrounding liquids will enter between the two eddies, which also produces transverse movement. In this case, the surrounding liquids enter the layer at y from both the upper and lower layers. According to the description of flow patterns above, it can be considered that the transverse fluctuating velocity u'_y and streamwise fluctuating velocity u'_x have the same order of magnitude, namely $u'_y \sim u'_x$, or it can be written as

$$u'_y = c_1 u'_x = c_1 l' \frac{d\bar{u}_x}{dy} \tag{4-37}$$

where c_1 is the coefficient.

Note that $\overline{u'_x u'_y}$ and $u'_x \cdot u'_y$ are not the same, but that they are related in a way,

$$\overline{u'_x u'_y} = c_2 u'_x u'_y = c_1 c_2 l'^2 \left(\frac{d\bar{u}_x}{dy}\right)^2$$

where c_2 is another coefficient. Let $l^2 = c_1 c_2 l'^2$, then

$$\overline{u'_x u'_y} = l^2 \left(\frac{d\bar{u}_x}{dy}\right)^2$$

Thus, Equation (4-34) can be written as

$$\overline{\tau_2} = -\rho \overline{u'_x u'_y} = \rho l^2 \left(\frac{d\bar{u}_x}{dy}\right)^2 \tag{4-38}$$

where l is a physical quantity, which is proportional to the length of the time-averaged free movement of eddy, also called the mixing length. Equation (4-38) is the expression of turbulent shear stress by the time-averaged velocity.

Letting $\varepsilon = l^2 \dfrac{d\bar{u}_x}{dy}$ or $\eta = \rho l^2 \dfrac{d\bar{u}_x}{dy}$, then Equation (4-38) becomes

$$\bar{\tau}_2 = \rho\varepsilon \dfrac{d\bar{u}_x}{dy} = \eta \dfrac{d\bar{u}_x}{dy} \qquad (4-39)$$

where ε is called the turbulent kinematic viscosity coefficient, which has the dimension of $[L^2 T^{-1}]$, and $\varepsilon \gg \nu$. η is called the turbulent dynamic viscosity coefficient or the eddy viscosity coefficient with the dimension of $[ML^{-1}T^{-1}]$, $\eta \gg \mu$. The value of η can be up to $10^4 \sim 10^5$ times of μ.

The time-averaged shear stress of turbulent flow is

$$\bar{\tau} = \bar{\tau}_1 + \bar{\tau}_2 = \mu \dfrac{d\bar{u}_x}{dy} + \rho l^2 \left(\dfrac{d\bar{u}_x}{dy}\right)^2$$

For turbulent flow, all flow parameters need to be taken time-averaged value. For simplicity, the time-averaged symbol is dropped,

$$\tau = \mu \dfrac{du_x}{dy} + \rho l^2 \left(\dfrac{du_x}{dy}\right)^2 \qquad (4-40)$$

The Prandtl mixing length theory describes that turbulent momentum exchange is an intermittent phenomenon. Every transversed movement is a process: mixing → transverse movement → mixing again. The transverse time-averaged velocity of eddy will keep the original value in the transverse process but the instantaneous momentum occurs during the mixing. In fact, the time-averaged velocity and turbulent diffusion are continuous. Therefore, this theory does not really reflect the actual motion of flow, so it is a semi-hypothesis theory. However, the turbulent velocity distribution that is derived based on the theory agrees well with the data.

4.5.4 The viscous sublayer and flow zone of turbulent flow

In fully developed turbulent flow, much attention should be paid to the flow near solid boundaries. Firstly, the turbulence is restricted by the boundary, so the transverse mixed motion is limited, i.e. $u'_y = 0$. Secondly, large velocity gradient $\dfrac{du_x}{dy}$ occurs near the boundary, where the viscous shear stress $\tau_1 = \mu \dfrac{du_x}{dy}$ predominates. Finally, the velocity u of fluid particle is very small near the boundary, where the laminar flow is dominant; this thin region is called the laminar sublayer or viscous sublayer. Although this layer is very thin (only a few hundredths of a millimeter in thickness), it has significant influence on the flow resistance. There exists a thin transition layer between the laminar sublayer and the core of turbulent flow, which is generally treated as the turbulence. Fig. 4-20 shows a sketch of the boundary layer of flow.

4.5.4.1 The thickness δ_0 of viscous sublayer

Althougth velocity distribution of viscous sublayer is parabolic, because the thickness δ_0 of the laminar sublayer is very small, the velocity distribution is approximately linear. Thus, $y = 0$,

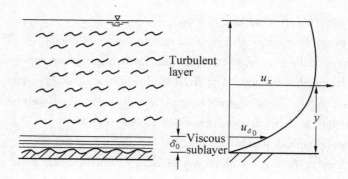

Fig. 4-20

$u_x = 0$ and $y = \delta_0$, $u_x = u_{\delta_0}$, where u_{δ_0} is the velocity at viscous sublayer boundary. According to $\dfrac{u_{\delta_0}}{\delta_0} = \dfrac{du_x}{dy}$, we can obtain

$$\tau_0 = \mu \frac{du_x}{dy} = \mu \frac{u_{\delta_0}}{\delta_0}$$

$$\delta_0 = \frac{\mu}{\tau_0} u_{\delta_0} = \frac{\frac{\mu}{\rho}}{\frac{\tau_0}{\rho}} u_{\delta_0} = \frac{\nu}{u_*^2} u_{\delta_0}$$

where $u_* \left(= \sqrt{\dfrac{\tau_0}{\rho}} = \sqrt{gRJ}\right)$ is the frictional velocity, also see Equation (4-16'). Writing u_*^2 in the form of $u_* \cdot u_*$, we can obtain

$$\delta_0 = \frac{u_{\delta_0}}{u_*} \cdot \frac{\nu}{u_*}$$

Because $\dfrac{u_{\delta_0}}{u_*}$ is a dimensionless number, denoted by N, according to the well-known Nikuradse experimental data, which show that $N = 11.6$, we can obtain

$$\delta_0 = \frac{N\nu}{u_*} = 11.6 \times \frac{\nu}{u_*} \qquad (4-41)$$

From Equation (4-18), $u_* = \sqrt{\dfrac{\lambda}{8}} v$, v is the average velocity of cross-section. Put it into Equation (4-41), then

$$\delta_0 = \frac{11.6\nu}{\sqrt{\dfrac{\lambda}{8}} v} = \frac{11.6\sqrt{8}d}{\sqrt{\lambda}\dfrac{vd}{\nu}} = \frac{32.8d}{\sqrt{\lambda} Re} \qquad (4-42)$$

Equation (4-42) shows that the thickness of laminar sublayer decreases with increasing Reynolds' number. The larger the Reynolds number, the stronger the turbulence intensity is, and the thickness of viscous sublayer will reduce.

Recent researches show that the actual laminar sublayer is even thinner than the value given by Equation (4-42). This thickness $\delta'_0 = (3.5 \sim 5)\dfrac{\nu}{u_*}$ or $\delta'_0 = (9.9 \sim 14.2)\dfrac{d}{\sqrt{\lambda} Re}$.

4.5.4.2 The flow region of turbulent flow

The surface of solid boundary always has a certain degree of roughness. The roughness height of any solid surface is called the absolute roughness, which is represented by Δ. For a surface of certain material, its roughness height Δ is fixed while the thickness δ_0 of laminar sublayer varies with the Reynolds number. Thus, δ_0 may be greater or less than Δ. Based on the relative ratio of Δ to δ_0, turbulent flow is divided into three different flow regions.

(1) Hydraulically smooth region

When Re is small, δ_0 can be much bigger than Δ. In this case, although the surface of boundary is rough, the height of roughness is immersed under the laminar sublayer, as shown in Fig. 4-21a. Therefore, the absolute roughness has no impact on the turbulence of flow. The flow resistance of boundary is mainly the viscous resistance of laminar sublayer. From the hydraulics point of view, this type of rough surface acts like a smooth surface, and it is called the hydraulically smooth region.

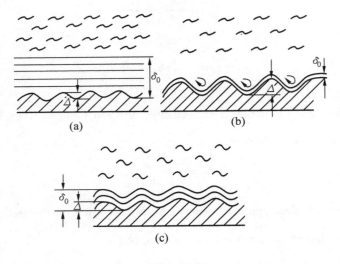

Fig. 4-21

(2) Hydraulically rough region

When Re is large, the laminar sublayer is thin and δ_0 is much smaller than Δ. In such a case, the roughness of boundary takes a major role in turbulent flow. Small eddies will be formed around the rough boundary, as seen in Fig. 4-21b. The flow resistance of boundary is mainly caused by these small eddies, while the viscous force of laminar sublayer has little impact. This type of rough surface is called the hydraulically rough region.

(3) Transitional rough region

There is also a situation falling in between the two cases above. The thickness of laminar sublayer is not sufficient to cover the effect of the roughness of boundary, as seen in Fig. 4-21c, in which the absolute roughness has a partial influence on the flow. This type of rough surface is called the transitional rough region.

(4) The criteria of turbulent flow region

There are two criteria for identifying the turbulent flow region. One is to use the ratio of absolute roughness (Δ) to the thickness of laminar sublayer (δ_0); the other one is the rough Reynolds number Re_*, which is defined as $\dfrac{u_* \Delta}{\nu}$. Their relationship is as follows.

$$\frac{\Delta}{\delta_0} = \frac{\Delta}{\dfrac{N\nu}{u_*}} = \frac{u_* \Delta}{11.6\nu} = \frac{Re_*}{11.6}$$

or

$$Re_* = 11.6 \frac{\Delta}{\delta_0}, \qquad (4-43)$$

Table 4-1 The turbulent flow region

Criterion	Turbulent flow region		
	Hydraulically smooth region	Transitional rough region	Hydraulically rough region
$\dfrac{\Delta}{\delta_0}$	$\dfrac{\Delta}{\delta_0} < 0.4$	$0.4 < \dfrac{\Delta}{\delta_0} < 6$	$\dfrac{\Delta}{\delta_0} > 6$
Re_*	$Re_* < 5$	$5 < Re_* < 70$	$Re_* > 70$

Finally, it is necessary to point out that the so-called hydraulically smooth region or hydraulically rough region is not entirely dependent on whether the solid boundary surface itself is smooth or rough, but on the relative ratio of $\dfrac{\Delta}{\delta_0}$. Even for the same solid boundary, under a certain Reynolds number it may be a "hydraulically smooth" while under another Reynolds number, it may be a "hydraulically rough".

4.5.5 The velocity distribution of turbulent flow

The velocity distribution in the laminar sublayer can be regarded to be a linear distribution. What is the velocity distribution in the turbulent core region? For consistency between results from pipe and open-channel flows, take y as a coordinate axis, as shown in Fig. 4-22.

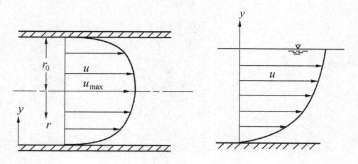

Fig. 4-22

The commonly used velocity distribution formulae of turbulent flow are the Prandtl-Karman logarithmic velocity distribution and Prandtl-Blasius exponential velocity distribution formula.

4.5.5.1 The logarithmic velocity distribution

The turbulent shear stress is

$$\tau = \mu \frac{du}{dy} + \rho l^2 \left(\frac{du}{dy}\right)^2$$

When Re is large, the turbulence shear stress $\tau_2 = \rho l^2 \left(\frac{du}{dy}\right)^2$ predominates. The influence of $\tau_1 = \mu \frac{du}{dy}$ on the flow can be ignored. Thus,

$$\tau = \rho l^2 \left(\frac{du}{dy}\right)^2 \qquad (4-44)$$

For pipe flow, it can be calculated by Equation (4-10), so

$$\tau = \frac{r}{r_0}\tau_0$$

Substitute $r = r_0 - y$ (see Fig. 4-22) into the above equation,

$$\tau = \frac{r_0 - y}{r_0}\tau_0 = \left(1 - \frac{y}{r_0}\right)\tau_0 \qquad (4-45a)$$

For open channel flow, Equation (4-11) gives

$$\tau = \frac{h - y}{h}\tau_0 = \left(1 - \frac{y}{h}\right)\tau_0 \qquad (4-45b)$$

Petkovic's study shows

$$l = \kappa y \sqrt{1 - \frac{y}{r_0}} \qquad (4-46)$$

where l is the mixing length, and κ is the von-Karman universal constant (usually $\kappa = 0.4$). Substitute Equations (4-45) and (4-46) into Equation (4-44),

$$\left(1 - \frac{y}{r_0}\right)\tau_0 = \rho \kappa^2 y^2 \left(1 - \frac{y}{r_0}\right)\left(\frac{du}{dy}\right)^2$$

$$\left(\frac{du}{dy}\right)^2 = \frac{\tau_0}{\rho \kappa^2 y^2} = \frac{u_*^2}{\kappa^2 y^2}$$

where $u_* = \sqrt{\frac{\tau_0}{\rho}}$ is the friction velocity.

$$\frac{du}{dy} = \frac{u_*}{\kappa y}$$

Separate the variables,

$$du = \frac{u_*}{\kappa y}dy \qquad (4-47)$$

By integration,

$$\frac{u}{u_*} = \frac{1}{\kappa}\ln y + C \qquad (4-48)$$

Substitute $\kappa = 0.4$ into Equation (4-48), and rewrite it in common logarithm,

$$\frac{u}{u_*} = 2.5 \times 2.3 \lg y + C$$

$$= 5.75 \lg y + C \qquad (4-49)$$

Equation (4-49) is the Prandtl-Karman logarithmic velocity distribution formula.

(1) In the hydraulically smooth region

The main factor that influences the flow motion is the motion state of flow rather than the absolute roughness of boundary. Therefore, it is important to consider the parameter of $\dfrac{u_* y}{\nu}$, which can reflect the flow pattern. Note that Equation (4-47): $du = \dfrac{u_*}{\kappa y} dy$, which can be written as

$$du = \frac{u_*}{\kappa} \frac{1}{\dfrac{u_* y}{\nu}} d\left(\frac{u_* y}{\nu}\right).$$

By integration,

$$\frac{u}{u_*} = \frac{1}{\kappa} \ln \frac{u_* y}{\nu} + C$$

Rewriting it in a common logarithm form,

$$\frac{u}{u_*} = 5.75 \lg \frac{u_* y}{\nu} + C \qquad (4-50)$$

According to the Nikuradse experimental data, the integral constant $C = 5.5$. Finally, the logarithmic velocity distribution in the hydraulically smooth region is

$$\frac{u}{u_*} = 5.75 \lg \frac{u_* y}{\nu} + 5.5 \qquad (4-51)$$

For flow of circular pipes, the average cross-sectional velocity is calculated by $v = \dfrac{\int_A u dA}{A}$ so that

$$\frac{v}{u_*} = 5.75 \lg \frac{r_0 u_*}{\nu} + 1.75 \qquad (4-52)$$

(2) In the hydraulically rough region

When the absolute roughness Δ of boundary is much larger than the thickness of laminar sublayer, the flow is fully developed turbulent flow. The main factor that influences the flow resistance and velocity distribution is the relative roughness $\dfrac{\Delta}{y}$, and its reciprocal is $\dfrac{y}{\Delta}$, which is called the relative smoothness. Namely, $\dfrac{u}{u_*} = f\left(\dfrac{y}{\Delta}\right)$, which indicates that the velocity should be a function of $\dfrac{\Delta}{y}$. Applying the similar analysis method of the hydraulically smooth region, and rewriting Equation (4-47) as

$$du = \frac{u_*}{\kappa \dfrac{y}{\Delta}} d\left(\frac{y}{\Delta}\right)$$

By integration,

$$\frac{u}{u_*} = \frac{1}{\kappa}\ln\frac{y}{\Delta} + C$$

Rewrite it in a common logarithm form,

$$\frac{u}{u_*} = 5.75\lg\frac{y}{\Delta} + C \tag{4-53}$$

According to the measured data by Nikuradse, when $y = \Delta$, $\frac{u}{u_*} = 8.5$, the integration constant $C = 8.5$. Finally, the logarithmic velocity distribution in the hydraulically rough region is

$$\frac{u}{u_*} = 5.75\lg\frac{y}{\Delta} + 8.5 \tag{4-54}$$

Integrating the above equation in light of continuity, the average cross-sectional velocity becomes

$$\frac{v}{u_*} = 5.75\lg\frac{r_0}{\Delta} + 4.75 \tag{4-55}$$

4.5.5.2 The exponential velocity distribution

In 1913, H. Blasius analyzed many data of pipe flow and found the exponential velocity distribution formula.

$$\frac{u}{u_{max}} = \left(\frac{y}{r_0}\right)^n \tag{4-56a}$$

where n is the index.

When $Re < 10^5$, $n = \frac{1}{7}$, which is called the seventh-power law of velocity distribution;

When $Re = 10^5 \sim 4 \times 10^5$, $n = \frac{1}{8}$;

When $Re = 1.1 \times 10^6$, $n = \frac{1}{9}$;

When $Re = (2 \sim 3.2) \times 10^6$, $n = \frac{1}{10}$.

As is seen from the above, the larger Re is, the smaller n is.

The exponential velocity distribution formula of the open channel is

$$\frac{u}{u_{max}} = \left(\frac{y}{h}\right)^n \tag{4-56b}$$

The exponential velocity distribution formula is simple, and is suited for both laminar flow and turbulent flow and flows in hydraulically smooth and rough regions. However, the value of velocity (u_{max}) is unknown, so the equation does not reflect the influence of Δ and u_* on the velocity.

Both the logarithmic velocity distribution and the exponential velocity distribution are more uniform than the parabolic velocity distribution for laminar flow, as shown in Fig. 4-23. The average cross-sectional velocity (v) of turbulent flow is $0.77 \sim 0.95$ times of the maximum velocity u_{max} of pipe. That is because the mixing and momentum exchange of turbulent eddy results in a more uniformed velocity distribution.

【Example 4-3】 Based on the turbulent logarithmic velocity distribution formula, calculate the position of average cross-sectional velocity in two-dimensional uniform flow in a

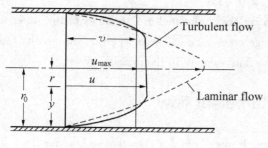
Fig. 4 − 23

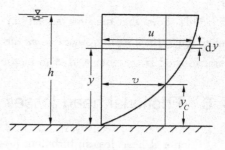
Fig. 4 − 24

rectangular open channel. See Fig. 4 − 24.

Solution: Most flows in open channel are in the hydraulically rough region and the turbulent logarithmic flow velocity distribution is

$$u = u_* \left(5.75 \lg \frac{y}{\Delta} + 8.5\right) = u_* \left(2.5 \ln \frac{y}{\Delta} + 8.5\right) \tag{A}$$

For rectangular open channel, the average cross-sectional velocity is $v = \frac{q}{h}$, where q is the discharge per unit width. Firstly, calculate q,

$$q = \int_0^h u \, dy = u_* \left(2.5 \int_0^h \ln \frac{y}{\Delta} dy + 8.5 \int_0^h dy\right) = u_* \left[2.5 \Delta \int_0^h \ln \frac{y}{\Delta} d\left(\frac{y}{\Delta}\right) + 8.5 h\right]$$

$$= u_* \left[2.5 \Delta \left(\frac{y}{\Delta} \ln \frac{y}{\Delta} - \frac{y}{\Delta}\right)\Big|_0^h + 8.5 h\right] = u_* \left[2.5 \left(h \ln \frac{h}{\Delta} - h\right) + 8.5 h\right]$$

$$= 2.5 u_* h \ln \frac{h}{\Delta} + 6 h u_* = u_* h \left(5.75 \lg \frac{h}{\Delta} + 6\right)$$

$$v = \frac{q}{h} = u_* \left(5.75 \lg \frac{y}{\Delta} + 6\right) \tag{B}$$

Since $u = v$, $y = y_c$, then substitute them into Equation A,

$$v = u_* \left(5.75 \lg \frac{y_c}{\Delta} + 8.5\right) \tag{C}$$

Since they both are the average cross-sectional velocity, they should be the same, namely Equation B = Equation C.

$$5.75 \lg \frac{y_c}{\Delta} + 8.5 = 5.75 \lg \frac{h}{\Delta} + 6$$

$$5.75 \left(\lg \frac{h}{\Delta} - \lg \frac{y_c}{\Delta}\right) = 8.5 - 6 = 2.5$$

$$\lg \frac{h}{y_c} = \frac{2.5}{5.75} = 0.435$$

$$\frac{h}{y_c} = 2.72$$

$$y_c = 0.367 h$$

or
$$h - y_c = 0.633h$$

This shows that the flow velocity at the point that is located at 0.6 times of the water depth below the surface can be regarded as the average cross-sectional velocity (v). This is why the one point method is sometimes used to measure the flow velocity in hydrologic survey.

4.6 Frictional head losses of turbulent flow

Frictional head loss in turbulent flow is much more complex than that in laminar flow. There are two methods to calculate frictional head loss. One method is through the experimental and theoretical analysis: find the frictional coefficient λ from Reynolds number and the relative smoothness or the relative roughness, and then calculate the frictional head loss h_f by using the Darcy-Weisbach formula. Another method is entirely based on empirical formula: calculate Chezy's coefficient C from the roughness coefficient n, then calculate the frictional coefficient from C, and finally calculate h_f by using the Darcy-Weisbach formula.

Frictional head loss is closely related to the roughness of solid boundary. The roughness of boundary can be measured by:

(1) Absolute roughness (denoted by symbol Δ): for artificial roughened surfaces, it is represented by the diameter d of uniform sand; for natural roughness, it is expressed by equivalent roughness.

(2) Relative roughness, which is the ratio of absolute roughness to sand diameter (or radius), denoted by Δ/d or Δ/r_0.

(3) Relative smoothness, which is the reciprocal of the relative roughness, denoted by d/Δ or r_0/Δ.

4.6.1 Experiment of frictional resistance coefficient

From the Darcy-Weisbach formula,

$$h_f = \lambda \frac{l}{d} \frac{v^2}{2g}$$

For laminar flow, $\lambda = 64/Re$ (pipe flow) or $\gamma = 24/Re$ (open-channel flow). For turbulent flow, the frictional coefficient λ was extensively studied by Nikuradse through the experiments of artificial sand rough pipes in 1931 – 1933. Goldstein studied it on artificial rough channel in 1938, and Colebrook carried out research on naturally rough commercial pipes in 1939. The Nikuradse experiment is mainly introduced in the following section.

4.6.1.1 Nikuradse's experiments

Nikuradse filled pipes with Japanese paint, and then drained off the paint, after half an hour, used rigorously screened uniform sands to fill test pipes, and removed the excess sands. The pipes were dried through the convection flow by electric bulb for 2～3 weeks, and then they were filled with Japanese paint once again and dried for 3～4 weeks before the experiment. Thus, the selected sands with paint stuck on the wall are uniformly distributed along the pipe. The sand size (d) is taken as the absolute roughness (Δ). Nikuradse did six group tests ($r_0/\Delta = 15, 30, 60, 126, 252, 507$, where $r_0 = d/2$) with the Reynolds number up to 10^6, which provide a wide range of data.

In the Nikuradse experiment, the piezometer tube was not installed on the wall. Instead, an elbow-typed tube with an outer diameter of 2 mm was used inside the pipe at $r_0/2$ position, as shown in Fig. 4 − 25. This is to avoid the impact of the variation of flow around the sands, which may cause positive or negative pressure.

Fig. 4 − 25

From Equation (4 − 14), we can obtain

$$\lambda = \frac{2gdh_f}{lv^2} \qquad (4-57)$$

where v is the average cross-sectional velocity, which can be obtained by $v = \dfrac{Q}{A}$ and $A = \dfrac{\pi d^2}{4}$. $Q = V/t$, where V is the volume of water, and t is the time. d, l, h_f (height difference in the piezometers) are all measured. Therefore, λ can be calculated from Equation (4 − 57). Meanwhile, the Reynolds number $Re = \dfrac{ud}{\nu}$ can also be obtained when the viscosity ν is given, based on the temperature of fluid.

Taking $\lg Re$ as the abscissa and $\lg(100\lambda)$ for the vertical axis, with a relative smoothness (r_0/Δ) as the reference parameter, the experimental data were plotted as shown in Fig. 4 − 26.

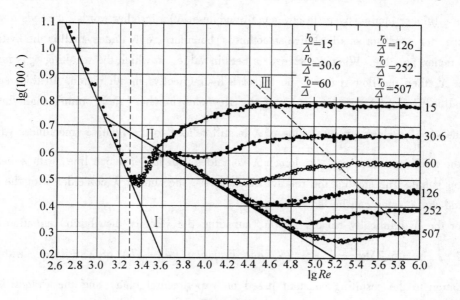

Fig. 4 − 26

4.6.1.2 Variation of frictional resistance coefficient

From the experimental results by Nikuradse (Fig. 4 − 26), the frictional coefficient (λ) has the following points:

(1) Laminar zone: $Re \leq 2000$, $\lg(Re) < 3.3$, represented by straight line I. All the data fall on this line, indicating that the relative smoothness in this region has no effect on λ, i.e. $\lambda \neq f\left(\dfrac{r_0}{\Delta}\right)$. λ is only a function of the Reynolds number, i.e. $\lambda = \dfrac{64}{Re}$. In laminar flow, the viscous friction predominates.

(2) The first transition zone: When $2000 < Re \leq 4000$, it is the transition zone from laminar flow to turbulent flow. λ is also only related to Re, but the relative smoothness has no impact on it. Since this zone is very narrow, i.e. a small range of Re, further research is needed.

(3) Turbulent zone: When $Re > 4000$, $\lg Re > 3.6$, it is the turbulent flow, and λ depends on the relative relation between the laminar sublayer thickness (δ_0) and the absolute roughness (Δ). Thus the flow is classified into the hydraulically smooth region, the second transition zone, and the hydraulically rough region.

① Hydraulically smooth region: denoted by straight line II (an envelope line). The data with different relative smoothness fit in a straight line initially, and then follow their own "rules". The smoother (or larger) $\dfrac{r_0}{\Delta}$, the longer the coincident segment is, because the larger $\dfrac{r_0}{\Delta}$ means smaller Δ, which does not require a thick laminar sublayer to cover it; while the laminar layer thickness (δ_0) is inversely proportional to the Reynolds number, that is to say, the thin sublayer in a large Re flow may also be immersed under the absolute rough (Δ) while maintaining the hydraulically smooth region. For a smaller relative smoothness ($\dfrac{r_0}{\Delta}$), it means a larger Δ, so it requires a thicker laminar layer to cover the roughness (Δ), in other words, it needs a small Re. Therefore, for rough (or small relative smoothness) boundary, only under small is the hydraulically smooth region possible. When Re exceeds a certain value, in which the resulting δ_0 is insufficient to cover Δ, then the flow region is no longer a hydraulically smooth region. In this region, the frictional coefficient (λ) is a function of the Reynolds number (Re), rather than the relative smoothness ($\dfrac{r_0}{\Delta}$). However, the value of λ is still related to the relative smoothness value. The larger the relative smoothness, the longer λ coincides with the envelope line, with λ being up to a smaller value. The smaller the relative smoothness, the shorter λ coincides with the envelope line, and the λ value can be larger.

For the hydraulically smooth region, substitue the velocity distribution equation Equation (4-52), $\dfrac{v}{u_*} = 5.75 \lg \dfrac{r_0 u_*}{v} + 1.75$, into Equation (4-17) of $\lambda = 8 \dfrac{v_*^2}{v^2}$, with a slight modification to the resulting constant based on experimental data, and the Prandtl-Nikuradse formula can then be obtained,

$$\dfrac{1}{\sqrt{\lambda}} = 2.0 \lg(Re \sqrt{\lambda}) - 0.8 \qquad (4-58)$$

Equation (4-58) shows that the frictional resistance coefficient λ of hydraulically smooth region is only related with the Reynolds number Re, but the absolute roughness has no impact on λ at all.

In the hydraulically smooth region, another commonly used formula for λ is Blasius' formula,

$$\lambda = \frac{0.3164}{Re^{1/4}} \qquad (4-59)$$

Equation (4-59) applies to $Re < 10^5$. It shows that λ is inversely proportional to $Re^{1/4}$, so the slope of the straight line II is smaller than that of the line I (λ is inversely proportional to $Re^{1.0}$). Substituting Equation (4-59) for λ in the Darcy-Weisbach formula (4-14), the head loss h_f is proportional to the velocity with the power of 1.75, i.e. $h_f \propto v^{1.75}$.

② In the second transition zone, namely in the region between the straight lines II and III in Fig. 4-26. This is the transition region of flow from the hydraulically smooth region to the rough region, which has a cluster of curves, shown as intially falling down and then gradually rising curves for these artificially sand roughened pipes. For the flow in the second transition zone, as the Reynolds number (Re) increases, the decreasing laminar sublayer thickness (δ_0) cannot completely cover the absolute roughness (Δ), so the roughness of boundary will now have an impact on λ. In this region, λ is a function of both Re and $\frac{r_0}{\Delta}$, i.e. $\lambda = f(Re, \frac{r_0}{\Delta})$. It follows that $h_f \propto v^{1.75 \sim 2.0}$, then the friction coefficient λ can be described by the Colebrook-White equation as

$$\frac{1}{\sqrt{\lambda}} = -2.0 \lg\left(\frac{2.51}{Re\sqrt{\lambda}} + \frac{\Delta}{3.7d}\right) \qquad (4-60)$$

which is valid for $2000 < Re < 10^6$.

③ The hydraulically rough region, also called a fully rough region or resistance square region, is the region to the right side of the dashed line III in Fig. 4-26. The resistance of flow shows a cluster of horizontal lines. λ is related only to $\frac{r_0}{\Delta}$, not Re, i.e. $\lambda = f(\frac{r_0}{\Delta})$. In this region, the flow has very large Re, the laminar sublayer thickness (δ_0) becomes very thin, which is well below the roughness height (Δ). Turbulent eddies form around the rough boundary, thus making a significant contribution to the frictional resistance. From Equation (4-14), $h_f \propto v^{2.0}$, this region is also called square region of resistance.

Substituting the average velocity equation Equation (4-55) for v in Equation (4-17) with the value of the corresponding constant that is slightly modified from experimental data, the Nikuradse's resistance formula becomes

$$\lambda = \frac{1}{\left(2\lg \frac{r_0}{\Delta} + 1.74\right)^2} = \frac{1}{\left[2\lg(3.7 \frac{d}{\Delta})\right]^2} \qquad (4-61)$$

which is valid for $Re > \frac{382}{\sqrt{\lambda}}(\frac{r_0}{\Delta})$.

Note that the dash line Ⅲ in Fig. 4-26 does not actually exist, and thus it is an imaginary dividing line between the second transition region and the resistance square region (a cluster of almost horizontal lines).

【Example 4-4】 In a pipe with a diameter of $d = 250$ mm, its inner wall is painted with 0.5 mm sand, and the water temperature is 10℃ (i.e. $\nu = 0.013$ cm²/s). What is the minimum flow rate for the hydraulically rough region? What is the maximum flow rate in the hydraulically smooth region? Calculate the corresponding wall frictional stress in the two cases.

Solution: (1) The maximum Re_* of flow in the hydraulically smooth region:

$$Re_{*,max} = \frac{u_* \Delta}{\nu} = 5$$

So

$$u_{*,max} = \frac{5\nu}{\Delta} = \frac{5 \times 0.013}{0.05} = 1.3 \text{ cm/s}$$

Since the average velocity for hydraulically smooth region is

$$\frac{v}{u_*} = 5.75 \lg \frac{r_0 u_*}{\nu} + 1.75$$

thus

$$v_{max} = u_{*,max}\left(5.75 \lg \frac{r_0 u_{*,max}}{\nu} + 1.75\right)$$

$$= 1.3 \times \left(5.75 \lg \frac{12.5 \times 1.3}{0.013} + 1.75\right) = 25.4 \text{ cm/s}$$

$$Q_{max} = A \cdot v_{max} = \frac{\pi \times 0.25^2}{4} \times 0.254 = 0.01244 \text{ m}^3/\text{s} = 12.44 \text{ L/s}$$

(2) Since the minimum Re for flow in the hydraulically rough region is $Re_{*,min} = 70$, i.e.

$$Re_{*,min} = \frac{u_* \Delta}{\nu} = 70$$

thus

$$u_{*,min} = \frac{70\nu}{\Delta} = \frac{70 \times 0.013}{0.05} = 18 \text{ cm/s}$$

Applying the average velocity in the hydraulically rough region,

$$v_{min} = u_{*,min}\left(5.75 \lg \frac{r_0}{\Delta} + 4.75\right)$$

$$= 18 \times \left(5.75 \lg \frac{12.5}{0.05} + 4.75\right) = 334 \text{ cm/s}$$

$$Q_{min} = A \cdot v_{min} = \frac{\pi \times 0.25^2}{4} \times 3.34 = 0.1637 \text{ m}^3/\text{s} = 163.7 \text{ L/s}$$

(3) Since the wall shear stress is $\tau_0 = \rho u_*^2$,

in the hydraulically smooth region $\tau_0 = \rho u_*^2 = 1000 \times 0.013^2 = 0.169 \text{ N/m}^2$;

in the hydraulically rough region $\tau_0 = \rho u_*^2 = 1000 \times 0.18^2 = 32.4 \text{ N/m}^2$.

Since the artificial coarse sand is used in this example, the solution can also be obtained by the Nikuladse experimental results: First determine the relative smoothness (r_0/Δ), then find the lgRe value, which corresponds to the intersections of r_0/Δ to straight lines Ⅱ and Ⅲ. Thus calculate Re and $v = \dfrac{\nu Re}{d}$ and finally calculate $Q = Av$. However, this method is not as accurate as the method above.

4.6.2 Frictional resistance coefficient of commercial pipes

The natural roughness of actual pipes is somewhat different from the roughness of artificial sands. The roughness of artificial sands is of relatively high density and regularity, where the sands are tightly adhered to the solid wall side by side and uniformly spread; the natural roughness of pipe is usually irregular in height, shape and distribution. Therefore, the concept of an "equivalent roughness" for actual pipes is introduced. First, find the resistance coefficient of the actual pipe made of various materials in the turbulent square region of resistance, which can be calculated from $\lambda = h_f \dfrac{2gd^2}{lv^2}$ by measuring h_f, d, l, and v, then the calculated Δ from

$\lambda = \dfrac{1}{\left[2\lg(3.7\dfrac{d}{\Delta})\right]^2}$ is the equivalent roughness for the pipe. Table 4-2 shows the equivalent roughness values of commonly used pipes.

Table 4-2 Equivalent roughness of various surface materials(Δ)

Wall Materials	Δ/mm	Wall Materials	Δ/mm
Glass, brass, transparent plastic, polyethylene, asbestos cement, aluminum	0.00015~0.015	Tin plate	0.15
Steel or wrought iron, welded steel pipe	0.046	Concrete, pink flat	0.3~0.8
Riveted steel	0.9~9	Concrete	1.0~2.0
Steel, painted	0.06	Reinforced concrete	0.3~3.0
Cast iron	0.3~1.0	Board, planed	0.2~0.9
Cast iron, bitumen coated surface	0.12~1.0	Wood, unplaned	1.0~2.5
Cast iron, painted	0.15	Fine brickwork	1.2~2.5
Cast iron, rusted	1.0~1.5	Fine stonework	1.5~3.0
Cast iron, fouling	1.5~3.0		

4.6.2.1 Moody diagram

Based on the Prandtl-Nikuradse, von Karman and Colebrook-White formulas, Moody(1944) made a diagram between λ and Δ/d, as shown in Fig. 4-27. The vertical coordinate is the value of $\log\lambda$, the horizontal coordinate is expressed in Re, and the values of different relative roughness (Δ/d) are represented by different curves. Using these curves, the frictional head loss of pipe flow can be obtained. For turbulent flow of non-circular pipes, the Moody diagram also suits, in which Δ/d should be replaced by $\Delta/(4R)$.

In the application of Moody diagram, first calculate Re, then determine the corresponding equivalent roughness of the pipe from Table 4 − 2, finally find λ by Δ/d and Re using the curves in Fig. 4 − 27. The frictional head loss h_f can be obtained from Equation (4 − 14) (Darcy-Weisbach formula). It is worth pointing out that both vertical and horizontal coordinates in the Moody diagram are expressed in logarithmic values. Usually interpolation is needed in the application of the Moody diagram.

The Moody diagram has five distinct regions:

Laminar region: $\quad\quad\quad\quad\quad\quad \lambda = f(Re)$, $h_f \propto v^{1.0}$

First transition region: How λ relates to Re and Δ/d is unclear.

Hydraulically smooth region: $\lambda = f(Re)$, $h_f \propto v^{1.75}$

Second transition region: $\lambda = f(Re, \frac{\Delta}{d})$, $h_f \propto v^{1.75 \sim 2.0}$

Hydraulically rough region: $\lambda = f(\frac{\Delta}{d})$, $h_f \propto v^{2.0}$

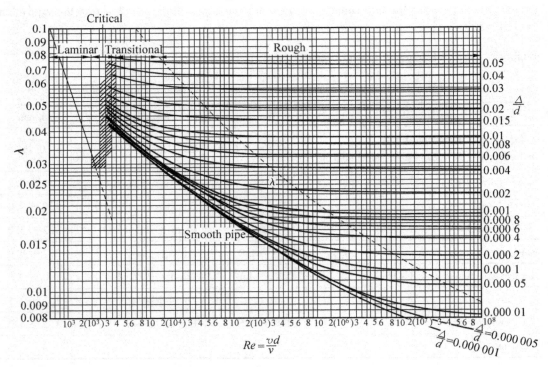

Fig. 4 − 27

Comparing the Nikuradse(Fig. 4 − 26) and Moody(Fig. 4 − 27) diagrams, it can be seen that the curves of the second transition region are not the same. For the Nikuradse diagram, λ drops first and then rises, while for the Moody diagram, λ decreases with increasing Re, and it continuously reduces to the horizontal curve, which is the value for the hydraulically rough region.

Although there is no satisfactory explanation about the difference above, it is generally believed that this is due to the different shape and distribution of the roughness. In practical applications, if the flow is in the second transition region, it is required to identify the difference between artificial roughness and the natural roughness.

The dashed lines in the Moody diagram is a dividing line between the second transition and hydraulically rough region, which can be expressed by $Re = 3500/(\frac{\Delta}{d})$.

Moody also gave an explicit formula for the λ in the second transition region.

$$\lambda = 0.0055\left[1 + \left(20000\frac{\Delta}{d} + \frac{10^6}{Re}\right)^{1/3}\right] \qquad (4-62)$$

which is valid in the range of $Re = 4000 \sim 10^7$, with the error to $\pm 5\%$, compared with the results calculated by the Colebrook-White formula $(4-60)$.

If the flow is in the completely turbulent rough zone, h_f is proportional to v^2, and λ is independent of Re. It is therefore straight forward to calculate h_f from the Moody diagram for a given Δ. Pipe flows in engineering are usually in the second transition zone, so the Reynolds number has to be calculated in order to use the diagram. If the flow velocity v is known, Re can be directly obtained. Otherwise, if the flow velocity v is unknown, Re is not known either, and thus h_f by the Darcy-Weisbach equation $(4-14)$ has two unknowns (λ and v). In such a case, the trial-and-error method has to be used: assume a λ value, calculate v and Re, and then find the corresponding λ value from the Moody diagram according to the obtained Re and Δ/d. If the λ value is not close to the assumed value, take another λ value to repeat the calculation procedure above, until the obtained λ equals the assumed λ. Since λ is typically between $0.02 \sim 0.06$, one or two times of iteration can achieve a satisfactory result.

【Example 4-5】 The pipe flow, in a new cast iron pipe with diameter $d = 600$ mm and length $l = 500$ m, has the mean cross-sectional velocity of $v = 1.7$ m/s. The water temperature is 15℃. Find the value of λ and the frictional head loss h_f.

Solution: At 15 ℃, the viscosity coefficient of water is $v = 0.0114$ cm²/s.

$$Re = \frac{vd}{\nu} = \frac{170 \times 60}{0.0114} = 895\,000$$

From Table 4-2, for the new cast iron pipe $\Delta \approx 0.6$ mm, so $\Delta/d = 0.001$.

From the Moody diagram, the flow is found in the second transition zone for $\Delta/d = 0.001$ and $Re = 895\,000$, and $\lambda = 0.020$.

$$h_f = \lambda\frac{l}{d}\frac{v^2}{2g} = 0.02 \times \frac{500}{0.6}\frac{1.7^2}{2 \times 9.8} = 2.46 \text{ m}$$

【Example 4-6】 The 20℃ water flows in a 50 cm welded steel pipe ($\Delta = 0.046$ m). Given the hydraulic gradient $J = 0.006$, determine the flow rate of the pipe.

Solution: To calculate the flow rate, the velocity has to be found by the formula $(4-14)$, $v = \sqrt{\frac{2gdJ}{\lambda}}$, which includes two unknowns in this formula: v and l. The trial-and-error method has to be used.

Firstly assume $\lambda = 0.03$, and substitute it into the formula above. So we can obtain $v = 1.4$ m/s. Thus,

$$Re = \frac{vd}{\nu} = \frac{1.4 \times 0.5}{1 \times 10^{-6}} = 7 \times 10^5$$

where the viscosity of water at 20℃ is 1×10^{-6} m²/s.

$$\frac{\Delta}{d} = \frac{0.046}{500} = 0.00009$$

For $Re = 7 \times 10^5$ and $\Delta/d = 0.00009$, find $\lambda = 0.0135$ from the Moody diagram, but it is not much close to the assumed value 0.03.

By assuming another $\lambda = 0.0135$,

$$v = \sqrt{\frac{2gdJ}{\lambda}} = 2.09 \text{ m/s}$$

$$Re = \frac{vd}{\nu} = \frac{2.09 \times 0.5}{1 \times 10^{-6}} = 1.0 \times 10^6$$

From the Moody diagram, for $Re = 1.0 \times 10^6$ and $\Delta/d = 0.00009$ we can obtain $\lambda = 0.0133$, which is close to the newly assumed value of 0.0135. Therefore, no further attempt is needed.

$$Q = Av = \frac{\pi \times 0.5^2}{4} \times 2.09 = 0.41 \text{ m}^3/\text{s}$$

【Example 4-7】 A clean, new wrought iron pipe ($\Delta = 0.046$ mm) conveys oil with the flow rate of $Q = 0.25$ m³/s. The viscosity coefficient of oil $\nu = 0.0929$ cm²/s; the pipe is 3000 m long and has the head loss $h_f = 23$ m. Calculate the diameter of pipe.

Solution: From Equation (4-14),

$$h_f = \lambda \frac{l}{d} \frac{v^2}{2g} = \frac{l}{d} \frac{Q^2}{2g \left(\frac{\pi d^2}{4}\right)^2} = \frac{8\lambda l Q^2}{g \pi^2 d^5}$$

$$d^5 = \frac{8\lambda l Q^2}{g \pi^2 h_f} \tag{4-63}$$

For known l, Q, h_f,

$$d^5 = \frac{8 \times 3000 \times 0.25^2}{9.8 \times 3.14^2 \times 23}\lambda = 0.675\lambda$$

$$Re = \frac{vd}{\nu} = \frac{4Q}{\nu \pi d} = \frac{4 \times 0.25}{0.0929 \times 10^{-4} \times 3.14 d} = \frac{34200}{d}$$

Assuming $\lambda = 0.02$ gives $d^5 = 0.0135$, i.e. $d = 0.423$ m, so $Re = 81000$. From the Moody diagram, $\lambda = 0.019$, which is not much close to the assumed value.

Assume another $\lambda = 0.019$, then $d^5 = 0.0128$, i.e. $d = 0.418$ m, so $Re = 82000$. From the Moody diagram, $\lambda = 0.019$, which is now close to the assumed value.

Select a pipe with the nominal diameter being the multiples of 50 mm, and take $d = 450$ mm.

4.6.2.2 Chevielev's formula

Chevielev proposed an expression of resistance coefficient λ for old steel and cast iron pipes in drainage systems.

(1) For the mean velocity $v < 1.2$ m/s (second transition zone),

$$\lambda = \frac{0.0179}{d^{0.3}} \left(1 + \frac{0.867}{v}\right)^{0.3} \quad (4-64)$$

(2) For the mean velocity $v \geqslant 1.2$ m/s (hydraulically rough zone),

$$\lambda = \frac{0.0210}{d^{0.3}} \quad (4-65)$$

In the application of new pipes, corrosion and fouling will occur, which will increase the resistance, so Equations (4-64) and (4-65) are widely used in design. Note that d and v are in meters in the formulae, and that the formulae were proposed for the water temperature at 10 ℃, which has the kinematic viscosity coefficient of $\nu = 1.3 \times 10^{-6}$ m²/s.

【Example 4-8】 An old steel pipe with diameter $d = 300$ mm carries water with a flow rate of $Q = 100$ L/s. The pipe length l is 1000 m, and the water temperature is 10℃. Calculate the frictional head loss.

Solution: For Chevielev's formula,

$$A = \frac{\pi \times d^2}{4} = \frac{3.14 \times 0.3^2}{4} = 0.07 \text{ m}^2$$

$$v = \frac{Q}{A} = \frac{0.1}{0.07} = 1.42 \text{m/s} > 1.2 \text{ m/s}$$

$$\lambda = \frac{0.0210}{d^{0.3}} = \frac{0.0210}{0.3^{0.3}} = 0.0301$$

$$h_f = \lambda \frac{l}{d} \frac{v^2}{2g} = 0.0301 \times \frac{1000}{0.3} \times \frac{1.42^2}{2 \times 9.8} = 10.3 \text{ m}$$

4.6.3 Empirical formulae for frictional head loss

In previous section, h_f is calculated by the Darcy-Weisbach formula, in which λ is obtained by the Moody diagram based on the relative roughness (Δ/d) and Re. However, in practical design in the early 1900s, the frictional head loss (h_f) was calculated based on the roughness coefficient (n), which is used to calculate Chezy coefficient C, thereby obtaining λ for h_f. Although the latter method lacks a theoretical basis, it is simple and the results can meet the requirements of practical engineering, and thereby it is widely used in practice.

4.6.3.1 Chezy formula

In 1775 the French engineer Chezy proposed a velocity calculation formula for uniform flow in open channel.

$$v = C\sqrt{RJ} = C\sqrt{Ri} \quad (4-66)$$

where v is the average velocity of cross-section, in m/s; R is the hydraulic radius, defined by A/χ, where A is the cross-sectional area and χ is the wetted perimeter; J is the hydraulic gradient, defined by h_f/l; for uniform flow, $J = i$, where i is the bed slope of channel; C is Chezy coefficient, a constant with dimension, $[C] = \frac{[v]}{[R^{1/2}]} = \frac{[LT^{-1}]}{[L^{1/2}]} = [L^{1/2}T^{-1}]$. Note that the C value is based on the unit systems of meter.

Chezy's formula is one of the oldest hydraulics formulae. However, it can be transformed into an expression consistent with the Darcy-Weisbach formula. Taking square on Equation (4-66),

$$v^2 = C^2 RJ = C^2 R \frac{h_f}{l}$$

$$h_f = \frac{8g}{C^2} \frac{l}{4R} \frac{v^2}{2g}$$

Let

$$\lambda = \frac{8g}{C^2}$$

$$h_f = \lambda \frac{l}{4R} \frac{v^2}{2g}$$

This is the same as Equation (4-13). As can be seen, the friction coefficient λ is related to C by

$$\lambda = \frac{8g}{C^2} \quad (4-67)$$

or

$$C = \sqrt{\frac{8g}{\lambda}} \quad (4-68)$$

The Darcy-Weisbach formula is applicable to laminar and turbulent flow, pipe and open-channel flow. It is a universal equation. It should be noted that the values of coefficient C are based on the data of flow in the square region of resistance, in other words, Chezy's equation applies only to the turbulent rough zone of either pipe or open-channel flows. Furthermore, coefficient C is a constant with dimension $[L^{1/2}/T]$, but the frictional coefficient λ is a dimensionless quantity, which theoretically seems more reasonable.

4.6.3.2 Empirical formula of Chezy coefficient C

(1) Manning's formula

In 1890 Manning proposed the following formula for coefficient C:

$$C = \frac{R^{1/6}}{n} \quad (4-69)$$

which is in meter and second units.

Manning's n is an overall roughness coefficient to take account for the impact of the irregularities and roughness of boundary on flow, called roughness or roughness factor. Its value is difficult to determine accurately. If a large n is chosen in the design, this will increase the cross-sectional area for a given discharge, resulting in an unnecessary cost of money; if a small n is chosen, this will lead to a large average velocity and a small cross-sectional area, which may not meet the requirements of the design flow. Choosing appropriate n values is always a difficult issue. Table 4-3 shows some typical n values.

Table 4-3 Roughness coefficient n value

No.	Boundary types and conditions	n	$1/n$
1	Carefully planed timber, cleanly new pig iron and cast iron pipes, laying flat with smooth seams	0.011	90.0
2	Unplaned timber but well connected, water supply pipes under normal circumstances, very clean drains, smooth concrete surface	0.012	83.3

Continued

No.	Boundary types and conditions	n	$1/n$
3	Drainage pipes in normal conditions, supply pipes with slight fouling, good brick work	0.013	76.9
4	Fouling water supply and drainage pipe, ordinary concrete surface, general brick work	0.014	71.4
5	Old brick work; rather rough concrete surface; smooth, carefully excavated rock surface	0.017	58.8
6	Compact clay canal; earth canal of loess with discontinuous silt layers; earth canal with sand and gravel, or large canal in good maintenance condition	0.0225	44.4
7	General large canal, small canal in good condition, extremely good natural rivers (clean straight, smooth flow, no riffle or pool)	0.025	40.0
8	Earth canal in bad condition (e.g. partchy weeds or gravel, embankment collapsed partially); natural rivers in good condition	0.030	33.3
9	Earth canal in very bad condition (e.g. irregular cross-sections, weeds, stone, no-smooth water flow); good natural river with only a few block stones and weeds	0.035	28.6
10	Extremely bad canal (e.g. pool or riffle, many weeds, large block stones in bed); natural rivers in not good condition (e.g. weeds, many block stones, irregular river bed with slightly meander, many collapsed banks or deep pools)	0.040	25.0

Inserting Equation (4-69) into (4-66) gives

$$v = C\sqrt{RJ} = \frac{1}{n}R^{2/3}J^{1/2} \qquad (4-70)$$

Because this formula is simple and can meet certain engineering requirements, it is widely used in engineering.

(2) Pavlovskii's formula

$$c = \frac{1}{n}R^y \qquad (4-71)$$

where
$$y = 2.5\sqrt{n} - 0.13 - 0.75\sqrt{R}(\sqrt{n} - 0.10) \qquad (4-72)$$

or y can approximately be calculated by

when $R < 1.0$ m, $y = 1.5\sqrt{n}$ \qquad (4-73a)

when $R > 1.0$ m, $y = 1.3\sqrt{n}$ \qquad (4-73b)

The application of the formula (4-73) is in the range of:

$0.1 \leqslant R \leqslant 3.0$ m, $0.011 \leqslant n \leqslant 0.035 \sim 0.04$

Since the index y is a variable, when $y = 1/6$, Pavlovskii's formula becomes Manning's formula. In other words, Manning's formula is a special case of Pavlovskii's formula. Therefore, Equation (4-71) has a wide range of application, and it can be used for both pipe and open-channel flows.

The dimension of the roughness n is still undecided. Based on Chezy's formula, $[n] = \dfrac{[R^{2/3}]}{[v]} = \dfrac{[L^{2/3}]}{[LT^{-1}]} = [L^{-1/3}T]$. Based on Pavlovskii's formula, $[n] = \dfrac{[R^y]}{[C]} = \dfrac{[L^y]}{[L^{1/2}T^{-1}]} = [L^{y-1/2}T]$, in which the dimension of n varies with y, so obviously it is not realistic. Meanwhile, it is also recommended by $[n] = [L^{1/6}]$, which is not appropriate either, because this will lead to $[C] = \left[\dfrac{1}{n}R^{1/6}\right] = \dfrac{[L^{1/6}]}{[L^{1/6}]} = [L^0] = [1]$, which is contradicted with the dimension of $[L^{-1/3}T]$.

In practice, it stipulates that when hydraulic radius R is used in unit of m, and regardless of the dimension of n, C is in $m^{1/2}/s$.

【Example 4-9】 A reinforced concrete tunnel has the diameter of 6 m, length of 1 km and carries the water with a flow rate of 400 m³/s, and the flow is in the hydraulically rough zone. Determine the frictional head loss.

Solution: (1) Calculate the hydraulic radius R.

Cross-sectional area:

$$A = \frac{\pi \times d^2}{4} = \frac{3.14 \times 6^2}{4} = 28.3 \text{ m}^2$$

Wetted perimeter:

$$\chi = \pi d = 3.14 \times 6 = 18.83 \text{ m}$$

Hydraulic radius:

$$R = \frac{A}{\chi} = \frac{28.3}{18.83} = 1.5 \text{ m}$$

(2) Calculate Chezy's coefficient C (based on Manning's n).

From Table 4-3, $n = 0.014$ for general reinforced concrete surface

$$C = \frac{R^{1/6}}{n} = \frac{1.5^{1/6}}{0.014} = 76.5 \text{ m}^{1/2}/\text{s}$$

(3) Calculate the head loss h_f.

From $v = C\sqrt{RJ}$, $J = \dfrac{h_f}{l} = \dfrac{v^2}{RC^2}$

$$v = \frac{Q}{A} = \frac{400}{28.3} = 14.1 \text{ m/s}$$

$$h_f = \frac{lv^2}{RC^2} = \frac{1000 \times 14.1^2}{1.5 \times 76.5^2} = 22.6 \text{ m}$$

or using $= \dfrac{8g}{C^2}$,

$$\lambda = \frac{8 \times 9.8}{76.5^2} = 0.0134$$

$$h_f = \lambda \frac{l}{d} \frac{v^2}{2g} = 0.0134 \times \frac{1000}{6} \times \frac{14.1^2}{2 \times 9.8} = 22.65 \text{ m}$$

The results are very close.

4.7 Local head loss

Local head loss is also called the form resistance, which is the head loss caused by the rapid changes of boundary in shape, size and direction. Examples of where local losses occur in practical engineering are: a change of pipe diameter, pipe bends, valves in pipelines; river bend, transition, hydraulic structures, and trash racks in open channels. In such cases, the changes of flow velocity and pressure can result in flow separation, vortex and the re-structuring of the velocity distribution, so this will increase the relative motion of fluid particles, thereby generating extra resistance locally.

Due to the complexity of solid boundaries, only a few simple cases can be approximated by a theoretical analysis, and in most cases the local losses have to be determined by experiments. The difficulty in any theoretical analysis lies in the calculation of hydrodynamic pressure for rapidly varied flows. Despite the complexity of boundary conditions, the local head loss can be expressed as a multiple of velocity head, namely

$$h_j = \zeta \frac{v^2}{2g} \qquad (4-74)$$

where ζ is called the local resistance coefficient or the local head loss coefficient. This will depend on local boundary conditions, and the ζ value is therefore usually determined by laboratory tests.

In the pipeline calculation, for simplicity of calculation, the concept of equal length (or equivalent length) is introduced and the local head loss in the pipeline is converted to the length of pipe by the frictional head loss, $h_f = \lambda \frac{L}{d} \frac{v^2}{2g}$. Let

$$\zeta = \lambda \frac{l}{d}$$

Thus,

$$l = \frac{\zeta d}{\lambda} \qquad (4-75)$$

where l is the equivalent length of local head loss. When all the local head losses are converted into corresponding equivalent lengths, the sum of the equivalent lengths of local losses is added to the actual length of the pipeline, which will then give the total resistance loss.

4.7.1 Local head loss of sudden expansion of pipe

Fig. 4-28 shows a sudden expansion of pipe, in which the cross-section A_1 is suddenly expanded to A_2. The circulation of flow around the corner will be created by the inertia influence of the flow, and the main flow gradually reaches the full velocity distribution after a distance of

$l = (5 \sim 8)d_2$. Streamlines at sections 1 – 1 and 2 – 2 tend to be parallel, so they are the cross-sections selected for gradually varied flow analysis. The energy equation can then be written as

$$z_1 + \frac{p_1}{\gamma} + \frac{\alpha_1 v_1^2}{2g} = z_2 + \frac{p_2}{\gamma} + \frac{\alpha_2 v_2^2}{2g} + h_{w1-2}$$

Since the distance between sections 1 – 1 and 2 – 2 is very short, the frictional head loss is negligible, so $h_{w_1-2} = h_j$, and the above equation can be written as

$$h_j = \left(z_1 + \frac{p_1}{\gamma}\right) - \left(z_2 + \frac{p_2}{\gamma}\right) + \left(\frac{\alpha_1 v_1^2}{2g} - \frac{\alpha_2 v_2^2}{2g}\right) \tag{4-76}$$

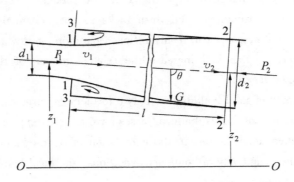

Fig. 4 – 28

Because the first two terms on the RHS of Equation (4 – 76) are still unknown, the momentum equation is required to solve h_j.

Taking the flow direction for the x axis, analyze the external forces on the flow between sections 1 – 1 and 2 – 2.

On sections 1 – 1 and 2 – 2, the hydrodynamic pressure forces are approximated by the hydrostatic pressure at the centroid points (i. e. average pressure) multiplied by the sectional area. For the annular section 1 – 3, the flow is in circulation region, its dynamic water pressure cannot be calculated; on sections 1 – 1, 2 – 2, the hydrodynamic pressure forces are assumed as $p_1 A_2$ and $p_2 A_2$ respectively, where p_1 and p_2 are the pressures at the centroids of sections 1 – 1 and 2 – 2, respectively.

The component of gravity in the direction of flow:

$$G_x = G\cos\theta = \gamma A_2 l \cos\theta = \gamma A_2 (z_1 - z_2)$$

where $\cos\theta = \dfrac{z_1 - z_2}{l}$.

Since the distance between sections 3 – 1 – 1 – 3 and 2 – 2 is very short, the friction resistance (T) on the boundary surface of flow section is small, and can be neglected.

Setting the flow rate Q, the momentum change in the flow direction is

$$\rho Q(\beta_2 v_2 - \beta_1 v_1)$$

According to the momentum equation,

$$p_1 A_2 - p_2 A_2 + \gamma A_2 (z_1 - z_2) = \frac{\gamma Q}{g}(\beta_2 v_2 - \beta_1 v_1)$$

Divided by γA_2, and rearranging the resulting equation gives

$$\left(z_1 + \frac{p_1}{\gamma}\right) - \left(z_2 + \frac{p_2}{\gamma}\right) = \frac{(\beta_2 v_2 - \beta_1 v_1) v_2}{g}$$

Substituting it into Equation (4-76),

$$h_j = \frac{(\beta_2 v_2 - \beta_1 v_1) v_2}{g} + \left(\frac{\alpha_1 v_1^2}{2g} - \frac{\alpha_2 v_2^2}{2g}\right)$$

Letting α_1, α_2, β_1, β_2 approximately equal 1, the above equation can be simplified to

$$h_j = \frac{(v_1 - v_2)^2}{2g} \tag{4-77}$$

Equation (4-77) is the theoretical formula for the local loss of sudden expansion of pipe, which is demonstrated to have sufficient accuracy based on experimental data.

From the continuity equation, substituting $v_1 = \frac{A_2 v_2}{A_1}$ into Equation (4-77),

$$h_j = \left(\frac{A_2}{A_1} - 1\right)^2 \frac{v_2^2}{2g} = \zeta_1 \frac{v_2^2}{2g} \tag{4-78}$$

or substituting $v_2 = \frac{A_1 v_1}{A_2}$ into Equation (4-77),

$$h_j = \left(1 - \frac{A_1}{A_2}\right)^2 \frac{v_1^2}{2g} = \zeta_2 \frac{v_1^2}{2g} \tag{4-79}$$

where $\zeta_1 = \left(\frac{A_2}{A_1} - 1\right)^2$ and $\zeta_2 = \left(1 - \frac{A_1}{A_2}\right)^2$. They are called the loss coefficients for a sudden pipe expansion. Note that the use of ζ_1 is associated with $\frac{v_2^2}{2g}$ and ζ_2 with $\frac{v_1^2}{2g}$.

4.7.2 Local head loss coefficient

Any local head loss can be expressed as $h_j = \zeta \frac{v^2}{2g}$, where the local loss coefficient ζ is usually determined by experiments. Thus, the key for calculating local head loss is to find the appropriate ζ values. Some commonly used ζ values are given in Table 4-4, for certain cases.

Table 4-4 Local head loss coefficients

(Local head loss $h_j = \zeta \frac{v^2}{2g}$, where v is shown in the figures)

Name	Figure	Local head loss coefficient (ζ)
Sudden expansion	$A_1 \rightarrow v \quad A_2$	$\zeta = \left(1 - \frac{A_1}{A_2}\right)^2$

Continued

Name	Figure	Local head loss coefficient (ζ)
Sudden contraction	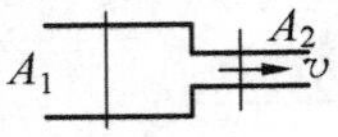	$\zeta = 0.5\left(1 - \dfrac{A_2}{A_1}\right)$
Entry	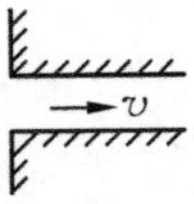	Right angle: $\zeta = 0.50$
		Slightly rounded corner: $\zeta = 0.20$ Bell-mouth shaped: $\zeta = 0.10$ Streamlined (no separated flow): $\zeta = 0.05 - 0.06$
		Cutting corner: $\zeta = 0.25$
Exit	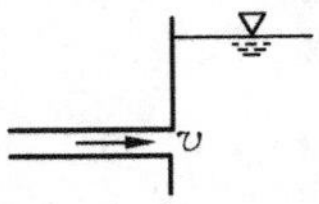	Exit to reservoir: $\zeta = 1.0$
		Exit to open channel: $\zeta = \left(1 - \dfrac{A_1}{A_2}\right)^2$
Diverging pipe	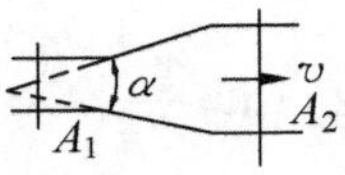	$\zeta = \kappa\left(\dfrac{A_2}{A_1} - 1\right)^2$

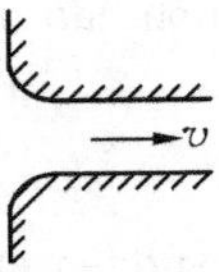

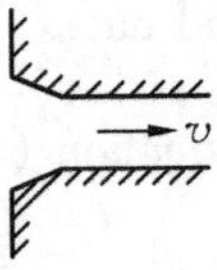

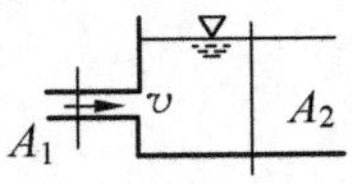

α	8°	10°	12°	15°	20°	25°
κ	0.14	0.16	0.22	0.30	0.42	0.62

Converging pipe

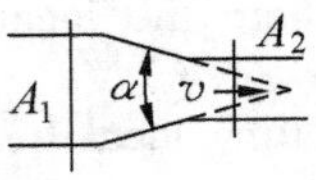

$\zeta = \kappa_1 \kappa_2$

α	10°	20°	40°	60°	80°	100°	140°
κ_1	0.40	0.25	0.20	0.20	0.30	0.40	0.60

A_2/A_1	0	0.10	0.20	0.30	0.40	0.50	0.60	0.70	0.80	0.90	1.0
κ_2	0.41	0.40	0.38	0.36	0.34	0.30	0.27	0.20	0.16	0.10	0

Name	Figure	Local head loss coefficient (ζ)
Rectangular to circular converging pipe	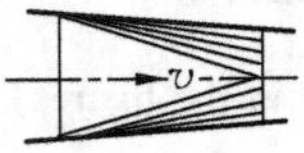	$\zeta = 0.05$ (velocity head of middle section)
Circular to rectangular converging pipe	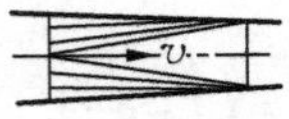	$\zeta = 0.10$ (velocity head of middle section)

Chapter 4 Types of flow and head loss

Continued

Name	Figure	Local head loss coefficient (ζ)						
Circular mild elbow		$\zeta = \left[0.131 + 0.1632\left(\dfrac{D}{R}\right)^{7/2} \right]\left(\dfrac{\theta°}{90°}\right)^{1/2}$						
Circular sharp elbow		$\zeta = 0.946\sin^2\left(\dfrac{\theta}{2}\right) + 2.05\sin^4\left(\dfrac{\theta}{2}\right)$						
		θ	15°	30°	45°	60°	90°	120°
		ζ	0.022	0.073	0.183	0.365	0.99	1.86
Pipe branch		$\zeta = 0.05$						
		$\zeta = 0.15$						
		$\zeta = 1.0$						
		$\zeta = 0.50$						
		$\zeta = 3.0$						
Tee pipe		$\zeta = 0.10$						
		$\zeta = 1.5$						
Right angle branch		$\zeta_{1-2} = 2,\ h_{j1-2} = 2\dfrac{v_2^2}{2g};\ h_{j1-3} = \dfrac{v_1^2 - v_3^2}{2g}$						
Plate gate		e/a	0.1~0.7	0.8	0.9			
		ζ	0.05	0.04	0.02			
		(velocity head at the contraction section)						
Gate slot		$\zeta = 0.05 - 0.20$ (usually 0.10)						

Name	Figure	Local head loss coefficient (ζ)
Curved gate		(chart showing ζ vs e/a for c/a = 1.2, 1.4, 1.6, 1.8 with $R/a \approx 2$) (velocity head at the contraction section)
Gate valve		Fully open ($a/d = 0$): see table below. Partially open (ζ values): see table below.
Disc valve		see table below
Check valve		$\zeta = 1.7$
Filter		$\zeta = 5 - 8$ (with bottom net) $\zeta = 2 - 3$ (without bottom net)
Screening bar		$\zeta = \beta \left(\dfrac{s}{b} \right)^{4/3} \sin\alpha$ s = bar width, b = gap of bar α = inclined angle, β = shape coefficient

Fully open ($a/d = 0$):

d/mm	15	20~50	80	100	150	200~250	300~450	500~800	900~1000
ζ	1.5	0.5	0.4	0.2	0.1	0.08	0.07	0.06	0.05

Partially open (ζ values):

d mm	in	1/8	1/4	3/8	1/2	3/4	1	d mm	in	1/8	1/4	3/8	1/2	3/4	1
12.5	1/2	450	60	22	11	2.2	1.0	50	2	140	20	6.5	3.0	0.68	0.16
19	3/4	310	40	12	5.5	1.1	0.28	100	4	91	16	5.6	2.6	0.55	0.14
25	1	230	32	9.0	4.2	0.90	0.23	150	6	74	14	5.3	2.4	0.49	0.12
								200	8	66	13	5.2	2.3	0.47	0.10
40	1 1/2	170	23	7.2	3.3	0.75	0.18	300	12	56	12	5.1	2.2	0.47	0.07

Disc valve:

α(°)	5	10	15	20	25	30	35	40
ζ	0.24	0.52	0.90	1.54	2.51	3.91	6.22	10.8
α(°)	45	50	55	60	65	70	90	
ζ	18.7	32.6	58.8	118	256	751	∞	

Screening bar β values:

No.	1	2	3	4	5	6	7
β	2.42	1.83	1.67	1.035	0.92	0.76	1.79

Chapter 4 Types of flow and head loss

Continued

Name	Figure	Local head loss coefficient(ζ)										
Plate orifice		$\frac{d}{D}$	0.30	0.40	0.45	0.50	0.55	0.60	0.65	0.70	0.75	0.80
		ζ	309	87	50.4	29.8	18.4	11.3	7.35	4.37	2.66	1.55
Standard orifice		$\frac{d}{D}$	0.30	0.40	0.45	0.50	0.55	0.60	0.65	0.70	0.75	0.80
		ζ	108.8	29.8	16.9	9.9	5.9	3.5	2.1	1.2	0.76	—
Venturi pipe		$\frac{d}{D}$	0.30	0.40	0.45	0.50	0.55	0.60	0.65	0.70	0.75	0.80
		ζ	19	5.3	3.06	1.9	1.15	0.69	0.42	0.26	—	—

【Example 4 – 10】 Water from a water tank is discharged to the atmosphere at a flow rate of $Q = 0.025$ m³/s through a pipeline with different diameter pipes and connections, as shown in Fig. 4 – 29. The given data are: $d_1 = 150$ mm, $l_1 = 25$ m, $\lambda_1 = 0.037$; $d_2 = 125$ mm, $l_2 = 10$ m, $\lambda_2 = 0.039$; the local loss coefficients: entry $\zeta_1 = 0.5$, converging transition $\zeta_2 = 0.15$, and valve $\zeta_3 = 2.0$. Calculate

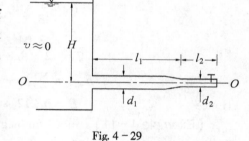

Fig. 4 – 29

(1) the total frictional head loss $\sum h_f$;
(2) the total local head loss $\sum h_j$;
(3) the head of the water tank H.

Solution: (1) Calculate the frictional head loss.
The first section:

$$v_1 = \frac{Q}{A_1} = \frac{4 \times 0.025}{3.14 \times 0.15^2} = 1.145 \text{ m/s}$$

$$h_{f1} = \lambda_1 \frac{l_1}{d_1} \frac{v_1^2}{2g} = 0.037 \times \frac{25}{0.15} \times \frac{1.145^2}{2 \times 9.8} = 0.63 \text{ m}$$

The second section:

$$v_2 = \frac{Q}{A_2} = \frac{4 \times 0.025}{3.14 \times 0.125^2} = 2.04 \text{ m/s}$$

$$h_{f2} = \lambda_2 \frac{l_2}{d_2} \frac{v_2^2}{2g} = 0.039 \times \frac{10}{0.125} \times \frac{2.04^2}{2 \times 9.8} = 0.663 \text{ m}$$

$$\sum h_f = h_{f1} + h_{f2} = 0.63 + 0.663 = 1.293 \text{ m}$$

(2) Calculate the local head loss.

The inlet head loss

$$h_{j1} = \zeta_1 \frac{v_1^2}{2g} = 0.5 \times \frac{1.145^2}{2 \times 9.8} = 0.051 \text{ m}$$

The head loss in the gradually converging transition

$$h_{j2} = \zeta_2 \frac{v_2^2}{2g} = 0.15 \times \frac{2.04^2}{2 \times 9.8} = 0.032 \text{ m}$$

The valve head loss

$$h_{j3} = \zeta_3 \frac{v_2^2}{2g} = 2 \times \frac{2.04^2}{2 \times 9.8} = 0.423 \text{ m}$$

$$\sum h_j = h_{j1} + h_{j2} + h_{j3} = 0.051 + 0.032 + 0.423 = 0.506 \text{ m}$$

(3) The static head of the water tank.

Taking the datum plane through the centroid of the exit cross-section, the energy equation between the water surface of the tank and the exit is written as

$$H + 0 + 0 = 0 + 0 + \frac{v_2^2}{2g} + h_{w1-2}$$

$$H = \frac{v_2^2}{2g} + h_{w1-2}$$

$$\frac{v_2^2}{2g} = \frac{2.04^2}{2 \times 9.8} = 0.212 \text{ m}$$

$$h_{w1-2} = \sum h_f + \sum h_j = 1.293 + 0.506 = 1.799 \text{ m}$$

Hence $\quad H = 0.212 + 1.799 = 2.011 \text{ m}$

【Example 4 – 11】 In a 10m long pipeline with a diameter of $d = 100$ mm, there exist two 90° bends ($d/R = 1.0$). The frictional coefficient of pipe is $\lambda = 0.037$. If the two bends are removed, assuming that both the length of pipe and the static head are unchanged, how much percentage of the flow is increased in the pipeline?

Solution: From Table 4 – 4, the loss coefficient of bend is calculated by

$$\zeta = \left[0.131 + 0.1632 \left(\frac{d}{R} \right)^{7/2} \right] \left(\frac{\theta°}{90°} \right)^{1/2}$$

where θ is the angle of bend, d is the pipe diameter, R is the curvature radius of the centerline of bend. In this case,

$$\zeta = 0.131 + 0.1632 = 0.2942$$

Before removing the bends, the head loss of the pipe is

$$h_w = \lambda \frac{l}{d} \frac{v_1^2}{2g} + 2 \frac{v_1^2}{2g} = \left(\lambda \frac{l}{d} + 2 \right) \frac{v_1^2}{2g}$$

$$= \left(0.037 \times \frac{10}{0.1} + 2 \times 0.294 \right) \frac{v_1^2}{2g} = 4.29 \frac{v_1^2}{2g}$$

The head loss, i.e. herein the frictional head loss of the pipe after the removal of the bends is

$$h_w = h_f = \left(0.037 \times \frac{10}{0.1}\right)\frac{v_2^2}{2g} = 3.7\frac{v_2^2}{2g}$$

If the total head remains the same, then

$$4.29\frac{v_1^2}{2g} = 3.7\frac{v_2^2}{2g}$$

where v_1, v_2 are the average velocity of pipe before and after the removal of the bends, respectively. Thus

$$\frac{v_2}{v_1} = \sqrt{\frac{4.29}{3.7}} = \sqrt{1.16} = 1.077$$

From the discharge $Q = Av$, where A remains the same,

$$\frac{Q_2}{Q_1} = \frac{v_2}{v_1} = 1.077$$

Thus, $\quad Q_2 = 1.077\ Q_1$, $\Delta Q = (1.077 - 1)Q_1 = 0.077\ Q_1$
which means an increase of 7.7% in discharge.

4.8 Basic concepts of boundary layer and flow resistance around an object

4.8.1 Basic concept of boundary layer

In 1904, a German scientist Prandtl proposed a theory of boundary layer. The basis of the theory is: in real fluid flow, when the fluid moves over a solid boundary, the fluid particles in contact with the boundary cling to the solid boundary and have no relative motion, i.e. their velocities are zero regardless of the Reynolds number. This is called the no slip condition. In the direction normal to the boundary, the flow velocity in the outer boundary increases rapidly from zero to the original incident velocity u_0 along the plate. Fig. 4-30 shows the simplest case of a flat plate boundary, where two-dimensional parallel flow is parallel to a flat plate, and only half of the flow is shown in the figure due to the symmetry. The velocity distribution is gradually developed from the boundary surface outwards to the outer layer edge where the flow is then uniform. The thickness of boundary layer should be infinite theoretically. However, at a certain distance away from the boundary surface, the flow velocity is actually very close to the incident velocity u_0. The height of boundary layer is currently defined as the vertical distance from the boundary surface to the layer that has the velocity of $u = 0.99\ u_0$, also called the thickness of the boundary layer, denoted by δ. No matter how large Re is, the boundary layer always exists, and Re will affect only the thickness of boundary layer. As a result, the actual flow can be seen to have two different types of flows:

(1) Boundary layer (inner flow zone): the impact of viscosity cannot be ignored in the region because a large velocity gradient exists;

(2) Outer boundary layer (outer flow zone): as the velocity gradient is very small, the

Fluid Mechanics in Civil Engineering

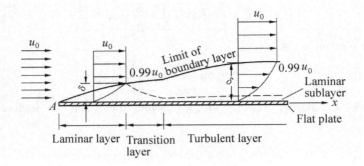

Fig. 4 – 30

viscous effect is negligible, and the flow can approximately be treated as ideal flow.

As the flow in the outer layer can be considered as an ideal fluid motion without internal frictions, the flow can be treated as potential flow to simplify the problems.

The thickness of boundary layer increases with the increasing x. By analysis of the order of magnitude in the equation of fluid mechanics, it is found that

$$\frac{\delta}{x} \propto \frac{1}{\sqrt{Re_x}} \qquad (4-80a)$$

where x is the longitudinal distance for the boundary layer thickness δ; $Re_x = \frac{u_0 x}{\nu}$ is the local Reynolds number of the boundary layer.

Equation (4 – 80a) can also be written as

$$\delta \propto \frac{x^{\frac{1}{2}} \nu^{\frac{1}{2}}}{u_0^{\frac{1}{2}}} = \sqrt{\frac{x\nu}{u_0}} \qquad (4-80b)$$

Under the condition of the certain fluid properties and flow velocity, the thickness of the boundary layer increases with the increasing x, and the velocity gradient in the boundary layer decreases with the increase of boundary layer thickness(δ).

Experimental results show that the flow regime in the boundary layer changes from laminar to turbulent flow when the boundary layer Reynolds number $Re_x = \frac{u_0 x}{\nu}$ increases and reaches a certain value. At the leading edge of the plate, the boundary layer(δ) is very thin, du/dy is very large, so the flow is laminar flow; with the increase of x, δ becomes thicker, the velocity gradient du/dy gets smaller, the viscous resistance is also smaller, so the flow in the boundary layer then gradually changes from laminar flow to turbulent flow. However, in the turbulent boundary layer the velocity gradient du/dy near the boundary surface is large, and the laminar sublayer still exits, as shown in Fig. 4 – 30.

From Equation (4 – 80b), the boundary layer thickness (δ) increases with $x^{1/2}$. If a pipe is long enough, the boundary layer can be filled with $\delta = d/2$, as shown in the entire pipe cross-section, that is, the whole cross-section is within the boundary layer, seen in Fig. 4 – 31. When the liquid flow enters the pipe, the parabolic or logarithmic (exponential) velocity distribution does not necessarily appear at the beginning, but they will appear and be maintained after a certain distance l (known as the development length, l, or the transition section), at which the flow is called fully developed boundary layer flow. For laminar pipe flow, $l = 0.065 dRe$, and for turbulent pipe flow, $l = (40 \sim 50)\, d$, where d is the diameter of pipe. The understanding of boundary layer transition length is important for the planning of experiments in laboratory tests. It is clear that the experiments should be conducted in the fully developed boundary layer flow region, in which the velocity and pressure distribution are stable.

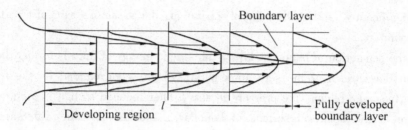

Fig. 4 − 31

4.8.2 Separation of boundary layer and flow resistance

Fig. 4 − 32 shows the flow pattern for real flow around a circular cylinder in a de-accelerating zone of velocity (the range of clockwise $90° \sim 180°$ from the stagnation point). The pressure in the boundary layer increases with the increase of pressure in the outer boundary layer along the flow. Within the boundary layer at point a, the kinetic energy of flow gradually decreases under the influence of both the increased pressure and the viscous resistance, leading to zero velocity near the boundary at point s, where the downstream pressure is higher than the pressure at point s. Under the adverse pressure gradient, the fluid particles from the downstream region move into the upstream region, thereby forming a circulation. By the continuity of flow, fluid particles from the upstream region will be forced to join the main flow, thus separating it from the boundary, and the downstream fluid will immediately fill the vacated area and form a wake region. This phenomenon is called the boundary layer separation or the boundary layer detachment. The point s, at which the boundary layer starts to be separated from the solid boundary, is called the

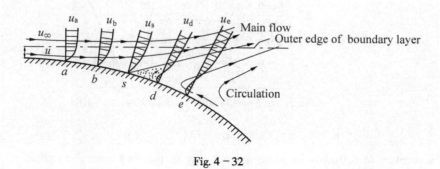

Fig. 4 − 32

separation point. The velocity distribution at the separation point is characterized by $\left(\frac{\partial u_x}{\partial y}\right)_{y=0} = 0$, that is, the velocity gradient on the boundary is zero; at the upstream point b of the separation, the velocity gradient is positive along the direction normal to boundary, i. e. $\left(\frac{\partial u_x}{\partial y}\right)_{y=0} > 0$; at the downstream point d, due to the return flow near the boundary, the velocity gradient is negative, i. e. $\left(\frac{\partial u_x}{\partial y}\right)_{y=0} < 0$. Because of the influence of the return flow, the thickness of boundary layer increases significantly, and at the point of separation, the streamline exits at a certain angle from the solid boundary.

After the separation of boundary layer, the kinetic energy of fluid is further dissipated in the wake region downstream of the separation point, which is surrounded by the fluid separation surface. In the wake region, the pressure is also further reduced so that it cannot be restored to the pressure at the upstream surface of boundary, and the pressure difference between the upstream and downstream sides of boundary forms a pressure resistance. The further downstream the separation point is, the smaller the wake zone becomes, with less kinetic energy and thereby creating a smaller pressure difference between the upstream and downstream of the object, so the pressure resistance is reduced. In engineering, streamlined objects are used in order to reduce the wake region, thereby reducing pressure resistance or form drag. In general, for high Re flows, the frictional resistance of boundary is much smaller than the pressure resistance, so the total flow resistance is greatly reduced if the pressure resistance can be reduced.

Fig. 4 – 33 shows the distribution of pressure resistance coefficient $C_p = \frac{p}{0.5\rho u_0^2}$ for flow around a cylinder. The cylindrical stagnation (stagnation point) on the upstream face is at the starting point of φ, i. e. the clockwise rotation is positive, the variation of pressure coefficient in $180° \sim 360°$ is symmetric with that in $0° \sim 180°$. The separation point for laminar flow is theoretically calculated at $\varphi = 110°$, but is actually around $\varphi = 80°$; the separation point of turbulent flow is at about $\varphi = 120°$.

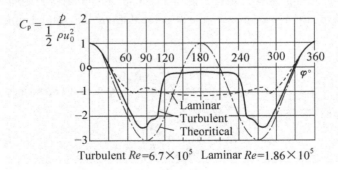

Fig. 4 – 33

The separation of boundary layer occurs not only in the flow around a cylinder, but also in

Chapter 4 Types of flow and head loss

the flows of poorly designed channels with a large diverging angle(>80°). The vortexes behind piers or sluice gates also belong to such types of separation.

The separation of the boundary layer is related to not only the shape of the object, but also the direction of flow relative to the object. For example, the flow around a thin plate, if the plate is placed parallel to the flow field, will not cause any separation; but if the plate is placed normal to the flow direction, a separation zone will be formed behind the plate, as shown in Fig. 4 – 34. Similarly, when even streamlined bodies are placed in the flow at a certain angle, boundary layer separation will also occur, as shown in Fig. 4 – 35.

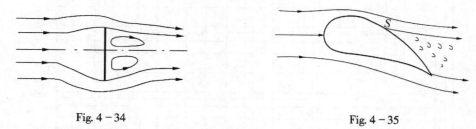

Fig. 4 – 34   Fig. 4 – 35

The flow resistance around an object includes both the frictional resistance and the pressure resistance. The total resistance of flow around an object can be expressed by

$$F_D = C_D \frac{\rho u_0^2}{2} A \qquad (4-81)$$

where u_0 is the undisturbed inflow velocity; A is the projection area of object on the plane which is normal to the flow direction, also known as the front area of the object. C_D is the drag coefficient of flow, and C_D values depend on the shape, the Reynolds number and the orientation of the object to the flow, along with the roughness of the surface of object. Fig. 4 – 36 shows the variation of drag coefficient (C_D) with the Reynolds number (Re) for different shaped objects. It is seen that: When Re is very small, the flow is in the laminar boundary layer region, so only frictional resistance exits, no eddies occurs, and C_D is inversely proportional to Re. As Re increases, vortices will form downstream of the object, and both the friction and pressure resistance exist. When Re is increased to 10^4, the pressure resistance predominates and the frictional resistance is relatively small, so the flow resistance is almost independent of Re. When Re is further increased to 3×10^5, the C_D values suddenly decrease. This is because the upstream boundary layer changes into turbulent flow, and as a consequence the separation point moves downwards and the wake region becomes smaller. The pressure resistance of turbulent flow is much smaller than that of laminar flow (see Fig. 4 – 36). The drag coefficient (C_D) and the drag force F_D of turbulent flow are smaller than those of laminar flow.

【Example 4 – 12】 An outdoor chimney is approximately cylindrical, with a height of 25 m and an average diameter of 1.3 m. The chimney is affected by a lateral wind of 40 km/h at the temperature of 10℃ in the air. Find the airflow thrust to the chimney.

Solution: Airflow through the chimney will produce a flow resistance F_D, which is the thrust of the airflow on the chimney. The flow drag coefficient C_D is related to the Reynolds number Re.

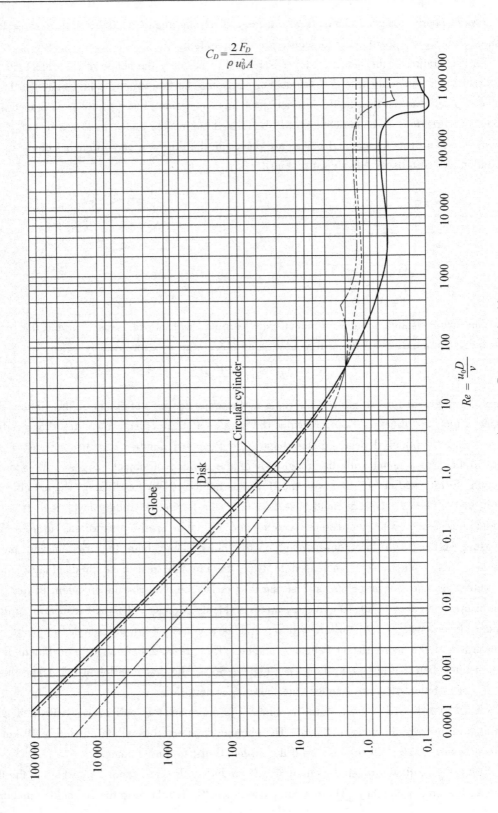

Fig. 4-36 C_D - Re relationship

The density of air at 10℃ is $\rho = 1.25$ kg/m^3, and the viscous coefficient is $\mu = 1.78 \times 10^{-5}$ N·s/m^2, so the flow Reynolds number is

$$Re = \frac{u_0 D}{\nu} = \frac{\rho u_0 D}{\mu} = \frac{1.25 \times 40/3.6 \times 1.3}{1.78 \times 10^{-5}} = 1.01 \times 10^6$$

where u_0 is the wind speed, and D is the diameter of the chimney. From Fig. 4-36, $C_D = 0.36$. According to Equation (4-81), the thrust of the chimney is

$$F_D = C_D \frac{\rho u_0^2}{2} DH = 0.36 \frac{1.25 \times 11.1^2}{2} \times 1.3 \times 25 = 901 \text{ N}$$

Chapter summary

In this chapter, the theoretical analysis, experiments (Reynolds, Nikuradse), semi-empirical theories of flow and their engineering applications have been introduced. The flow resistances, the flow regimes and the law of head losses are also described.

1. According to the boundary conditions of flow, the flow resistance and the associated head losses are classified.

$$h_w = h_f + h_j = \lambda \frac{l}{d} \frac{v^2}{2g} + \frac{v^2}{2g}$$

2. There are two types of flow: laminar flow and turbulent flow, which are identified by a dimensionless number, the Reynolds number.

Laminar flow: $Re = \frac{vd}{\nu} < 2000$, or $Re_R = \frac{vR}{\nu} < 500$, $h_f \propto v^{1.0}$

Turbulent flow: $Re = \frac{vd}{\nu} > 2000$, or $Re_R = \frac{vR}{\nu} < 500$, $h_f \propto v^{1.75 \sim 2.0}$

3. The relationship between the flow frictional head loss and the shear stress is $\tau_0 = \gamma R J$. The shear stress is significantly different in different flow regimes and the head losses of laminar and turbulent flow are different too.

4. For laminar pipe flow, $\tau = \mu \frac{du}{dy}$, the velocity distribution is parabolic, and $\lambda = \frac{64}{Re}$.

5. Turbulent flow is characterized by the mixing of fluid particles and the fluctuating of flow parameters. The shear stress of turbulent flow includes the viscous shear stress and turbulent eddy shear stress:

$$\tau = \tau_1 + \tau_2 = \mu \frac{du_x}{dy} - \rho \overline{u'_x u'_y}$$

The velocity distribution of turbulent flow complies with the logarithmic law.

$$\frac{u}{u_*} = \frac{1}{\kappa} \ln y + C$$

6. Turbulent flow has different zones of resistance (due to the existence of a viscous sublayer). Nikuradse's experiment reveals: in hydraulically smooth region, $\lambda = f(Re)$; in turbulent transition region, $\lambda = f(Re, \frac{r_0}{\Delta})$; in hydraulically rough region (or resistance square zone), $\lambda = f(\frac{r_0}{\Delta})$.

7. The main cause of the local head loss is the separation of flow over boundary, the formation of vortex, and the re-structuring of velocity distribution. The local head loss coefficient (ζ) depends on the boundary conditions, and is usually determined experimentally.

8. The resistance of flow around an object, including the frictional resistance and the pressure resistance, can be expressed as

$$F_D = C_D \frac{\rho u_0^2}{2} A$$

where the drag coefficient C_D depends on the Reynolds number, the shape of object, and the surface roughness of the object.

Multiple-choice questions (one option)

4-1 The Reynolds number is defined as the ratio of _____.
(A) the pressure force to the viscous force
(B) the viscous force to the gravity
(C) the inertial force to the viscous force
(D) the viscous force and viscous shear stress

4-2 For flow of real fluid in a pipe, the critical Reynolds number is related to _____.
(A) the diameter of the pipe
(B) the kinematic viscosity coefficient of the fluid
(C) the wall roughness of the pipe
(D) none of above, but it is the same

4-3 When the discharge and temperature of fluid remain constant, if the ratio of two sections of a pipe flow is 2, then the ratio of their Reynolds numbers is _____.
(A) 2   (B) 1/2   (C) 4   (D) 1/4

4-4 In uniform flow, the equation $\tau = \gamma RJ$ applies to _____.
(A) only the laminar flow
(B) only the turbulent flow
(C) both laminar and turbulent flow
(D) neither laminar nor turbulent flow

4-5 The shear stress distribution of uniform flow over a circular pipe is _____.
(A) uniform
(B) parabolic
(C) linear, with zero at the center of pipe and the maximum on the wall
(D) linear, with the maximum at the center of pipe and zero on the wall

4-6 For uniform laminar flow in a pipe, the velocity distribution is _____.
(A) uniform   (B) exponential   (C) parabolic   (D) logarithmic

4-7 For turbulent flow in a pipe, the velocity distribution is _____.
(A) uniform   (B) linear   (C) parabolic   (D) logarithmic

4-8 For the uniform laminar flow in a pipe, the flow velocity is 0.8 m/s, and the average flow velocity is _____.

(A) 0.6 m/s   (B) 0.5 m/s   (C) 0.4 m/s   (D) 0.3 m/s

4-9   The eddy shear stress of turbulent flow is due to the momentum exchange caused by _____.

(A) fluid molecules   (B) fluctuating velocity
(C) time-averaged velocity   (D) fluctuating pressure

4-10 The kinetic energy correction factor of turbulent flow in a pipe _____ that of laminar flow.

(A) is greater than   (B) is less than
(C) is equal to   (D) cannot be determined, compared with

4-11 In turbulent flow, when the thickness of viscous sublayer is much larger than the absolute roughness height, the wall boundary is called _____.

(A) hydraulically smooth surface   (B) hydraulically transition rough surface
(C) hydraulically rough surface   (D) none of the above

4-12 In turbulent rough region, the relationship between the head loss and velocity is _____.

(A) linear   (B) quadratic
(C) of $1.75^{th} \sim 2^{nd}$ power   (D) logarithmic

4-13 If the frictional coefficient λ of flow is known to be only related to the roughness of boundary, the flow is in _____.

(A) laminar zone   (B) turbulent smooth zone
(C) turbulent transition zone (D) turbulent rough zone

4-14 The frictional coefficient λ of laminar flow is only related to _____.

(A) the Reynolds number
(B) the relative roughness
(C) both the flow length and the hydraulic radius
(D) both the Reynolds number and the relative roughness

4-15 For laminar flow of pipe, the frictional head loss is _____ the velocity.

(A) proportional to the squared root of
(B) inversely linear to
(C) proportional to the 1.75^{th} power of
(D) proportional to the square of

4-16 Regarding the Chezy formula, it is _____.

(A) completely different from the Darcy-Weisbach formula
(B) an empirical formula for calculating the frictional head loss
(C) consistent with the Darcy-Weisbach formula after the transformation
(D) only applicable to the square resistance region when the Manning formula is used for the Chezy coefficient

4-17 The boundary layer is _____.

(A) the flow layer with uniform velocity distribution
(B) the viscous sublayer
(C) the flow layer significantly affected by the viscosity of fluid
(D) the flow layer without the influence of viscosity of fluid

4 − 18 The flow resistance around an object _____.
(A) includes the frictional resistance and pressure resistance
(B) is the frictional resistance only
(C) has no relation with the wake flow
(D) is proportional to the surface area of the object

Problems

4 − 1 Water flows through a converging transition pipe, where the small pipe has the diameter d_1, and the diameter of the large pipe is d_2, and $d_2/d_1 = 2$. Which section has a larger the Reynolds number? What is the ratio of the Reynolds numbers of the two sections?

4 − 2 A 1 cm circular water pipe of 1 cm diameter has the average flow velocity of 0.25 m/s. The water temperature is 10℃. ① What type of flow is the flow in the pipe? ② If the diameter of pipe is 2.5 cm, and both the flow rate and the water temperature are kept the same, what type of flow will it be? ③ For the 2.5 cm pipe, what is the flow rate of laminar flow? ④ For the same temperature, what maximum average velocity of cross-section is allowed for the 2.5 cm pipe to transport oil ($v = 0.4$ cm²/s) in laminar flow?

4 − 3 In a ventilation system, the diameter of pipe is $d = 500$ mm, the average flow velocity is $v = 15$ m/s, and the air temperature is 40℃ ($\nu = 0.18$ cm²/s). What type of air flow is in the pipe? If the diameter of pipe is changed to 200 mm, what is the maximum average velocity of the pipe to remain a laminar flow?

4 − 4 A 100 m long pipe has the hydraulic gradient of $J = 0.8\%$, and the diameter of pipe is 0.2 m. Determine the wall shear stress τ_0 and the frictional head loss of pipe h_f.

4 − 5 The average velocity of water flow in a 2.5 cm pipe is 2.5 m/s, the frictional coefficient $\lambda = 0.044$, and the viscosity coefficient of water is $\nu = 0.01$ cm²/s. Determine the friction velocity u_* and the wall shear stress τ_0.

4 − 6 A mineral oil is flowing in a 6 mm circular pipe with laminar flow, and a mercury differential manometer is installed at two sections of 2 m apart (Fig. 4 − 37). When $h = 12$ cm, and the flow rate is 7.2 cm³/s, what is the dynamic viscosity coefficient of the mineral oil (the density of oil is 8.33 kN/m³)?

4 − 7 For laminar flow in a circular pipe, what is the location (the distance away from the center of pipe) at which the velocity of flow (u) equals the average velocity of the pipe (v)?

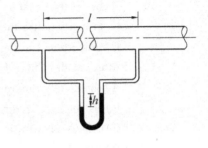

Fig. 4 − 37

4-8 A 10 cm diameter circular pipe conveys water at the average flow velocity of 2 m/s, the water temperature is 20℃, and the frictional coefficient $\lambda = 0.03$. What is the thickness of the laminar sublayer?

4-9 In a pipe flow, the water temperature is 15℃, the average velocity $v = 3.5$ m/s, and the roughness coefficient $\lambda = 0.015$. What is the height of the viscous sublayer? How about it if the average flow velocity is increased to 6 m/s?

4-10 Some heavy oil is transported in a pipeline with the diameter of $d = 75$ mm, the specific weight of the oil is $\gamma_{oil} = 8.83$ kN/m^3, the kinematic viscosity coefficient of the oil $\nu_{oil} = 0.9$ cm^2/s, and the height difference in the mercury differential manometer is $\Delta h = 20$ mm, as shown in Fig. 4-38. What is the mass flow rate of heavy oil per hour?

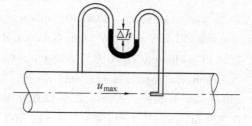

Fig. 4-38

4-11 Fig. 4-39 shows a device for measuring the frictional coefficient λ of pipe section AB. The length (l) of section AB is 10 m, and the diameter of pipe d is 50 mm. The measured data are: the head difference between A and B is 0.8 m; the water flow rate is 0.247 m^3 volume rise in the tank per 90 seconds. What is the frictional coefficient λ?

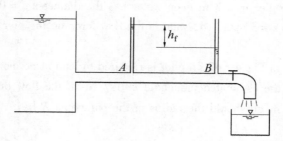

Fig. 4-39

4-12 A cast iron pipe ($\Delta = 0.3$ mm) has the diameter of $d = 300$ mm, the length $l = 1000$ m, the flow rate $Q = 100$ L/s, and the water temperature is 10℃. Determine the value of λ using the Moody diagram and calculate the frictional head loss h_f. Find the pressure difference (Δp) between the two ends if the pipe is horizontally placed.

4-13 15℃ water is flowing in a riveted steel pipe of $d = 30$ cm. The 300 m long pipe has the head loss of $h_f = 6$ m and the roughness height $\Delta = 3$ mm. What is the flow rate in the pipe?

4-14 In a test of head loss, the pipe has the diameter $d = 1.5$ cm, the length of the measuring section of $l = 4$ m, and the water temperature of 4℃. If the flow rate is 0.02 L/s, determine: ① whether the flow of pipe is laminar or turbulent flow; ② the frictional coefficient of the pipe; ③ the head loss of the measuring section; ④ the maximum water head difference in the measuring section if a laminar flow remains in the pipe.

4-15 A 5-meter long pressurized pipe has the equivalent roughness of $\Delta = 0.3$ mm, and the water temperature is 15℃. Determine the head loss of pipe, when: ① the diameter is 1 cm, and the flow rate is 0.05 L/s; ② the diameter is changed to 7.5 cm and the flow rate is the same as ①; ③ the diameter is 7.5 cm, but the flow is increased to 20 L/s.

4-16 For a pipe of $d = 50$ mm in diameter, which has the friction coefficient $\lambda = 0.04$, how much is the Chezy coefficient C? If the Manning formula is used, what is the roughness coefficient n?

4-17 Fig. 4-40 shows the device for determining the local loss coefficient of a 90° elbow. Given the length of segment AB is $l = 10$ m, the diameter d is 50 mm, and the friction coefficient $\lambda = 0.03$, determine the local loss coefficient ζ, when the measurement data are: ① the piezometric head difference between A and B is 0.629 m; ② the amount of 0.329 m³ water entered into the tank in 2 minutes.

4-18 As shown in Fig. 4-41, a 25 m long water pipe is linked between tanks A and B, the diameter of pipe is $d = 25$ mm, the friction coefficient $\lambda = 0.03$, and the pipeline has two 90° bends ($d/R = 1$) and a gate valve ($a/d = 0.5$). When the water level difference between the two tanks is $H = 1$ m, find the flow rate through the pipe.

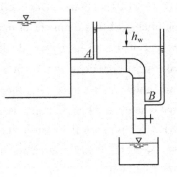

Fig. 4-40

4-19 A circular pier in a 3 m deep water has the diameter $d = 0.4$ m, and the average flow velocity of water is $v = 3$ m/s, so, what is the flow force on the pier? The water temperature is 20℃.

4-20 As shown in Fig. 4-42, a pilot is required to land at no more than 4 m/s. The total weight of the pilot and the associated parachute is 687 N. If the flow drag coefficient C_D of the umbrella is 1.33, how much should the radius of the parachute R be?

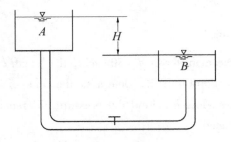

Fig. 4-41

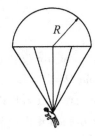

Fig. 4-42

Chapter 5 Steady orifice, nozzle and pipe flow

5.1 Introduction

An orifice is an opening in the side wall or base of a tank or reservoir, through which fluid is discharged. This type of flow is called orifice flow. When the shape of an orifice remains unchanged and the upstream liquid level in the tank is constant, the flow is a steady flow. If the flow through an orifice is discharged into the air in the form of a jet, it is a free orifice flow. If the flow is discharged into a tank containing a liquid with its level higher than the orifice, it is a submerged orifice flow. In a free orifice flow, the flow streamlines near the entrance of an orifice are bent due to the impact of the wall. Although the streamlines are smooth curved lines, the flow passing the orifice will be further contracted until at section $C-C$ because of the effect of inertia, and then the flow is diverged and gradually falls down due to gravity (Fig. 5-1). Section $C-C$ is the contracted section that has the minimum area, also termed the "vena contracta", where the streamlines are approximately parallel straight and the pressure distribution within the jet could be assumed to be nearly uniform. For a free orifice flow, the pressure at the vena contracta therefore equals that of the atmosphere surrounding the jet. If the orifice is in a thin wall and sharp-edged, as shown in Fig. 5-1, the thickness of wall has little

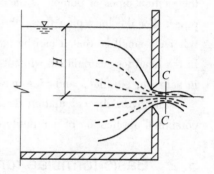

Fig. 5-1

influence on the flow due to its minimal contact with the fluid. Generally, since the upper and lower edges of the orifice are at different depths below the free surface, the flow through the upper part of orifice differs from that through the lower part. However, when the diameter d or height e of the orifice is very small compared with the liquid head H above the center of the orifice, the difference can be negligible and the heads at each point on the orifice section are assumed to be the same. According to the value of d/H or e/H, the orifice can be classified as small orifice with $d/H \leqslant 1/10$ or large orifice with $d/H > 1/10$.

If the thickness of the orifice wall or the length of a short tube connected to the orifice is $3 \sim 4$ times the diameter of the orifice, such a short tube is seen as a nozzle, through which the liquid will fill the outlet section and be discharged. This type of flow is called nozzle flow, which can also be classified as a free flow or a submerged flow. Likewise, a vena contracta is formed by the flow in nozzle, but the flow may be detached from the side wall where a vortex zone exists.

Afterwards the flow diverges and eventually extends to the whole cross-section as it leaves the nozzle.

Examples of orifice flow are: water intake, drain hole, perforated water distribution sprinklers, some flow measurement device such as orifices; pressurized short pipe, short discharge pipe in dam, fire water gun, and nozzles for water guns for hydraulically mechanized construction. In general, for both orifice flow and nozzle flow, the frictional head losses could be negligible and thus the total head loss is primarily due to minor losses.

The flow full of pipe cross-section is called pressurized pipe flow or pipe flow for short. It is one of the most common types of flow in engineering as pipes are mainly used to transport liquids and gases. A pipe flow has the following features: no free surfaces and the pressure on the pipe wall is usually not equal to the atmospheric pressure. For a certain length pipeline, the head losses of flow will be both frictional head losses and local head losses. In engineering for simplicity of calculation, the pipes with full flow are classified into two types: hydraulically short pipe and hydraulically long pipe, depending on the relative weighting of these two kinds of losses in the total head losses. For the short pipe flow, both the frictional head loss and the local head loss are important so that they both should be included in the head loss calculation. For example, the suction pipes of pumps, siphon pipes, pipes for culvert, and industrial vent tubes are short pipes. For the long pipe flow, the frictional head loss is dominant where the local head loss is relatively small so that it can be ignored or estimated as a certain percentage of the frictional loss in the head loss calculation without causing much error. For example, a simple pipeline, pipelines in series or parallel, and pipe networks in water supply system are long pipes.

The hydraulic calculation for steady flow through orifice, nozzle or pressurized pipe is the combined application of continuity equation, energy equation, flow friction law and head loss.

5.2　Basic formulae for steady flow through orifice and nozzle

5.2.1　Steady flow through thin-wall orifice

5.2.1.1　Free flow in small orifice

As shown in Fig. 5 - 2, in order to derive the basic discharge formula for free orifice flow, taking the GVF cross-sections 1 - 1 in the vessel and $C - C$ at the vena contracta, with the horizontal plane $O - O$ through the center of the orifice as the datum level, and applying the energy equation gives

$$H + 0 + \frac{\alpha_1 v_1^2}{2g} = 0 + 0 + \frac{\alpha_c v_c^2}{2g} + h_{w_{1-c}}$$

(5 - 1)

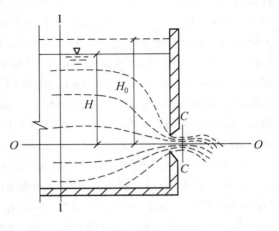

Fig. 5 - 2

where v_1 = the approaching velocity;

v_C = the averaged velocity at the vena contracta;

α_1, α_C = the kinetic correction factors for the two sections respectively;

h_{w_1-C} = the total head loss between two sections.

Since the frictional head losses in the vessel are very small and thus negligible, h_{w_1-C} is only the local head loss, which is due to contraction when the flow passes through the orifice. That is

$$h_{w_1-C} = h_j = \zeta \frac{v_C^2}{2g}$$

where ζ is the local head loss coefficient due to the contraction.

Letting $H_0 = H + \frac{\alpha_1 v_1^2}{2g}$, where H_0 is termed as the total head and H the hydrostatic head, substituting it into Equation (5-1), and rearranging the resulting equation gives

$$v_C = \frac{1}{\sqrt{\alpha_C + \zeta}} \sqrt{2gH_0} = \varphi \sqrt{2gH_0} \qquad (5-2)$$

where H_0 is the total head or the acting head;

φ is the velocity coefficient, $\varphi = \frac{1}{\sqrt{\alpha_C + \zeta}} \approx \frac{1}{\sqrt{1 + \zeta}}$.

If A represents the area of the orifice and A_C is the area of the vena contracta, $\varepsilon = \frac{A_C}{A}$ is termed as contraction coefficient. Then, the flow rate through an orifice is

$$Q = A_C v_C = \varepsilon \cdot A \cdot \varphi \sqrt{2gH_0} = \mu A \sqrt{2gH_0} \qquad (5-3)$$

where μ is the discharge coefficient of orifice, $\mu = \varepsilon \varphi$.

Equations (5-2) and (5-3) are the basic equations for free orifice flow.

Note that ε, φ and μ are different coefficients. Experimental tests show that $\varepsilon = 0.60 \sim 0.64$, usually $\varepsilon = 0.62$, $\varphi = 0.97 \sim 0.98$, and $\mu = 0.60 \sim 0.62$ for the thin-wall, sharp-edged circular orifice flows.

5.2.1.2 Submerged flow in small orifice

As shown in Fig. 5-3, unlike the free orifice flow, the submerged outflow is discharged into a downstream water tank, rather than the atmosphere. Through the orifice, the submerged outflow is first contracted at the vena contracta due to the effect of inertia and then expanded due to the resistance.

Similarly, taking the GVF cross-sections 1-1 and 2-2, with the horizontal datum plane $O-O$ passing through the center of the orifice, and applying the energy equation gives

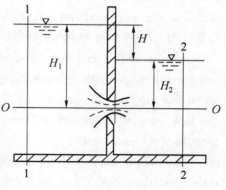

Fig. 5-3

$$H_1 + 0 + \frac{\alpha_1 v_1^2}{2g} = H_2 + 0 + \frac{\alpha_2 v_2^2}{2g} + h_{w_1-2} \qquad (5-4)$$

Fluid Mechanics in Civil Engineering

The head losses of the submerged orifice flow include two parts: the local head loss due to flow contraction before the vena contracta and the local head loss due to flow expansion afterwards. Then,

$$h_{w1-2} = \sum h_j = (\zeta + \zeta_s) \frac{v_C^2}{2g}$$

where ζ is the local head coefficient due to the contraction;

ζ_s is the local head coefficient due to the sudden expansion after the vena contracta. Since $A_2 \gg A_C$, $\zeta_s \approx 1$.

Let $H_0 = (H_1 + \frac{\alpha_1 v_1^2}{2g}) - (H_2 + \frac{\alpha_2 v_2^2}{2g})$. If the areas of the cross-section are much larger than that of the orifice, it is reasonably to assume that $v_1 \approx v_2 \approx 0$, and then $H_0 \approx H$. Substituting H_0 and the head loss into Equation (5-4), and arranging the resulting equation, gives

$$H_0 = (\zeta + 1) \frac{v_C^2}{2g}$$

Further arranging the above equation, we can obtain the basic equations for submerged orifice flow:

$$v_C = \frac{1}{\sqrt{1+\zeta}} \sqrt{2gH_0} = \varphi \sqrt{2gH_0} \tag{5-5}$$

$$Q = \varphi \varepsilon A \sqrt{2gH_0} = \mu A \sqrt{2gH_0} \tag{5-6}$$

Equations (5-5) and (5-6) are the same in form as those for free orifice flow, and the experiments also show that their coefficients are quite similar to those of the free orifice flow, so that the same coefficient values can be used for the free or submerged orifice flow in the calculations. However, we notice that the acting head H_0 has a different meaning: H_0 is the difference of water level between the upstream and the downstream tanks for the submerged orifice flow, while H_0 is the elevation distance between the center of orifice and the upstream surface of tank for the free orifice flow. Furthermore, in the submerged outflow, there is no distinction between a large orifice and a small orifice because the acting heads H_0 for every point on the orifice section are equal and independent on the depth of the orifice below the free surfaces.

5.2.1.3 Free flow in large orifice

A large orifice can be decomposed into many small orifices with different acting heads, so its flow rate can be obtained by integrating the flow rates of small orifices based on the formula of the small orifice flow.

As shown in Fig. 5-4, a rectangular large orifice has the width b and the height e, and the acting heads at the upper and lower edge of the orifice are H_1 and H_2, respectively. Take a small element of height, dh, for analysis, and

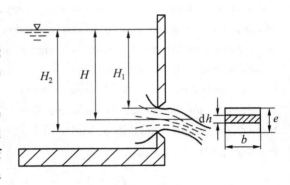

Fig. 5-4

its acting head is h, so

$$dQ = \mu \sqrt{2gh}dA = \mu b \sqrt{2gh}dh$$

Assuming that the value of μ does not change in the e direction, and the velocity distribution of the contraction section is uniform, we have

$$Q = \int_A dQ = \int_{H_1}^{H_2} \mu b \sqrt{2gh}dh = \mu b \sqrt{2g}\int_{H_1}^{H_2} \sqrt{h}dh$$

$$= \frac{2}{3}\mu b \sqrt{2g}(H_2^{3/2} - H_1^{3/2}) \quad (5-7)$$

Let H be the acting head at the centroid of the orifice, then $H_1 = H - e/2$, $H_2 = H + e/2$, and taking the Taylor series expensions of $H_2^{3/2}$ and $H_1^{3/2}$ with the linear term considered, we have

$$H_2^{3/2} - H_1^{3/2} = \left(H_2^{3/2} + \frac{3}{2} \cdot H^{1/2} \cdot \frac{e}{2}\right) - \left(H_1^{3/2} + \frac{3}{2} \cdot H^{1/2} \cdot \frac{e}{2}\right)$$

$$= \frac{3}{2} \cdot H^{1/2} e$$

Substituting it into Equation (5-7) gives

$$Q = \mu A \sqrt{2gh} \quad (5-8)$$

It can be seen that the calculation formula of Q for the small orifice is also applicable to the large orifice, demonstrated by this calculation. However, the discharge coefficient μ in the large orifice is larger than that in the small orifice because the flow is not fully contracted in the large orifice. Table 5-1 shows the common values of μ in the large orifices.

Table 5-1 μ values of large orifices

Contraction of flow	μ
Not fully contracted	0.70
No contraction on bottom, but a moderate side contraction	0.65~0.70
No contraction on bottom, but a small side contraction	0.70~0.75
No contraction on bottom, but a very small side contraction	0.80~0.90

5.2.2 Steady flow through nozzle

As shown in Fig. 5-5, there are generally at least five types of nozzle commonly used in engineering practice: ① cylindrical external; ② cylindrical internal; ③ conical convergent; ④ conical divergent; ⑤ streamlined external.

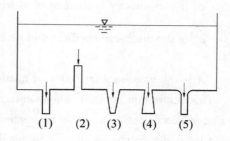

Fig. 5-5

5.2.2.1 Free flow through cylindrical external nozzle

This type flow is schematically shown in Fig. 5-6. Like orifice flow, the streamlines are bent and converged to pass the nozzle due to the restriction of boundaries and then contracted at a minimum section at the vena contracta $C-C$ due to the effect of inertia; subsequently, the flow is expanded to the full cross-section and then discharged. Taking the GVF sections $1-1$ and $2-2$ with a horizontal datum plane $O-O$ passing through the center of the nozzle, the energy equation is written as

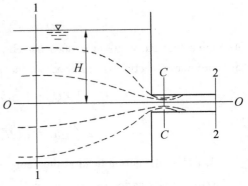

Fig. 5-6

$$H + \frac{\alpha_1 v_1^2}{2g} + 0 = 0 + 0 + \frac{\alpha_2 v_2^2}{2g} + h_{w_{1-2}} \qquad (5-9)$$

Since the frictional head losses in both the vessel and the nozzle are so small, they can be ignored. $h_{w_{1-2}}$ refers only to the local head losses, which are the entry loss before $C-C$ and the expansion loss after $C-C$, and each of them is equivalent to the right-angled entry loss of a tube, i.e.

$$h_{w_{1-2}} = \zeta_n \frac{v_2^2}{2g}$$

Letting $H_0 = H + \dfrac{\alpha_1 v_1^2}{2g}$, substituting it into Equation (5-9), and rearranging the resulting equation gives

$$v_2 = \frac{1}{\sqrt{\alpha + \zeta_n}} \sqrt{2gH_0} = \varphi_n \sqrt{2gH_0} \qquad (5-10)$$

$$Q = \varphi_n A \sqrt{2gH_0} = \mu_n A \sqrt{2gH_0} \qquad (5-11)$$

where v_2 is the velocity at the outlet of nozzle;

Q is the flow rate or the discharge of nozzle;

ζ_n is the local head loss coefficient of nozzle, i.e. the local loss coefficient for the right-angled entrance of tube, and $\zeta_n = 0.5$ in general;

φ_n is the velocity coefficient of nozzle, $\varphi_n = \dfrac{1}{\sqrt{\alpha + \zeta_n}} \approx \dfrac{1}{\sqrt{1 + 0.5}} = 0.82$;

μ_n is the discharge coefficient of nozzle, $\mu_n = \varphi_n = 0.82$, because the contraction coefficient is 1;

A is the cross-sectional area of nozzle.

This shows that the basic equation forms for nozzle and orifice flow are identical. However the discharge coefficient of a nozzle is greater than that of an orifice, which means under the same acting head the discharge through a nozzle is greater than that for an equal-sized orifice, so nozzles are usually used for drain pipes in engineering.

5.2.2.2 The vacuum at vena contracta in cylindrical external nozzle

When a nozzle is externally connected to an opening, although the head losses are increased, the flow rate will be increased; this is due to the effect of a vacuum, which occurs in the vena contracta of the nozzle.

Applying the energy equation to $1-1$ and $C-C$ in Fig. 5-6 gives

$$H + 0 + \frac{\alpha_1 v_1^2}{2g} = 0 + \frac{p_C}{\gamma} + \frac{\alpha_C v_C^2}{2g} + \zeta \frac{v_C^2}{2g} \qquad (5-12)$$

where ζ is the same as that of the free orifice flow.

Letting $H_0 = H + \frac{\alpha_1 v_1^2}{2g}$ and substituting it into Equation (5-12), then

$$v_C = \varphi \sqrt{2g\left(H_0 - \frac{p_C}{\gamma}\right)} \qquad (5-13)$$

where $\varphi = \dfrac{1}{\sqrt{\alpha_C + \zeta}}$ is approximately the same as that of the orifice flow.

Thus, the flow rate is

$$Q = v_C A_C = v_C \varepsilon A = \mu A \sqrt{2g\left(H_0 - \frac{p_C}{\gamma}\right)} \qquad (5-14)$$

where ε and μ have the same values as those of the orifice flow.

Comparing Equation (5-14) with the discharge equation of orifice flow, when cross-sections are the same in the two cases, μ and A are the same for the two equations but the effective head for the nozzle flow is $(H_0 - \frac{p_C}{\gamma})$. Since the flow rate of nozzle is larger than that of the orifice, it implies that a vacuum will occur at section $C-C$.

To calculate $\frac{p_C}{\gamma}$, applying the energy equation to sections $C-C$ and $2-2$, gives

$$0 + \frac{p_C}{\gamma} + \frac{\alpha_C v_C^2}{2g} = 0 + 0 + \frac{\alpha_2 v_2^2}{2g} + h_{w_{C-2}} \qquad (5-15)$$

Since $v_C = \dfrac{A}{A_C} v_2 = \dfrac{1}{\varepsilon} v_2$, and the only head loss is the local head loss due to the enlargement, i.e. $h_{w_{C-2}} = \zeta_s \dfrac{v_2^2}{2g}$; substitute these into Equation (5-15), then

$$\frac{p_C}{\gamma} = (\alpha_2 + \zeta_s) \frac{v_2^2}{2g} - \frac{\alpha_C}{\varepsilon^2} \frac{v_2^2}{2g}$$

where $v_2 = \varphi_n \sqrt{2gH_0}$, i.e. $\dfrac{v_2^2}{2g} = \varphi_n^2 H_0$, and $\zeta_s = (\dfrac{A}{A_C} - 1)^2$, which is the local head loss given by Equation (4-78) for an abrupt enlargement of circular pipe. Substituting them into the above equation gives

$$\frac{p_C}{\gamma} = \left[\alpha_2 + \left(\frac{1}{\varepsilon} - 1\right)^2 - \frac{\alpha_C}{\varepsilon^2}\right] \varphi_n^2 H_0$$

Let $\alpha_2 = \alpha_C = 1$, $\varepsilon = 0.64$ and $\varphi_n = 0.82$, then

$$\frac{p_C}{\gamma} = -0.756H_0 \qquad (5-16)$$

By substituting it into Equation (5 – 14), we find that the equivalent acting head of nozzle is 1.756 times that of a geometrically same orifice and consequently the multiple of discharge is $\sqrt{1.756} = 1.325$.

5.2.2.3 Working conditions of nozzle

The vacuum pressure in the vena contracta of a nozzle is proportional to the acting head H_0. When the vacuum pressure is more than 7 m of water column, vaporization (cavitation) will take place or the vacuum can cause damage due to air being sucked in the outlet. Hence, owing to the restriction of vacuum pressure at the vena contracta, the acting head of a nozzle flow normally has a limit, which is

$$H_0 \leq \frac{7}{0.756} \approx 9 \text{ m}$$

In addition, a nozzle has a requirement in length. If the nozzle is too short, there is not enough room for the flow to expand into the whole section after the contraction, which will then not generate a vacuum zone. On the other hand, if the nozzle is too long, the frictional head loss will be increased and it cannot be ignored any more, thus resulting in a decrease in flow rate. In general, the length of nozzle should be $3 \sim 4$ times its diameter.

Thus, the normal working conditions of a cylindrical external nozzle are: ① the acting head $H_0 \leq 9$ m; ② the nozzle length $l = (3 \sim 4)d$.

5.2.2.4 Submerged flow through nozzle

The same method of anlysis applies to free flow in a nozzle, and the associated coefficients are the same as those for free nozzle flow. However, the pressure p_C at vena contracta may not be a vacuum pressure but less than the pressure p_e at the outlet, and $\frac{p_e}{\gamma} - \frac{p_C}{\gamma} = 0.756H_0$. Thus the restriction of acting head to the depth of outlet could be lessened to some extent.

The hydraulic characteristics of some common orifices and nozzles are listed in Table 5 – 2.

Table 5 – 2 Hydraulic parameters of orifice and nozzle

Type	Small thin-wall orifice	Cylindrical internal nozzle	Cylindrical external nozzle	Conical diverging nozzle ($\theta = 5° \sim 7°$)	Conical converging nozzle ($\theta = 5° \sim 7°$)	Streamlined nozzle
Shape						

Chapter 5 Steady orifice, nozzle and pipe flow

Continued

Type	Small thin-wall orifice	Cylindrical internal nozzle	Cylindrical external nozzle	Conical diverging nozzle ($\theta = 5° \sim 7°$)	Conical converging nozzle ($\theta = 5° \sim 7°$)	Streamlined nozzle
Local frictional coefficient	0.06	0.98	0.50	3.0~4.0	0.09	0.04
Contraction coefficient	0.64	1.00	1.00	1.00	0.98	1.00
Velocity coefficient	0.97	0.71	0.82	0.45~0.50	0.96	0.98
Discharge coefficient	0.62	0.71	0.82	0.45~0.50	0.94	0.98

5.3 Steady flow in pressurized pipes

The flow in a pressurized pipe in which both boundary and incoming flow conditions do not change is steady flow.

5.3.1 Hydraulic calculation of hydraulically short pipes

A hydraulically short pipe refers to the pipeline where both frictional head losses and local head losses cannot be ignored in calculation. The hydraulic calculation for such pipes can be classified into two types: the free and the submerged outflow.

5.3.1.1 Basic formulae for the free outflow

As shown in Fig. 5 - 7, the water is discharged into the atmosphere through a short pipe connected with a water tank (reservoir) and has a pipe length l and a constant diameter d. Taking the GVF section 1 - 1 in the tank and GVF section 2 - 2 at the exit of the pipe, with the horizontal reference datum plane passing the center of pipe exit, the energy equation can be written as

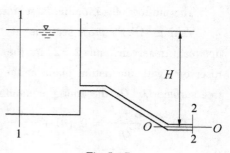

Fig. 5 - 7

$$0 + H + \frac{\alpha_1 v_1^2}{2g} = 0 + 0 + \frac{\alpha_2 v_2^2}{2g} + h_{w1-2} \tag{5-17}$$

Letting $H_0 = H + \frac{\alpha_1 v_1^2}{2g}$, then $H_0 = \frac{\alpha_2 v_2^2}{2g} + h_{w1-2}$

where v_1 is the flow velocity of section 1 - 1, which is also termed as approaching velocity;

v_2 is the mean cross-sectional velocity of the pipe;

H_0 is the total acting head of the flow.

The head loss $h_{w_{1-2}}$ is the sum of all the frictional losses and the minor (or local) losses in the pipeline.

$$h_{w_{1-2}} = \sum h_{f_{1-2}} + \sum h_{j_{1-2}} = \lambda \frac{\sum l}{d} \frac{v_2^2}{2g} + \sum \zeta \cdot \frac{v_2^2}{2g}$$

where $\sum \zeta$ represents the sum of all minor loss coefficients of the pipeline. Then

$$H_0 = \left(\alpha_2 + \lambda \frac{\sum l}{d} + \sum \zeta\right) \frac{v_2^2}{2g}$$

Letting $\alpha_2 = 1$, the flow velocity in pipe becomes

$$v_2 = \frac{1}{\sqrt{1 + \lambda \frac{\sum l}{d} + \sum \zeta}} \sqrt{2gH_0} \qquad (5-18)$$

and the discharge is

$$Q = v_2 \cdot A = \frac{1}{\sqrt{1 + \lambda \frac{\sum l}{d} + \sum \zeta}} A\sqrt{2gH_0} = \mu_c A \sqrt{2gH_0} \qquad (5-19)$$

where A is the cross-section area, $A = \frac{1}{4}\pi d^2$; μ_c is the discharge coefficient.

The approaching velocity v_1 and the corresponding velocity head $\frac{\alpha v_1^2}{2g}$ are usually so small that they are negligible, and thus Equation (5-19) can be simplified to be as

$$Q = \mu_c A \sqrt{2gH} \qquad (5-20)$$

where H is the static head.

5.3.1.2 Basic formulae for the submerged outflow

The outflow of a pipeline is submerged, as shown in Fig. 5-8.

Taking GVF cross-sections $1-1$ in the upstream reservoir and $2-2$ in the downstream reservoir, with the datum plane at the downstream free surface $O-O$, the energy equation is then

$$H + 0 + \frac{\alpha_1 v_1^2}{2g} = 0 + 0 + \frac{\alpha_2 v_2^2}{2g} + h_{w_{1-2}}$$

$$(5-21)$$

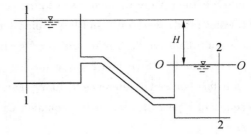

Fig. 5-8

Letting $H_0 = H + \frac{\alpha_1 v_1^2}{2g}$ yields

$$H_0 = \frac{\alpha_2 v_2^2}{2g} + h_{w_{1-2}} \qquad (5-22)$$

The head loss $h_{w_{1-2}}$ is the sum of all the frictional losses and minor losses of the pipeline, i.e.

$$h_{w_{1-2}} = \sum h_{f_{1-2}} + \sum h_{j_{1-2}} = \left(\lambda \frac{\sum l}{d} + \sum \zeta\right)\frac{v^2}{2g}$$

where $\sum \zeta$ is the sum of all minor loss coefficients in the pipeline including the sudden expansion loss coefficient ζ_s at the pipe outlet, which is an additional loss coefficient relative to the free outflow case; $\zeta_s = 1.0$ when $A_2 \gg \frac{1}{4}\pi d^2$; v is the mean flow velocity in the pipe.

Thus, Equation (5-22) becomes

$$H_0 = \frac{\alpha_2 v_2^2}{2g} + \left(\lambda \frac{\sum l}{d} + \sum \zeta\right)\frac{v^2}{2g}$$

Since $A_2 \gg \frac{1}{4}\pi d^2$, taking $v_2 = 0$ and rearranging the resulting equation gives

$$v = \frac{1}{\sqrt{\lambda \dfrac{\sum l}{d} + \sum \zeta}} \sqrt{2gH_0} \qquad (5-23)$$

Consequently, the discharge of the pipe is

$$Q = vA = \frac{1}{\sqrt{\lambda \dfrac{\sum l}{d} + \sum \zeta}} A\sqrt{2gH_0} = \mu_c A \sqrt{2gH_0} \qquad (5-24)$$

where μ_c is the discharge coefficient of the short pipe, $\mu_c = \dfrac{1}{\sqrt{\lambda \dfrac{\sum l}{d} + \sum \zeta}}$.

Similarly, the approaching velocity v_1 and the corresponding velocity head $\dfrac{\alpha v_1^2}{2g}$ are usually small so that they can be ignored, so Equation (5-24) can be simplified to be as

$$Q = \mu_c A \sqrt{2gH} \qquad (5-25)$$

From the basic formulae for both free and submerged outflows in a hydraulically short pipe, it can be seen that they are in the same form, and that their coefficients actually take the same values, despite some differences in expressions used. The main difference lies in the acting heads: for submerged outflow the acting head is the level difference between the free surfaces upstream and downstream, whereas for free outflow the acting head is the height from the center of the pipe exit to the free surface of the tank (reservoir). In addition, the pressures in the pipelines are not the same in the two outflow conditions.

It should be noted that Equations (5-20) and (5-25) are derived for a uniform pipeline only in the flow direction. If the pipe sizes change along the flow, the equations need to be revised by solving the energy equation and continuity equation again, along with all the associated head losses.

5.3.1.3 Hydraulic calculation about short pipes

For the steady flow in a short pipe, if the length, diameter, wall material, roughness, and type of minor loss are known, the typical types of hydraulic calculation are as follows:

(1) Given the pipe length l, diameter d, friction coefficient λ, minor losses and the acting head H, determine the discharge Q through the pipe.

For this kind of problem, one can directly apply the aforementioned basic formulae for the simple pipeline case or solve the energy equation and continuity equation simultaneously with all head losses accounted for in the calculation for more complex pipelines.

【Example 5-1】 A short pipe shown in Fig. 5-9, is 200 m long, with a diameter of 400 mm and the acting head of $H = 10$ m. The minor loss coefficient of the two equivalent bends is 0.25, and the loss coefficient of the fully-opened valve is 0.12. The friction coefficient is $\lambda = 0.03$. Find the flow rate Q when the valve is fully opened.

Solution: Firstly calculate the discharge coefficient,

$$\mu_c = \frac{1}{\sqrt{1 + \lambda \sum \frac{l}{d} + \sum \zeta}}$$

The minor losses include one right-angle sudden enlargement ($\zeta_{inlet} = 0.5$), two bends ($\zeta_{bending} = 0.25$) and one control valve ($\zeta_{valve} = 0.12$). Substituting them into the above expression gives

$$\mu_c = \frac{1}{\sqrt{1 + 0.03 \times \frac{200}{0.4} + 0.5 + 2 \times 0.25 + 0.12}} = 0.2417$$

According to Equation (5-20), ignoring the approaching velocity in the reservoir, the flow rate is

$$Q = \mu_c A \sqrt{2gH} = 0.2417 \times \frac{1}{4} \times 3.14 \times 0.4^2 \times \sqrt{2 \times 9.8 \times 10} = 0.4254 \text{ m}^3/\text{s}$$

(2) Given the discharge Q, pipe length l, diameter d, frictional head loss coefficient λ, and minor losses, determine the acting head H.

For this kind of problem, one can firstly calculate the velocity from the continuity equation, then the head losses and the acting head from the energy equation.

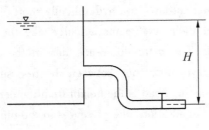
Fig. 5-9

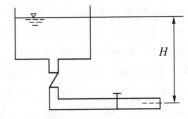
Fig. 5-10

【Example 5-2】 As shown in Fig. 5-10, the water supply pipe from a tank is 20 m long, with a diameter of 40 mm. The friction coefficient is 0.03 and the total minor loss coefficient is 15. Determine the acting head H when the flowrate is 2.75 L/s.

Solution: The flow velocity in the pipe is

$$v = \frac{Q}{\frac{1}{4}\pi d^2} = \frac{2.75 \times 10^{-3}}{0.25 \times 3.14 \times 0.04^2} = 2.188 \text{ m/s}$$

Then the acting head is calculated by

$$H = \left(1 + \lambda \frac{l}{d} + \sum \zeta\right) \frac{v^2}{2g} = \left(1 + 0.03 \times \frac{20}{0.04} + 15\right) \times \frac{2.188^2}{2 \times 9.8} = 5.126 \text{ m}$$

(3) Given the discharge Q, acting head H, pipe length l, frictional head loss coefficient λ, and minor losses, determine the pipe diameter d.

【Example 5-3】 As shown in Fig. 5-11, the pressurized circular culvert is 20 m long and the height difference between the free surfaces of upstream and downstream is 3 m. The friction coefficient is $\lambda = 0.03$ and the minor loss coefficients are: $\zeta_1 = 0.5$ for the inlet, $\zeta_2 = 0.71$ for the first bend, $\zeta_3 = 0.65$ for the second bend, and $\zeta_4 = 1.0$ for the outlet. If the flow rate through the culvert is required to be 3 m³/s, what diameter, d, of the culvert is required?

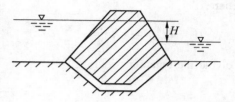

Fig. 5-11

Solution: The water flow in the pressurized circular culvert can actually be taken as the submerged outflow for a short pipe. From $Q = \mu_c A \sqrt{2gH}$, we have

$$Q = \frac{1}{\sqrt{\lambda \frac{\sum l}{d} + \zeta_1 + \zeta_2 + \zeta_3 + \zeta_4}} \cdot \frac{\pi d^2}{4} \sqrt{2gH}$$

Putting the known values to it and arranging the resulting equation, gives

$$d^5 - 0.7089d - 0.3718 = 0$$

Using a trial-on-error method, one can solve the equation to obtain $d = 1.018$ m. Thus the nominal diameter of pipe is taken to be 1.0 m in practice, and consequently the actual discharge will be slightly less than 3 m³/s.

(4) Hydraulic calculation for siphon and pump

Since part of siphon is higher than the free surface, there must exist a vacuum somewhere inside the siphon. However, if value of the vacuum is too large, the fluid continuity will not remain due to the effect of vaporization. Hence, in engineering, the maximum vacuum allowed is usually 7 m of water column. The advantage of a siphon is that it can be placed over a higher terrain thereby reducing the cost of excavation.

【Example 5-4】 As shown in Fig. 5-12, a siphon pipe is used for carrying water over the hill. The pipe length is $l = l_{AB} + l_{BC}$, where $l_{AB} = 20$ m and $l_{BC} = 30$ m, with a constant diameter of $d = 200$ mm. The water level difference of the two reservoirs is $H = 1.2$ m. If the friction coefficient is $\lambda = 0.03$ and the minor loss coefficients are $\zeta_e = 0.5$ for the entrance, $\zeta_s = 1.0$ for the exit, $\zeta_1 = 0.2$ for the first bend, $\zeta_2 = \zeta_3 = 0.4$ for the second and the third, $\zeta_4 = 0.3$ for the fourth. Part B is 4.5 m above the upstream free surface. Find the discharge and the maximum vacuum value in the siphon.

Solution: Taking cross-sections 1 – 1 and 2 – 2, choosing the level $O - O$ as the reference datum, and applying the energy equation (ignoring the approaching velocity and assuming $\alpha = 1.0$), we have

$$H + 0 + 0 = 0 + 0 + 0 + h_{w_{1-2}}$$

Then,

$$H = h_{w_{1-2}} = \left(\lambda \frac{l}{d} + \sum \zeta\right) \frac{v^2}{2g}$$

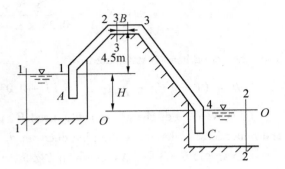

Fig. 5 – 12

Solving the equation yields

$$v = \frac{1}{\sqrt{\lambda \dfrac{l}{d} + \sum \zeta}} \sqrt{2gH}$$

where

$$\sum \zeta = 0.5 + 0.2 + 2 \times 0.4 + 0.3 + 1.0 = 2.8$$

Hence

$$v = \frac{1}{\sqrt{0.03 \times \dfrac{50}{0.2} + 2.8}} \times \sqrt{2 \times 9.81 \times 1.2} = 1.51 \text{ m/s}$$

Thus

$$Q = Av = \frac{1}{4}\pi d^2 v = \frac{1}{4}\pi \times 0.2^2 \times 1.51 = 0.0475 \text{ m}^3/\text{s}$$

The maximum vacuum most likely occurs at a higher place where the pressure energy is converted to the potential energy and the head losses are the majority of the total head. So, applying the energy equation to section 2 – 2 (the downstream) and section 3 – 3, which is immediately after the third bend,

$$(H + 4.5) + \frac{p_B}{\gamma} + \frac{\alpha v^2}{2g} = 0 + 0 + 0 + h_{w_{3-2}}$$

where $\alpha = 1$, $h_{w_{3-2}} = \left(\lambda \dfrac{l_{BC}}{d} + \zeta_4 + \zeta_s\right)\dfrac{v^2}{2g}$

Substituting them into the above equation and rearranging the resulting equation gives

$$\frac{p_B}{\gamma} = \left(\lambda \frac{l_{BC}}{d} + \zeta_4 + \zeta_s - 1\right)\frac{v^2}{2g} - (H + 4.5)$$

$$= \left(0.03 \times \frac{30}{0.2} + 0.3 + 1 - 1\right) \times \frac{1.51^2}{2 \times 9.81} - (1.2 + 4.5)$$

$$= -5.1 \text{ m}$$

When the pump in working, the impeller, which is turning with a high speed and driven by an electric motor, carries the water out of the pressurized pipe. Consequently a vacuum will occur inside the pump and the water in the reservoir will be sucked into the pump chamber through the inlet pipe in order to maintain fluid continuity.

The vacuum height at the entrance of pump should be limited to a certain value to prevent any cavitation damage due to the break-up and collapse of generated vapour bubbles. The actual limit value can be used to determine the installation elevation of the pump.

The working principle of a pump is to convert the mechanical energy of motor into the input energy of the fluid, which is then able to be lifted. The total lift head of a pump, which can directly be used in the energy equation, refers to the energy of the fluid per unit weight provided by the pump motor. This includes the static head and all the losses that might occur in the flow system, usually in meters of water height. The effective power of a pump, i.e. hydraulic power N_e, is defined as the actual mechanical energy per unit time of the fluid obtained from the pump.

$$N_e = \gamma Q H \qquad (5-26)$$

where γ is the specific weight of the liquid in kN/m^3;

Q is the flow rate through the pump in m^3/s;

H is the total lift head of the pump in m, water height;

N_e is the effective power in kW.

The shaft power is the input power which the motor transmits to the pump, generally in kW. The efficiency of a pump is the ratio of the effective power of fluid to the shaft input power, which is usually in the range of 70%~90%.

【Example 5-5】 Shown in Fig. 5-13, the pump is lifting water from the lower reservoir to the higher reservoir, with the level difference $z = 45$ m. Both the inlet or sucking pipe and pressurized pipe have the same diameter $d = 500$ mm. The pump shaft is $h = 2$ m higher than the upstream water surface, and the sucking pipe is $l_1 = 10$ m long while the pressurized pipe $l_2 = 90$ m. The friction coefficient is $\lambda = 0.03$, and the minor loss coefficients are: $\zeta_1 = 3.0$ for the sucking inlet, $\zeta_2 = \zeta_3 = 0.3$ for the two 90° bends, and $\zeta_4 = 1.0$ for the pipe outlet. If the flow rate is $Q = 0.4$ m^3/s, find the total lift head H_p, the shaft power at a hydraulic efficiency of 80% and the vacuum $\dfrac{p_v}{\gamma}$ at the inlet pipe of the pump.

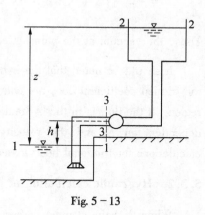

Fig. 5-13

Solution: Take cross-sections 1-1 and 2-2, choose the level $O-O$ as the reference datum, and then write the energy equation (ignoring the approaching velocity) as

$$0 + 0 + 0 + H_p = z + 0 + 0 + h_{w_{1-2}}$$

where $h_{w_{1-2}} = \left(\lambda \dfrac{l}{d} + \sum \zeta\right)\dfrac{v^2}{2g} = \left(\lambda \dfrac{l}{d} + \sum \zeta\right)\dfrac{1}{2g}\left(\dfrac{Q}{\frac{1}{4}\pi d^2}\right)^2$

Substituting it into the energy equation gives

$$H_p = z + \left(\lambda \dfrac{l}{d} + \sum \zeta\right)\dfrac{16Q^2}{2g\pi^2 d^4}$$

$$= 45 + \left(0.03 \times \dfrac{100}{0.5} + 3 + 2 \times 0.3 + 1\right) \times \dfrac{1}{2 \times 9.81} \times \dfrac{16}{\pi^2 \times 0.5^4} \times 0.4^2$$

$$= 47.23 \text{ m}$$

Then the shaft power is

$$N_s = \dfrac{\gamma Q H_p}{\eta} = \dfrac{9.81 \times 0.4 \times 47.26}{0.8} = 231.8 \text{ kW}$$

Applying the energy equation for cross-sections $1-1$ and $3-3$ gives

$$0 + 0 + 0 = h + \dfrac{p_3}{\gamma} + \dfrac{v^2}{2g} + h_{w_{1-3}}$$

Since $h_{w_{1-3}} = \left(\zeta_1 + \zeta_2 + \lambda \dfrac{l_1}{d}\right)\dfrac{v^2}{2g}$, substituting it into the above equation yields

$$\dfrac{p_3}{\gamma} = -h - \dfrac{v^2}{2g} - h_{w_{1-3}}$$

$$= -2 - \left(3 + 0.3 + 0.03 \times \dfrac{10}{0.5} + 1\right) \times \dfrac{1}{2 \times 9.81} \times \dfrac{16}{\pi^2 \times 0.5^4} \times 0.4^2 = -3.04 \text{ m}$$

Thus, the vacuum at the pump inlet is $\dfrac{p_v}{\gamma} = 3.04$ m.

It should be noted that the hydraulic calculation of short pipes is based on the premise that the friction coefficient does not change with velocity, i.e. the flow is in the hydraulically rough region. If the flow is in the hydraulically smooth or transitional region, the friction coefficients are dependent on Re and the velocity. Therefore, strictly speaking, in such cases the above calculations for frictional head losses should be reviewed and verified by trial-on-errors.

5.3.2 Hydraulic calculation of hydraulically long pipes

When the sum of minor losses and velocity heads are so small compared with the friction loss that they can be ignored in hydraulic calculations, the pipes in the system are classified as the hydraulically long pipes or long pipes in short, such as occur in simple pipelines, pipelines in series, pipelines in parallel, and pipeline networks.

5.3.2.1 Simple pipeline

A simple pipeline refers to the pipe that has constant diameter and flow rate without any diversion.

As shown in Fig. 5-14, a simple pipeline from a reservoir has the length l, diameter d and friction coefficient λ. The vertical height from the upstream free surface to the center of the exit of pipeline is H, i.e. the static head of the flow system.

Taking the GVF cross-section $1-1$ in the reservoir and $2-2$ at the exit, with the horizontal datum level $O-O$ at the center of section $2-2$, applying the energy equation gives

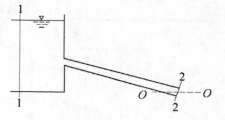

$$H + 0 + \frac{\alpha_1 v_1^2}{2g} = 0 + 0 + \frac{\alpha_2 v_2^2}{2g} + h_{w_{1-2}} \quad (5-27)$$

Fig. 5-14

For the long pipe, the minor head losses and velocity heads are both negligible, i.e.

$$H = h_{w_{1-2}} = h_{f_{1-2}} \quad (5-28)$$

which indicates the acting head is almost the same as the friction losses.

Since $\quad H = h_f = \lambda \dfrac{l}{d} \dfrac{v_2^2}{2g} = \lambda \dfrac{l}{d} \dfrac{1}{2g} \left(\dfrac{4Q}{\pi d^2}\right)^2 = \dfrac{8\lambda}{g \pi^2 d^5} l Q^2$

Letting
$$a = \frac{8\lambda}{g \pi^2 d^5} \quad (5-29)$$

gives
$$H = alQ^2 \quad (5-30)$$

If let $s = al$, then
$$H = sQ^2 \quad (5-31)$$

where a is the specific resistance, which refers to the head loss per unit length of pipe at unit flow rate, and it depends on λ and d; s is the frictional resistance referring to the corresponding head losses per unit flow rate in the pipe, and it depends on the specific resistance and length of pipe.

In engineering, the following formulae are commonly used for the long pipe systems, e.g. Chevielev's and Pavlovskii's are widely used in China, while Hazen-Williams's and Colebrook-White's are preferred in Western countries.

(1) Chevielev's formula

For the used steel or cast-iron pipes, when water temperature is 10 °C:

$$\left. \begin{array}{l} v \geqslant 1.2 \text{ m/s}, \quad a = \dfrac{0.001\,736}{d^{5.3}} \\[2mm] v < 1.2 \text{ m/s}, \quad a = 0.852 \times \left(1 + \dfrac{0.867}{v}\right)^{0.3} \left(\dfrac{0.001\,736}{d^{5.3}}\right) = Ka \end{array} \right\} \quad (5-32)$$

where K is the correction factor, $K = 0.852 \left(1 + \dfrac{0.867}{v}\right)^{0.3}$.

The values of K at 10 °C are given in Table 5-3.

Table 5-3 The correction factor K

v /m·s^{-1}	K	v /m·s^{-1}	K	v /m·s^{-1}	K
0.2	1.41	0.50	1.15	0.80	1.06
0.25	1.33	0.55	1.13	0.85	1.05
0.30	1.28	0.60	1.115	0.90	1.04
0.35	1.24	0.65	1.10	1.0	1.03
0.40	1.20	0.70	1.085	1.1	1.015
0.45	1.175	0.75	1.07	≥1.2	1.00

For the new steel and cast-iron pipes, when the velocity $v \geqslant 1.2$ m/s, the values of specific resistance are shown in Tables 5-4 and 5-5.

Table 5-4 Specific resistance a for steel pipes (s^2/m^6)

Nominal diameter /mm	a (Q in m^3/s)	Nominal diameter /mm	a (Q in m^3/s)	Nominal diameter /mm	a (Q in m^3/s)
15	8.809×10^6	150	44.95	450	0.1089
20	1.643×10^6	175	18.96	500	0.06222
25	436.7×10^3	200	9.273	600	0.02384
32	93.86×10^3	225	4.822	700	0.01150
40	44.53×10^3	250	2.583	800	0.005665
50	11.08×10^3	275	1.535	900	0.003034
70	2.893×10^3	300	0.9392	1000	0.001736
80	1.168×10^3	325	0.6088	1200	0.0006605
100	267.4	350	0.4078	1300	0.0004322
125	106.2	400	0.2062	1400	0.0002918

Table 5-5 Specific resistance a for cast iron pipes (s^2/m^6)

Nominal diameter /mm	a (Q in m^3/s)	Nominal diameter /mm	a (Q in m^3/s)
50	15190	400	0.2232
75	1709	450	0.1195
100	365.3	500	0.06839
150	41.85	600	0.02602
200	9.029	700	0.01150
250	2.752	800	0.005665
300	1.025	900	0.003034
350	0.4529	1000	0.001736

(2) Pavlovskii's formula

For concrete or reinforced concrete pipes,

$$\left. \begin{array}{l} a = 0.001743 \dfrac{1}{d^{5.33}} \text{ when } n = 0.013 \\ a = 0.002021 \dfrac{1}{d^{5.33}} \text{ when } n = 0.014 \end{array} \right\} \quad (5-33)$$

where n is the roughness of pipe.

Table 5-6 also gives detailed reference values for the specific resistance of concrete or reinforced concrete pipes.

Table 5-6 Specific resistance a for concrete pipes (s^2/m^6)

Inner diameter /mm	a ($n=0.013$, Q in m³/s)	a ($n=0.014$, Q in m³/s)	Inner diameter /mm	a ($n=0.013$, Q in m³/s)	a ($n=0.014$, Q in m³/s)
100	373	432	500	0.0701	0.0813
150	42.9	49.8	600	0.02653	0.03076
200	9.26	10.7	700	0.01167	0.01353
250	2.82	3.27	800	0.00573	0.00664
300	1.07	1.24	900	0.00306	0.00354
400	0.23	0.267	1000	0.00174	0.00202

(3) Hazen-Williams' formula

$$h_f = \frac{10.67 Q^{1.852} l}{C^{1.852} d^{4.87}} \quad (5-34)$$

where C is the factor, whose value can be found in Table 5-7; l is the pipe length in m; Q is the flow rate in m³/s; d is the pipe diameter in m.

Table 5-7 C values in the Hazen-Williams formula

Pipe types	Values of C
Plastic	150
New cast iron, bituminous/cement coating cast iron	130
Concrete, welded steel	120
Old cast iron or steel	100

(4) Colebrook-White's formula

$$\frac{1}{\sqrt{\lambda}} = -2\lg\left(\frac{\Delta}{3.7d} + \frac{2.51}{Re\sqrt{\lambda}}\right) \quad (5-35)$$

where λ is the major loss coefficient; Δ is the equivalent roughness in mm, see Table 5-8; d is the pipe diameter in mm; Re is the Reynolds number.

Table 5-8 Equivalent roughness Δ

Pipe types	Values of Δ/mm
Cast iron pipe with bituminous coating	0.05~0.125
Cast iron pipe with cement coating	0.50
Steel pipe with bituminous coating	0.05
Galvanized steel pipe	0.125
Asbestos cement pipe	0.03~0.04
Reinforced concrete pipe	0.04~0.25
Plastic pipe	0.01~0.03

【Example 5-6】 As shown in Fig. 5-15, the free surface in the water-supply tower is 30 m higher than the outlet (the user supply point), i.e. $z = 30$ m. The used steel pipe is 3000 m long and 200 mm in diameter. The pressure at the user supply point is required to be 20 m in water column. Find the flow rate in the pipe by using Chevielev's formula and Hazen-Williams' formula, respectively.

Solution:

Firstly, the calculation is carried out by Chevielev's formula.

From Table 5-4, for $d = 200$ mm, the corresponding specific resistance is $a = 9.273$ s^2/m^6. The acting head is $H = z - 20 = 30 - 20 = 10$ m. Substituting them into $H = alQ^2$ gives

$$Q = \sqrt{\frac{H}{al}} = \sqrt{\frac{10}{9.273 \times 3000}} = 0.01896 \text{ m}^3/\text{s}$$

Then the velocity is

$$v = \frac{Q}{\frac{\pi}{4}d^2} = \frac{0.01896}{\frac{\pi}{4} \times 0.2^2} = 0.60 \text{ m/s} < 1.2 \text{ m/s}$$

which shows that the flow is transient turbulent, so the value of a needs to be amended. Therefore, from Table 5-3 we can find $K = 1.115$. Repeat the calculation, thus

$$Q = \sqrt{\frac{10}{9.273 \times 1.115 \times 3000}} = 0.01796 \text{ m}^3/\text{s}$$

Secondly, the Hazen-Williams formula is applied to the same question.

From Table 5-7, it can be found that $C = 100$ for the used steel pipe. Since

$$H = \frac{10.67 Q^{1.852} l}{C^{1.852} d^{4.87}}$$

thus

$$Q = \left(\frac{HC^{1.852}d^{4.87}}{10.67l}\right)^{\frac{1}{1.852}} = \left(\frac{10 \times 100^{1.852} \times 0.2^{4.87}}{10.67 \times 3000}\right)^{\frac{1}{1.852}} = 0.01859 \text{ m}^3/\text{s}$$

In this example, the calculated results may have a small difference because of the different formulae used.

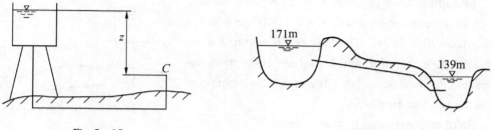

Fig. 5 – 15   Fig. 5 – 16

【Example 5 – 7】 A long distance water distribution pipeline shown in Fig. 5 – 16 is a reinforced concrete conduit, which is connected to two reservoirs with the water level being 171 m and 139 m respectively. The total pipe length is $l = 10$ km, and the wall roughness of the pipe is $n = 0.014$. If a flow rate of about 0.7 m³/s is required to carry out, determine the diameter of the pipe.

Solution: Since $H = alQ^2$, where $Q = 0.7$ m³/s and $H = 171 - 139 = 32$ m, then

$$a = \frac{H}{lQ^2} = \frac{32}{10\,000 \times 0.7^2} = 0.006\,531$$

From Table 5 – 6, when $n = 0.014$ and $d = 0.8$ m, the corresponding $a = 0.006\,64$ s²/m⁶, which is the closest value to the calculated result above; thus an 800 mm diameter pipe will be selected for the system. The corresponding flow rate will be

$$Q = \sqrt{\frac{H}{al}} = \sqrt{\frac{32}{0.006\,64 \times 10\,000}} = 0.694 \text{ m}^3/\text{s}$$

which is slightly less than 0.7 m³/s but acceptable in practice.

5.3.2.2 Pipelines in series

Pipes that are of different diameters and with various flow rates, connected to each other successively, are referred to as a pipeline in series. Generally, each pipe in the system is different from each other in length, diameter, flow rate and velocity of flow, so the head losses should be taken into account seperately for each individual pipe length. Assuming the pipe length, diameter, flow rate and specific resistance of a pipeline in series as l, d, Q and a, respectively, the total head loss for the pipeline in series is

$$h_w = \sum_{i=1}^{n} h_{fi} = \sum_{i=1}^{n} a_i l_i Q_i^2 \tag{5-36}$$

where n is the total number of the sections.

If the flow rate in the pipeline is constant, the total friction resistance of the pipeline is

$$s = \sum_{i=1}^{n} s_i$$

The flow rate in the pipeline should meet the continuity equation. The intersection between any two adjacent simple pipes is termed as nodal point or node. Thus the continuity equation can be interpreted as the inflow rate is equal to the outflow rate at any node, i. e.

$$Q_i = q_i + Q_{i+1} \qquad (5-37)$$

where q_i is the flow rate leaving the pipeline system at the end of i-th section.

【Example 5-8】 Shown in Fig. 5-17, the total length of a water supply pipeline is $l = 3000$ m with the acting head $H = 28$ m. The flow rate is required to be $Q = 160$ L/s. In order to make full use of the head and guarantee the demand of flow rate, design the pipeline in series by two cast iron pipes.

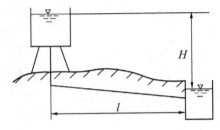

Fig. 5-17

Solution: Applying $H = alQ^2$ gives

$$a = \frac{H}{lQ^2} = \frac{28}{3000 \times 0.16^2} = 0.3646 \text{ s}^2/\text{m}^6$$

By referring to Table 5-5, the value above lies in one between an inner diameter 350 mm and 400 mm of cast iron pipe. Hence the pipes of $d_1 = 350$ mm and $d_2 = 400$ mm are chosen as a pipeline in series with the lengths l_1 and l_2 respectively. From Table 5-5, it is found that $a_1 = 0.4529 \text{ s}^2/\text{m}^6$ and $a_2 = 0.2232 \text{ s}^2/\text{m}^6$. Then

$$\begin{cases} H = a_1 l_1 Q^2 + a_2 l_2 Q^2 = (a_1 l_1 + a_2 l_2) Q^2 \\ l_1 + l_2 = l \end{cases}$$

Substituting the given data into the above equations,

$$\begin{cases} 28 = (0.4529 l_1 + 0.2232 l_2) \times 0.16^2 \\ l_1 + l_2 = 3000 \end{cases}$$

and solving the equations gives

$$l_1 = 1848.28 \text{ m } (d_1 = 350 \text{ mm})$$
$$l_2 = 1151.72 \text{ m } (d_1 = 400 \text{ mm})$$

5.3.2.3 Pipelines in parallel

When two or more simple pipelines are connected at the same starting point and ending point, where the flow is split at the inlet and subsequently joined together at the outlet, the pipelines are known as in parallel.

The hydraulic characteristic of such pipeline system is that the head loss of each individual pipeline is the same, i. e.

$$h_{f_1} = h_{f_2} = \ldots = h_{f_n} \qquad (5-38)$$

where n denotes the total number of the pipelines in parallel. It is also written as

$$a_1 l_1 Q_1^2 = a_2 l_2 Q_2^2 = \ldots = a_n l_n Q_n^2 \qquad (5-39)$$

Supposing the overall discharge through the system is represented by Q and the frictional resistance s, then

$$\begin{cases} Q = Q_1 + Q_2 + \ldots + Q_n \\ sQ^2 = s_1 Q_1^2 = s_2 Q_2^2 = \ldots = s_n Q_n^2 \end{cases}$$

It can easily be deduced that

$$\frac{1}{\sqrt{s}} = \frac{1}{\sqrt{s_1}} + \frac{1}{\sqrt{s_2}} + \ldots + \frac{1}{\sqrt{s_n}} \qquad (5-40)$$

【Example 5 – 9】 Three concrete pipes in parallel, as shown in Fig. 5 – 18, have the same roughness coefficient $n = 0.013$, and $d_1 = 300$ mm, $d_2 = 250$ mm and $d_3 = 200$ mm respectively. $l_1 = l_3 = 1000$ m and $l_2 = 800$ m. If the total flow is $Q = 0.32$ m³/s, find the flow rates in each pipe.

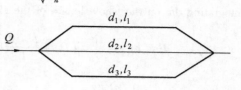

Fig. 5 – 18

Solution: From Table 5 – 6, for $n = 0.013$, $a = 1.07$ s²/m⁶ at $d_1 = 300$ mm; $a = 2.82$ s²/m⁶ at $d_1 = 250$ mm; $a = 1.07$ s²/m⁶ at $d_1 = 300$ mm.

For given values, $a_1 l_1 Q_1^2 = a_2 l_2 Q_2^2 = a_3 l_3 Q_3^2$ becomes

$$1.07 \times 1000 \times Q_1^2 = 2.82 \times 800 \times Q_2^2 = 9.26 \times 1000 \times Q_3^2$$

Thus

$$Q_2 = 0.6887 Q_1, \quad Q_3 = 0.3399 Q_1$$

From the continuity equation, one has

$$Q = Q_1 + Q_2 + Q_3 = (0.6887 + 0.3399 + 1) Q_1$$

Then

$$Q_1 = \frac{1}{0.6887 + 0.3399 + 1} \times 0.32 = 0.1577 \text{ m}^3/\text{s}$$

$$Q_2 = 0.6887 Q_1 = 0.1086 \text{ m}^3/\text{s}$$

$$Q_3 = Q - Q_1 - Q_2 = 0.0537 \text{ m}^3/\text{s}$$

5.3.2.4 Pipeline with uniform lateral flow drawn-off

The laterally discharged pipeline refers to a discharging pipe or distributing pipe, e. g. artificial rainfall pipe, perforated distribution pipe, and backwashing pipes in a filter chamber. Usually the lateral discharge from the pipeline is not uniform. If a pipeline has been designed to discharge the same per unit length, the pipeline is called lateral uniformly discharging pipeline, which will be discussed here in this section.

Shown in Fig. 5 – 19, it is supposed that pipe AB has the length l, the specific resistance a, the acting head H, the outflow rate Q_t at the end, and the total uniformly lateral flow Q_l. At the cross-section that is at a distance x away from the starting point A, the flow rate is

$$Q_x = Q_t + \frac{l - x}{l} Q_l$$

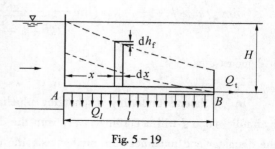

Fig. 5 – 19

Thus in an elemental segment dx, the frictional head loss

$$dh_f = aQ_x^2 dx = a\left(Q_t + \frac{l-x}{l}Q_1\right)^2 dx$$

Integrating dh_f on the whole length of the pipe gives the total frictional head loss of line AB

$$H = h_{f_{AB}} = \int_0^l a\left(Q_t + \frac{l-x}{l}Q_1\right)^2 dx = \int_0^l d\left[a\frac{\left(Q_t + \frac{l-x}{l}Q_1\right)^3}{3}\left(\frac{-l}{Q_1}\right)\right] \quad (5-41)$$

$$= \frac{1}{3}\frac{al}{Q_1}[(Q_t + Q_1)^3 - Q_t^3] = al\left(Q_t^2 + Q_tQ_1 + \frac{1}{3}Q_1^2\right)$$

Thus $h_{f_{AB}} = alQ_t^2$ when $Q_1 = 0$, and $h_{f_{AB}} = \frac{1}{3}alQ_1^2$ when $Q_t = 0$. Except for these two special cases, an equivalent flow rate, denoted by Q_r, is introduced in order to have an equivalent head loss $h_{f_{AB}}$.

$$Q_r = Q_t + \alpha Q_1$$

where α is the equivalent factor.

The head loss with the equivalent flow

$$H = alQ_r^2 = al(Q_t + \alpha Q_1)^2$$

Substituting it into Equation (5-41) gives

$$\alpha = \left(\sqrt{Q_t^2 + Q_tQ_1 + \frac{1}{3}Q_1^2} - Q_t\right)/Q_1 \quad (5-42)$$

Letting $k = \frac{Q_t}{Q_1}$ gives

$$\alpha = \sqrt{k^2 + k + \frac{1}{3}} - k \quad (5-43)$$

which shows that α is only dependent on k.

Since $\dfrac{d\alpha}{dk} = \dfrac{k + \frac{1}{2}}{\sqrt{k^2 + k + \frac{1}{3}}} - 1 < 0$, α always decreases with k.

If $Q_t \ll Q_1$ or $Q_t = 0$, $k = 0$ and $\alpha = \sqrt{\frac{1}{3}} = 0.577$. If $Q_t \gg Q_1$ or $Q_1 = 0$, $k \to \infty$. Then

$$\lim_{k \to \infty}\sqrt{k^2 + k + \frac{1}{3}} - k = \lim_{k \to \infty}\frac{k + \frac{1}{3}}{\sqrt{k^2 + k + \frac{1}{3}} + k} = \lim_{k \to \infty}\frac{1 + \frac{1}{3k}}{\sqrt{1 + \frac{1}{k} + \frac{1}{3k^2}} + 1} = \frac{1}{2}$$

It demonstrates

$$\alpha \in (0.5, 0.577) \quad (5-44)$$

In the lateral uniformly discharging pipeline, as the velocity varies along the flow, the hydraulic slope J will accordingly vary with the distance. If the minor losses are negligible, and the diameter and roughness are unchanged, the total energy line and hydraulic grade line would be as illustrated in Fig. 5-19.

5.3.3 Hydraulic calculation for pipeline networks

5.3.3.1 Introduction

The pipeline network for urban water distribution is generally composed of a number of pipelines connected in a complex way, which individually can be regarded to be hydraulically long pipes.

The pipeline network is classified as branch network (Fig. 5 – 20a) or loop network (Fig. 5 – 20b) according to its connection formation.

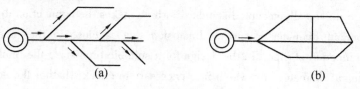

Fig. 5 – 20

The hydraulic calculations for pipeline networks are also classified into two kinds, as described as follows.

(1) Design computation

In the planning of new networks or pipeline extensions, we usually know the structure and terrain of the network system, the length of each pipe, the flow rate through the pipe as well as the designed water consumption and minimum head requirement at each node, so it is required to determine the pipe diameters and the acting heads (usually the total lift head of pump or the height of water tower) for the system.

Diameter d of a pipe is dependent on the flow rate Q and velocity v, as given by

$$d = \sqrt{\frac{4Q}{\pi v}}$$

When Q is constant, a smaller d resulted for a larger v implies the lower cost of the pipes and construction. However, larger velocities would result in higher head losses and consequently an increase in the power consumption of pump operating or the height of water tower, thus increasing the capital expenditure.

Therefore, while deciding the pipe diameters, the cost-and-benefit analysis on the project should also be taken. As shown in Fig. 5 – 21, the velocity that has a minimum total cost to the flow system (including the cost of the pipe system and water tower construction, and pump station running) is termed as the economic velocity v_e.

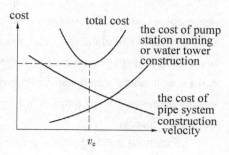

Fig. 5 – 21

Many factors affect the economic velocity and they are dependent on the location and time. Usually in the design handbooks, detailed analysis for the velocity is given in a tabulated format, e. g.

when $d = 100 \sim 200$ mm, $v_e = 0.6 \sim 1.0$ m/s;

when $d = 200 \sim 400$ mm, $v_e = 1.0 \sim 1.4$ m/s.

In the design computation of a pipeline network, the general procedure is as follows: the flow rate in each pipe is calculated first, then the pipe diameter of each pipe is determined according to the flow rate and the economic velocity, and finally the head losses and the acting head are worked out.

Besides, for convenience of manufacture and maintenance, various nominal diameters of pipe are stipulated by a universally accepted standard, which makes the connection dimensions of pipes unified. The nominal diameter is just a dimension that is close to the inner diameter, and represents a specification of pipe. In the design for a suitable pipe size, the final pipe needs to be expressed in nominal diameter, for which it is necessary to check whether the actual flow velocity through the selected pipe is in the prescribed range.

(2) Check computation

This task is usually to examine whether the free pressure head of each node meets the operating requirement, for given the network structure and system terrain, the length and diameter of each pipe, the water consumption of users, and the acting head of the network system.

The steps for such a check computation include calculations: of flow rate in each pipe firstly, then for the head losses, and finally for the pressure or free head at every node.

5.3.3.2 Hydraulic calculation for branched network

The overall flow direction in a branched network is determined, i. e. always from the inlet to the exit, so that we can directly calculate the flow rate in each pipe from the continuity equation. Afterwards, we need to select the control point and main branches. The control point, also known as the worst water supply point, refers to the point among all the nodes that has the highest terrain elevation together with a free head and the head losses from the headwater to the supply point. It is usually at the place furthest from the water source and on the higher ground; however, if difficult to estimate, some initial computations and comparisons are needed to help the decision. The main branch refers to the connecting pipeline from the headwater to the control point. One can designate a diameter for each pipe on the main branch according to its flow rate and the economic velocity, then calculate the head losses, and finally work out the total head loss in the main branch or the acting head for the system. Thereby, the head of each node along the main branch can be estimated, which can be viewed as the acting head of the other branches. Furthermore, the diameters of branch pipes could be determined and the corresponding node heads can then be calculated finally.

【Example 5-10】 Shown in Fig. 5-22, the topographical elevation of every node is all at 0.00 m, with a required minimum free head 16 m for each node. The water demands at each node are: $q_1 = 5$ L/s, $q_2 = 12$ L/s, $q_3 = 31$ L/s, $q_4 = 16$ L/s, and $q_5 = 12$ L/s.

(1) Determine diameters of the cast iron pipes and the acting head of the pipe system at the water

tower. (2) If node 5 needs an additional discharge 10 L/s with a minimum working free head 10 m for fire protection, does the acting head of tower satisfy the requirement?

Solution: (1) From the given data shown in Fig. 5-22, we can deduce that the control point has to be either Node 4 or Node 5, so the head losses in lines 3-4 and 3-5 should be compared. Consider the flow rate 16 L/s of section 3-4 and the limitation of economic

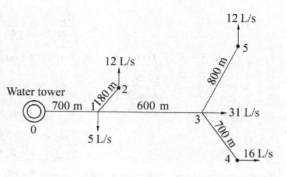

Fig. 5-22

velocity, the nominal diameter of section 3-4 could be 150 mm with the corresponding velocity v_{3-4} = 0.905 m/s. So does section 3-5, of 150 mm diameter but with the corresponding velocity v_{3-5} = 0.679 m/s. From Tables 5-3 and 5-5, the specific resistances for sections 3-4 and 3-5 after correction are as follows:

$$a_{3-4} = 41.85 \times 1.04 = 43.524 \text{ s}^2/\text{m}^6,$$
$$a_{3-5} = 41.85 \times 1.09 = 45.616 \text{ s}^2/\text{m}^6$$

Thus the head losses are,

$$h_{3-4} = a_{3-4} l_{3-4} q_{3-4}^2 = 43.524 \times 700 \times 0.016^2 = 7.799 \text{ m}$$
$$h_{3-5} = a_{3-5} l_{3-5} q_{3-5}^2 = 45.6165 \times 800 \times 0.12^2 = 5.26 \text{ m}$$

Since $h_{3-4} > h_{3-5}$, and the terrain elevations and the required free heads at Nodes 4 and 5 are identical, so Node 4 is chosen to be the control point. Thus, line 0-1-3-4 is selected to be the controlling trunk line and its hydraulic computation process is tabulated as in Table 5-9.

Table 5-9 Hydraulic computation along line 0-1-3-4

Pipe section	Length /m	Discharge /L·s^{-1}	Diameter /mm	Velocity /m·s^{-1}	Specific resistance /s^2·m^{-6}	Correction factor k	Head loss /m
0-1	700	76	350	0.79	0.4529	1.061	1.943
1-3	600	59	300	0.835	1.025	1.053	2.254
3-4	700	16	150	0.905	41.85	1.04	7.799

The head of Node 4 is 16 + 0 = 16 m

The head of Node 3 is 16 + h_{3-4} = 16 + 7.799 = 23.799 m

The head of Node 1 is 23.799 + h_{1-3} = 23.799 + 2.254 = 26.053 m

The head of Node 0 is 26.053 + h_{0-1} = 26.053 + 1.943 = 27.996 m

Hence, the acting head of the water tower is 27.996 m.

As for designing of the branch pipeline 1-2, we start to fully use the head of Node 1 rather than to calculate the economic velocity.

$$a_{1-2} = \frac{H_{1-2}}{l_{1-2}Q_{1-2}^2} = \frac{26.053 - 16}{180 \times 0.012^2} = 387.85$$

From Table 5−5, a 100 mm pipe diameter could be selected, with checking the actual velocity

$$v_{1-2} = \frac{4Q}{\pi d_{1-2}^2} = \frac{4 \times 12 \times 10^{-3}}{3.14 \times 0.1^2} = 1.24 \text{ m/s}$$, which is within the range of economic velocity.

Similarly,

$$a_{3-5} = \frac{H_{3-5}}{l_{3-5}Q_{3-5}^2} = \frac{23.799 - 16}{800 \times 0.012^2} = 67.7$$

The pipe diameter 100 mm is then selected with the actual velocity of 0.82 m/s.

Therefore, the diameters for each pipe are $d_{0-1} = 350$ mm, $d_{1-2} = 100$ mm, $d_{1-3} = 300$ mm, $d_{3-4} = 150$ mm and $d_{3-5} = 100$ mm, and the acting head of the water tower is 27.996 m.

(2) If a flow of 10 L/s is added to Node 5, select line 0 − 1 − 3 − 5 for the hydraulic calculation, as shown in Table 5 − 10.

Table 5 − 10. Hydraulic computation along line 0 − 1 − 3 − 5

	Length /m	Discharge /L·s^{-1}	Diameter /mm	Velocity m·s^{-1}	Specific resistance /s^2·m^{-6}	Correction factor k	Head loss /m
0 − 1	700	86	350	0.894	0.4529	1.04	2.439
1 − 3	600	69	300	0.976	1.025	1.038	3.039
3 − 5	800	22	150	1.24	41.85	1.0	16.204

Calculate the free head of Node 5 from the tower by the acting head,

$$27.996 - h_{0-1} - h_{1-3} - h_{3-5} - 0 = 27.996 - 2.439 - 3.039 - 16.204$$
$$= 6.314 \text{ m} < 10 \text{ m}$$

which is unable to satisfy the requirement for fire protection. Therefore, increase the pipe size of line 3 − 5 to 200 mm in diameter and then calculate again combined with Tables 5 − 3 and 5 − 5,

$$h_{3-5} = 9.029 \times 1.085 \times 800 \times 0.022^2 = 3.793 \text{ m}$$

Thus the free head at Node 5 is

$$27.996 - 2.439 - 3.039 - 3.793 - 0 = 18.721 \text{ m}$$

which now has satisfied the requirement.

5.3.3.3 Hydraulic calculation for looped networks

The flow direction of a pipe in a looped network has many possibilities so that the flow rate of each pipe cannot be calculated directly. If the flow rate of pipe is known, the pipe diameter can be determined by the economic velocity and the head loss is then obtained. Therefore, given the network structure and terrain, the length of each pipe and the water consumption at each node, the unknown variable would be the flow rate of each pipe.

Assuming P is the number of pipe sections, L the number of loops and J the number of nodes, then the relationship among above three numbers is

$$P = J + L - 1$$

Hydraulic calculation for a looped network also needs to apply both continuity equation and energy equation.

(1) Continuity equation

For any Node i, the flow rate going towards the node is equal to that leaving from the node. Consider the flow rate leaving from the node is positive and that approaching the node is negative, then

$$q_i + \sum Q_{ij} = 0 \qquad (5-45)$$

where q_i is the water consumption or supply of Node i; Q_{ij} is the flow rate in pipe ij which is connected to Node i.

There are $J-1$ independent continuity equations.

(2) Energy equation

For any loop, considering that the head loss of clockwise flow is positive and that of anticlockwise flow is negative, the algebraic sum of each head loss in the loop is zero.

$$\sum_i h_{ij} = \sum_i a_{ij} l_{ij} Q_{ij}^2 = 0 \qquad (5-46)$$

There are L independent energy equations.

The continuity equations and the energy equations are coupled together to give $J-1+L=P$ independent equations, whose number is exactly the same as the number of unknown discharge. So there is a certain solution to the equations. In fact, the hydraulic calculation for a looped network is in fact the process of solving a set of equations.

There are lots of methods for the hydraulic calculation of a looped network. Herein, only the Hardy-cross method is introduced.

Firstly, allocate a preliminary discharge for each pipe according to the continuity equation and then compute the head loss of each pipe. Substituting them into the energy equations, we have

$$\Delta h_i = \sum_i a_{ij} l_{ij} Q_{ij}^2 = \sum_i h_{ij}$$

where Δh_i is termed as the closing error. If $\Delta h_i \neq 0$, it indicates the tentative flow rates need to be adjusted. Assuming that the correction discharges for each loop are denoted by ΔQ_i, then

$$\sum_i a_{ij} l_{ij} (Q_{ij} + \Delta Q_i)^2 = \sum_i a_{ij} l_{ij} Q_{ij}^2 \left(1 + \frac{\Delta Q_i}{Q_{ij}}\right)^2 = 0$$

As $\frac{\Delta Q_i}{Q_{ij}}$ is a small quantity, take Taylor series expansion for the term of $\left(1 + \frac{\Delta Q_i}{Q_{ij}}\right)^2$ and retain the linear items, and then we have

$$\sum_i a_{ij} l_{ij} Q_{ij}^2 \left(1 + 2\frac{\Delta Q_i}{Q_{ij}}\right) = \sum_i a_{ij} l_{ij} Q_{ij}^2 + \sum_i 2 a_{ij} l_{ij} Q_{ij} \Delta Q_i = 0$$

Thus

$$\Delta Q_i = -\frac{\sum_i a_{ij} l_{ij} Q_{ij}^2}{2 \sum_i a_{ij} l_{ij} Q_{ij}} = -\frac{\sum_i h_{ij}}{2 \sum_i \frac{a_{ij} l_{ij} Q_{ij}^2}{Q_{ij}}}$$

$$= -\frac{\Delta h_i}{2\sum_i \dfrac{h_{ij}}{Q_{ij}}} \qquad (5-47)$$

Adding ΔQ_i to the first tentative discharge allocation of each pipe will result in the second tentative discharge allocation. Repeat the same step until the results satisfy the accuracy requirement (Δh is close to zero).

For a large looped network, there may be thousands of pipes and loops, so it is impossible for manual calculation, but can be done using computers. The relevant software can be found in some reference books.

Chapter summary

In this chapter, the application of fundamenal principles of fluid mechanics on pressurized pipe flow has been described, with focuses on the application of the continuity equation, energy equation and head loss of total flow.

1. The characteristics of orifice, nozzle and pressurized pipe flow:
   - For orifice and nozzle flow, only local head loss is considered, $h_w = h_j$;
   - For flow of short pipe, both the frictional and local head losses are needed to take into account, $h_w = h_f + h_j$;
   - For flow of long pipe, the local head loss and velocity head are negligible compared with the frictional head loss, $h_w = h_f$.

2. The orifice flow and the nozzle flow have the same formula of discharge:

$$Q = \mu A \sqrt{2gH_0}$$

but they have different discharge coefficients, and usually the μ for nozzle flow is larger.

The normal operation conditions of nozzle flow are: H_0 (acting head) < 9 m; l (length) $= (3 \sim 4)\, d$. Both orifice and nozzle flows can be free outflow or submerged outflow, which has the same discharge coefficients and formula expression. The difference of discharge in two conditions of outflow is only reflected in the calculation of acting head.

3. In hydraulic calculation of short pipe, the total head loss should be taken into account with the continuity and energy equations.

4. For hydraulic calculation of long pipe:

Simple pipeline: $H = h_f = a\, l\, Q^2$

Pipes in series: $H = \sum h_{fi} = \sum a_i l_i Q_i^2$

Pipes in parallel: $H = a_1 l_1 Q_1^2 = a_2 l_2 Q_2^2 = \cdots = a_n l_n Q_n^2$, where n is the number of pipe.

5. In hydraulic calculation of branched pipe network, according to the flow continuity, calculate the discharge for each pipe, and then choose the control points and main pipeline based on the water demands in the network. Based on the requirements of economic flow velocity and discharge in each pipe, choose the diameter of pipe, and then calculate the head loss and the total head loss of trunk pipe, which is used for the design of acting head in the network. The diameter of branch pipe can be determined by the head at its junction with the main pipeline.

Chapter 5 Steady orifice, nozzle and pipe flow

Review questions

5-1 Qualitatively analyze the impact of the contraction coefficient on the velocity and flow rate of orifice and nozzle flow.

5-2 For the same sized orifice and nozzle, under the same acting head, what is their relationship for the velocity and discharge of outflow?

5-3 Under the same acting head, what are the similarities and differences between the free and submerged outflow in orifice or nozzle flow?

5-4 Why does not the size of orifice matter in the condition of submerged outflow?

5-5 What is the difference between short and long pipes?

5-6 Under the same acting head, analyze the relationship of discharge and pressure between the free and submerged outflow of short pipe.

5-7 Fig. 5-23 shows two identical short pipes 1 and 2. If the upstream water level remains constant, and the downstream water level is at A, B and C respectively, what is the relationship between the discharges of two pipes under above three conditions? How about the relationship of pressure?

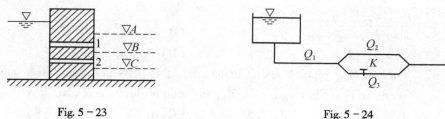

Fig. 5-23   Fig. 5-24

5-8 Analyze the conditions of two parallel pipes having the same discharge and hydraulic gradient, respectively.

5-9 As shown in Fig. 5-24, the water tank level is constant, and the flow rate at the normal operation is Q_1, Q_2, Q_3; if the valve K is closed down a bit, how will Q_1, Q_2 and Q_3 be changed?

5-10 Is the control point of a pipe network allocated at the highest terrain or farthest from the water supply source?

Multiple-choice questions (one option)

5-1 Under normal operating conditions, for the same water head and diameter, the discharge of small orifice flow is _____ that of cylindrical external nozzle flow.
  (A) greater than   (B) less than
  (C) equal to   (D) unknown compared with

5-2 Orifice, nozzle and pipe flow all have the free outflow and submerged outflow, in which _____.
  (A) the basic formula of discharge is different
  (B) the discharge coefficient is different

(C) the velocity coefficient is different

(D) the acting head is different

5-3 For the free outflow of cylindrical external nozzle, _____.

(A) its head loss is smaller than that of orifice flow under the same conditions

(B) its acting head in normal operating conditions is greater than 9 m

(C) the length of nozzle is generally equal to the diameter of the nozzle

(D) vacuum pressure occurs in the contraction section

5-4 The head loss (h_w) of pressurized pipe flow is classified into the frictional head loss (h_f) and the local head loss (h_j). In the hydraulic calculation of short pipe flow, _____ can be ignored.

(A) the frictional loss   (B) the local loss

(C) the velocity head   (D) none of the above

5-5 The total head line of long pipe flow and its piezometric head line are _____.

(A) the same   (B) parallel, straight lines

(C) parallel, in a stepwise form (D) none of the above

5-6 For parallel pipelines 1, 2, 3, the head loss between nodes A (start) and B (end) h_{fAB} is _____.

(A) $h_{f1} + h_{f2} + h_{f3}$ (B) $h_{f1} + h_{f2}$

(C) $h_{f2} + h_{f3}$ (D) $h_{f1} = h_{f2} = h_{f3}$

5-7 For two long pipes 1 and 2 in parallel, they have the same diameter and friction coefficient; if the length $l_2 = 3l_1$, the ratio of discharge Q_1 to Q_2 is _____.

(A) 1   (B) 1.5   (C) 1.73   (D) 3

5-8 Long pipe is the pipe for which _____.

(A) the length is more than 100 m

(B) the sum of local head loss and velocity head is small and negligible compared with the frictional head loss

(C) the frictional coefficient is very large

(D) the frictional head loss is greater than 5 m

5-9 The water in a tank passes through a pipeline to be discharged to the atmosphere; in the application of the energy equation for calculating the discharge, the section should be chosen at _____.

(A) any cross section of the pipe

(B) the inlet and outlet of the pipe

(C) the water surface of the tank and the outlet of the pipe

(D) the water surface of the tank and the inlet of the pipe

5-10 The pressure at the highest position of a siphon is _____ the atmospheric pressure.

(A) >   (B) <   (C) =   (D) none of the above

5-11 The main pipeline of a branched pipe network has _____.

(A) the largest discharge   (B) the smallest head loss
(C) the largest acting head   (D) the largest head loss

5-12 In the design of pipe network, the economic velocity is _____.
(A) the maximum flow velocity of pipeline
(B) the minimum flow velocity of pipeline
(C) the actual operating flow velocity
(D) the velocity for the lowest total cost

Problems

5-1 The sidewall of a container has a circular hole of 20 mm diameter, and under the constant water head of $H = 0.5$ m, the measured flow rate is 0.772 L/s, determine the discharge coefficient of the orifice flow.

5-2 As shown in Fig. 5-25, the flat barge ship has the height $h = 1$ m, the weight 9.81 kN, and the horizontal cross-sectional area 8 m². Its bottom suddenly has a hole of 100 mm diameter, and the orifice discharge coefficient was 0.60. How long will the ship take to sink? (Neglecting the ship's wall thickness)

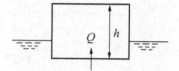

Fig. 5-25

5-3 As shown in Fig. 5-26, in order to make the water evenly enter the sedimentation tank, the perforated wall is installed at the entrance of the tank. The wall is perforated with 14 square holes with each having the side length of 80 mm, and the total flow rate is 110 L/s, regardless of the wall thickness and the influence between the holes. If the discharge coefficient is 0.62, what is the water level difference before and after the wall being perforated?

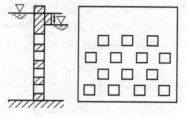

Fig. 5-26

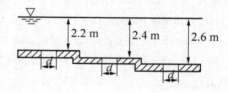

Fig. 5-27

5-4 As shown in Fig. 5-27, the reservoir has three horizontal bottom outlets, whose diameter is $d = 0.2$ m; the water depth of the orifices is 2.2 m, 2.4 m and 2.6 m respectively, and the discharge coefficient is 0.60. Find the total discharge (Neglecting the thickness of outlet).

5-5 With a constant head of 2 m, a 20 mm-diameter thin-walled orifice has the discharge coefficient of 0.62; under the same conditions the orifice is connected to a cylindrical external nozzle of 20 mm diameter, which has the discharge coefficient of 0.8. Determine the orifice flow rate Q_H, the nozzle flow rate Q_P, and the vacuum degree (p/γ) in the nozzle for free outflow.

5-6 If the acting head of a nozzle is 2.8m, the diameter is 0.5m, and the discharge is

1.368 m³/s, determine the discharge coefficient of nozzle flow.

5-7 If the acting head of a nozzle is 6 m, the discharge of outflow is 12 m³/s, and the discharge coefficient is 0.98, find the nozzle diameter.

5-8 As shown in Fig. 5-28, the water tank is divided into chambers A and B with a thin clapboard, which has an opening of 40 mm diameter. On the bottom of chamber B, there is a cylindrical external nozzle of a diameter of 30 mm. If $h = 0.5$ m, $H = 3$ m, the water flow is steady flow, the discharge coefficients of the nozzle and orifice are 0.82 and 0.62, respectively. Find the level difference (ΔH) of the two chambers A and B, and the flow rate at the outlet of tank.

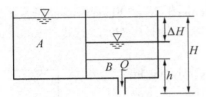
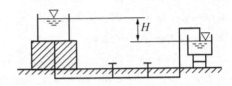

Fig. 5-28   Fig. 5-29

5-9 As shown in Fig. 5-29, the water supply pipe has the diameter $d = 100$ mm, the length $l = 80$ m, two fully open valves (the loss coefficient $\zeta = 0.12$), four identical 90° elbows ($\zeta = 0.48$), the entry loss coefficient $\zeta = 0.5$, and the frictional coefficient of pipe $\lambda = 0.03$. The acting head of the water tower $H = 2.6$ m, and the effective volume of the tank is 10 m³. For steady flow, how long is required to fulfill the tank?

5-10 As shown in Fig. 5-30, a pipe with the length of $l_1 = 20$ m and the diameter of $d_1 = 100$ mm is used to convey the water from the reservoir to the tank, from which the water is then discharged to the atmosphere through another pipe (the length $l_2 = 100$ m, diameter $d_2 = 50$ mm). If $H = 10$ m, the loss coefficient of the valve is $\zeta = 3$, the entry $\zeta = 0.5$, the outlet to tank $\zeta = 1.0$, and the friction coefficient $\lambda = 0.03$. Find the flow rate and the water level difference (ΔH) between the reservoir and the tank.

5-11 As shown in Fig. 5-31, the length of circular culvert is 10 m, the diameter 1 m, the total local loss coefficient $\sum \zeta = 1.5$ (including the outlet), the flow rate $Q = 4.3$ m³/s, and the frictional coefficient $\lambda = 0.02$. Find the water level difference H.

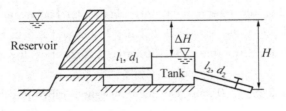
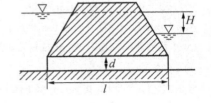

Fig. 5-30   Fig. 5-31

5-12 As shown in Fig. 5-32, the water is extracted from reservoir A to B by a pump with the lift head $H = 10$ m. The pipe diameter is $d = 100$ mm, the total length is 100 m (the length

BC is 60 m), the frictional coefficient $\lambda = 0.03$, and the local loss coefficients: ζ(entry) = 0.8, ζ(outlet) = 1, ζ(elbow) = 0.2. The water level difference between the two reservoirs is 7.5 m, and point C is 3 m above the water level of reservoir B. Determine the flow rate of pipe and the pressure head at point C.

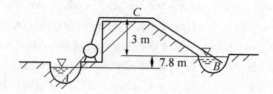

Fig. 5 – 32

5 – 13 As shown in Fig. 5 – 33, the water in the sump is pumped into the water tower. The length of suction pipe is 20 m, the diameter is $d = 200$ mm; the pressurized pipe length is 45 m, and the diameter d is 150 mm. The frictional coefficient $\lambda = 0.03$, the local loss coefficients: ζ(pump entry) = 7.0, ζ(90° elbow) = 0.05, ζ(outlet) = 1.0. The water level difference between the sump and tower is $H_o = 30$ m, and the flow rate is $Q = 30$ L/s. Determine the lift head of the pump and its shaft power at the efficiency of 80%.

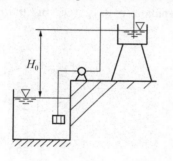

Fig. 5 – 33

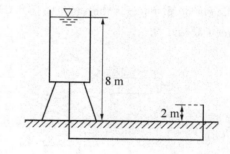

Fig. 5 – 34

5 – 14 As shown in Fig. 5 – 34, the water surface of tower is 8 m above the ground, the length of the old cast iron pipe is $l = 1000$ m, the diameter $d = 400$ mm. If the point for the demand of water is at 2 m above the ground and requires a free head of 2 m, how much is the flow rate of pipe?

5 – 15 Fig. 5 – 35 shows the pipeline in series: $d_1 = 300$ mm, $d_2 = 200$ mm, $d_3 = 150$ mm; the length $l_1 = l_2 = l_3 = 300$ m; the roughness coefficient of concrete pipe $n = 0.013$. How much of water head H is required for $Q_1 = Q_2 = Q_3 = 25$ L/s?

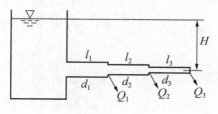

Fig. 5 – 35

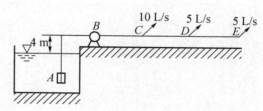

Fig. 5 – 36

5-16 Fig. 5-36 shows the water supply pipeline in series. They are the old cast iron pipes, for which $l_{AB} = 30$ m, $d_{AB} = 200$ mm; $l_{BC} = 240$ m, $d_{BC} = 150$ mm; $l_{CD} = 200$ m, $d_{CD} = 100$ mm; $l_{DE} = 100$ m, $d_{DE} = 100$ mn. At point E it requires a free head of 5 m. The pump flow is 20 L/s, the local loss coefficients are: $\zeta(\text{elbow}) = 0.5$, $\zeta(\text{pump entry}) = 6$. Find the pump lift head H.

5-17 Three parallel pipes connected at two ends have the diameter d, $2d$, $3d$ respectively, and they have the same length and roughness coefficient. If the pipe of diameter d has the flow rate $Q_1 = 30$ L/s, how much is the flow rate of the other two pipes (Q_2, Q_3)?

5-18. As shown in Fig. 5-37, two identical pipes in parallel are replaced by n segments of short pipes of the same diameter, which are connected in parallel. In normal operation, the working pressure remains constant, which corresponds to a total water flow rate Q. At the same working pressure, what is the total flow rate Q' if one of the segments is closed?

Fig. 5-37

5-19 Fig. 5-38 shows the pipeline system: the flow rate $Q = 120$ L/s, the pipe lengths: $l_A = 1000$ m, $l_B = 900$ m, $l_C = 300$ m; $d_A = 250$ mm, $d_B = 300$ mm, $d_C = 250$ mm. The friction coefficient of all pipes is the same ($\lambda = 0.03$). Determine Q_A, Q_B, Q_C, and the head loss h_{MN} between M and N.

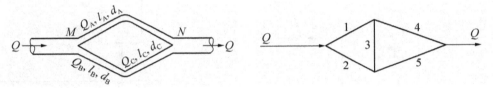

Fig. 5-38   Fig. 5-39

5-20 As shown in Fig. 5-39, a looped network has five pipes: the resistance coefficient $a_1 = a_2 = a_3 = a$, $a_4 = a_5 = 2a$. The steady water flow rate is Q. Find the corresponding flow rate of each pipe: Q_1, Q_2, Q_3, Q_4, Q_5. If the flow rate of the 4th pipe is reduced, qualitatively analyze how the flow rate ($Q_1 \sim Q_5$) will change. (Assuming the water flow Q remains unchanged)

Chapter 6　Steady flow in an open channel

Open-channel flow refers to the water flow in natural or man-made canals and channels, which have a free surface in contact with the atmosphere. Since the relative pressure at the free surface is zero, it is also called non-pressurized flow. It is usually seen wherever water flows in man-made canals, rivers, pipes and tunnels (partially filled), as shown in Fig. 6 - 1. Due to the existence of the free surface, open-channel flow is different from pressurized flow or pipe flow and the flow in closed conduits. In addition, the boundary conditions of open-channel flow are affected by many factors. Thus, it is much more complicated than pipe flow and consequently the related hydraulic calculations will also be different.

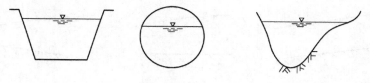

Fig. 6 - 1

Like pipe flow, open-channel flow can be classified into steady flow and unsteady flow according to whether the flow quantities vary with time or not, and uniform flow and non-uniform flow according to whether flow quantities vary with space or not. Non-uniform flows in open channel are further classified as gradually varied flow (GVF) or rapidly varied flow (RVF). Since open-channel flows in engineering are usually turbulent flow, they are close to or within the resistance squared zone of turbulent flow. In the following sections, discussions are limited to this kind of flow and also primarily restricted to one-dimensional flow analysis.

6.1　Geometry of open channel

The main function of an open channel is to carry water. Engineering practices show that geometric features including bed slope, shape and size of cross-section have an important impact on flow regime and carrying capacity.

6.1.1　Longitudinal bed slope of open channel

The longitudinal bed slope is the elevation difference per unit length of channel bed, denoted by i, which represents the gradient of channel bed. Shown in Fig. 6 - 2, the bed slope of channel is

$$i = \frac{z_{01} - z_{02}}{\Delta l} = -\frac{\Delta z_0}{\Delta l} = \sin\theta \quad (6-1)$$

where z_0 denotes the bed elevation of channel, subscripts 1 and 2 denote cross-sections 1 – 1 and 2 – 2 respectively, Δl is the distance between the two sections, and θ is the angle of channel bed to the horizontal. In practice, if θ is small ($\theta < 6°$), for convenience of measurement and calculation, the horizontal distance is usually replaced for the inclined length of bed, i.e. $i \approx \tan\theta$.

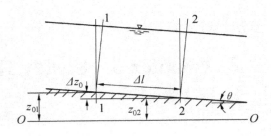

Fig. 6 – 2

As seen in Fig. 6 – 3, there are three possible bed slopes: positive slope or downhill slope, i.e. $i > 0$, if the bed elevation decreases stream-wise; zero slope or horizontal slope, i.e. $i = 0$, if bed elevation remains unchanged (horizontal) along the flow; and negative slope or uphill slope, i.e. $i < 0$, if bed elevation increases stream-wise. Most channels have a positive slope while a few may have either zero or negative slopes.

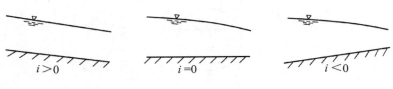

Fig. 6 – 3

The longitudinal channel bed of natural river is often an irregular slope, so the bed slope of a river bed refers to the average gradient over a certain river reach.

6.1.2 Cross-section of open channel

The cross-section of open channel can be of various shapes. The man-made channels usually have regular shapes such as trapezoidal for earth channels, circular or sometimes U-shaped for culvert pipes and tunnels, and rectangular or semi-circular for concrete channels and aqueducts; while the natural river cross-sections are usually irregular, and cross-sectional shapes or sizes may be significantly different even in the same river.

It should be noticed that the cross-section of an open channel is different from the wetted perimeter; the former refers to the whole transverse boundary of the channel and the latter is defined as the cross-section part that is in direct contact with the fluid.

6.1.3 Geometric parameters of flow cross-section

In many calculations of open-channel flow, the geometric parameters of the flow cross-section often need to be computed. Taking a trapezoidal section as an example (Fig. 6 – 4), the basic geometric parameters include:

b is the base width;

h is the water height, which refers to the vertical distance from the lowest point on the cross-section to the free surface. If the inclined angle θ of channel (or bed slope) is small, the flow cross-section can often be replaced by the vertical cross-section and consequently the vertical water depth is taken as the depth of the flow cross-section.

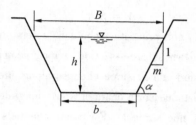

Fig. 6-4

m is the side slope coefficient, which reflects the gradient of both side slopes and is defined as cotangent of the side slope angle α, i. e. $m \approx \cot\alpha$. Its magnitude will depend on the soil properties or the armor characteristics of the side walls. For instance, Table 6-1 shows some reference values.

Table 6-1 Side slope coefficient for trapezoidal channels

Soil type	Side slope coefficient /m
Fine sand	3.0~3.5
Sandy or loose soil	2.0~2.5
Compact sandy soil or light clay	1.5~2.0
Gravel	1.5
Heavy clay, compact loess, clay	1.0~1.5
Compact heavy clay	1.0
Weathered rock	0.25~0.5
Non-weathered rock	0~0.25

The derived geometric quantities include

$$\left.\begin{array}{l} \text{the free surface width: } B = b + 2mh; \\ \text{the area of wetted cross-section: } A = (b + mh)h; \\ \text{the wetted perimeter: } \chi = b + 2h\sqrt{1 + m^2}; \\ \text{and the hydraulic radius: } R = \dfrac{A}{\chi}. \end{array}\right\} \quad (6-2)$$

The derived quantities of other shaped cross-sections can be obtained by the corresponding geometric relationship.

6.1.4 Prismatic and non-prismatic channel

Geometrically, open channels can be classified into prismatic channels and non-prismatic channels, as shown in Fig. 6-5. The straight channel whose cross-section shape and size as well as bed slope do not change along the distance is termed prismatic channel, for example, in a prismatic trapezoidal channel, the base width b, the side slope m and the bed slope i are the same for every channel cross-section. For a prismatic channel, the area of a flow cross-section

varies only with the water depth, i. e.
$$A = f(h)$$

A non-prismatic channel refers to a channel whose cross-section shape, geometric dimension or bed slope changes along streamwise distance, or a channel whose longitudinal axis is not straight. For such channels the cross-sections vary at different locations along the channel, so the flow cross-sectional area varies with both the water depth and the location, i. e.

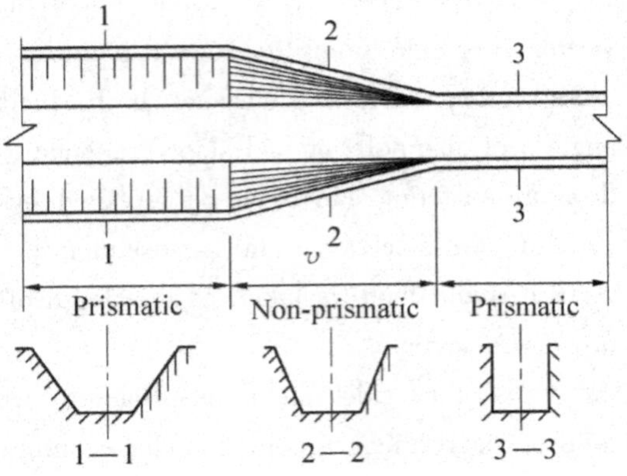

$$A = f(h, x)$$

Fig. 6 - 5

where x denotes the flow distance. In general, the transition reach in an open channel is typically a non-prismatic channel. Due to the irregularity in cross-section and bed, natural rivers are generally non-prismatic channels.

6.2 Uniform flow in open channel

In uniform open-channel flow, the streamlines are parallel and straight. Since there are free surfaces, it is impossible for uniform open-channel flow to be unsteady, that is to say, uniform open channel flow is, by definition, always steady flow.

Uniform flow is the simplest flow in open-channel flows, so the theory of uniform flow is the starting basis of hydraulic analysis and computation for open channels.

6.2.1 Characteristics and conditions of uniform open-channel flow

According to the definition of uniform open-channel flow, it has the following characteristics:

(1) The shape and size of flow cross-section do not change along the flow.

(2) The depth and velocity distribution of flow cross-section do not change along the flow direction either. Thus the flow rate, mean cross-sectional velocity, kinetic correction factor, momentum correction factor and velocity head are the same along the flow. The head losses of flow only include frictional losses, not minor losses.

(3) Since the free surface line is also the piezometric head line, the three lines of the total head line, the piezometric head line and the bed line are all parallel, i. e. the hydraulic gradient J, the slope (J_p) of piezometric head line or free surface line, and the bed slope i, are the same.

$$J = J_p = i \qquad (6-3)$$

which is shown in Fig. 6 - 6.

Because uniform open-channel flow is steady, there is no acceleration, so all external forces on the water body are in equilibrium. In the flow illustrated in Fig. 6 - 6, take the water body

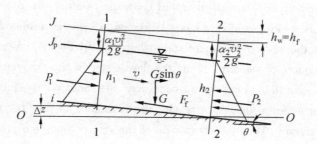

Fig. 6-6

between cross-sections 1-1 and 2-2 as the control volume for analysis. The forces exerted on the control volume are the dynamic pressure forces P_1 and P_2 on the cross-sections, the gravity G, and the frictional force F_f on the channel boundary. The equilibrium equation of force in the flow direction is

$$P_1 + G\sin\theta - F_f - P_2 = 0 \qquad (6-4)$$

Since cross-section 1-1 is the same as cross-section 2-2, on which the dynamic pressure distributions are the same but in opposite directions and comply with the hydrostatic pressure distribution, i.e. P_1 and P_2 have the same values but in opposite directions. Therefore, Equation (6-4) can also be written as

$$G\sin\theta = F_f$$

which demonstrates that uniform open-channel flow is such a flow that the component of gravity in direction of flow is balanced by the frictional force.

Take the horizontal plane through the bottom of cross-section 2-2 as the reference datum ($O-O$) and write the energy equation between 1-1 and 2-2

$$(h_1 + \Delta z) + \frac{\alpha_1 v_1^2}{2g} = h_2 + \frac{\alpha_2 v_2^2}{2g} + h_{w_{1-2}}$$

Since $h_1 = h_2, v_1 = v_2, \alpha_1 = \alpha_2, h_{w_{1-2}} = h_{f_{1-2}}$

Thus
$$\Delta z = h_{f_{1-2}} \qquad (6-5)$$

Therefore, from the perspective of energy, in uniform open-channel flow, the work done by the gravity for unit weight of water is equal to that by the frictional force. In other words, due to the same kinetic energy, the potential energy decreases as the water surface declines along the channel, with the decreased amount exactly equal to the lost energy that is dissipated to overcome the friction. That is the physical meaning of Equation (6-5).

From the above analysis, uniform open-channel flow happens only when the following conditions are satisfied: the flow is steady with constant flow rate, there is no tributary along the channel, and the channel is a long and straight downhill prismatic channel, which has constant roughness and no local disturbances caused by construction along the channel. Therefore uniform flow will not occur in horizontal or uphill channels, or non-prismatic channels and natural rivers.

Generally, with the restriction of other various factors, it is difficult for a flow in an open channel to meet the aforementioned conditions and thus most flows in open channels are non-uniform flows. However, for a straight downhill prismatic channel, as long as the channel is long enough, when flow is steady, it always has a tendency to become a uniform flow. Open-channel flows are usually in the resistance square zone where the frictional resistance (force) of flow is proportional to the square of velocity. When the flow is steady and the water depth is greater than the depth of the uniform flow in equivalent condition, the mean cross-sectional velocity would be less than that of the uniform flow. As a result, the resistance of flow becomes relatively smaller compared with the gravity. Thus the component of the gravity along the flow would become larger than the resistance force of flow, which will result in an accelerated movement of water. As the velocity increases, the flow depth decreases continuously in order to meet the continuity of flow and meanwhile the resistance is increasing. In this way, after a certain distance, the gravity component will eventually balance with the resistance to achieve an equilibrium in which a uniform flow is attained, and vice versa. In general, any non-uniform open-channel flow caused by various reasons always tends to become a uniform flow provided that the channel is long enough and has no other disturbance.

Man-made channels are usually built as straight as possible, and also the cross-sectional shape and size, wall roughness and bottom slope are usually maintained constant. Such channels therefore can basically meet the conditions required by uniform open-channel flow. Therefore, the canal or aqueduct in practical engineering is designed in the light of uniform flow.

6.2.2 Basic equations for uniform open-channel flow

Chezy's formula, an empirical expression for the head loss in the previous chapter, is
$$v = C \sqrt{RJ}$$
which is also suitable for uniform open-channel flow. Since in such flows $J = i$, thus
$$v = C \sqrt{Ri} \tag{6-6}$$
$$Q = Av = AC \sqrt{Ri} = K\sqrt{i} \tag{6-7}$$

Equations (6-6) and (6-7) are called as the basic equations of uniform open-channel flow, where $K = AC\sqrt{R}$ is termed as the discharge conveyance because $Q = K$ when $i = 1$. The discharge conveyance K depends on the cross-sectional shape and size, roughness, and water depth, and reflects the cross-sectional impact on the flow rate. The Chezy coefficient C can also be calculated by the commonly used Manning formula, $C = \dfrac{1}{n} R^{1/6}$, in which the coefficient C is related to the hydraulic radius R and the channel roughness n. In fact, R represents the cross-sectional size and shape whilst n denotes the overall roughness of channel. Engineering practice shows that the values of both R and n have direct impact on hydraulic calculation, especially with n values having much greater effect than R. Thus, choosing an appropriate n value is crucial to open-channel design and calculation. It should also be noted that when the flow is not in the turbulent resistance zone, the Manning coefficient n is not suitable to be used for calculating Chezy coefficient C because the Manning formula is independent of the Renolds number.

Typical n values for various channel materials have been obtained based on decades of observations and practical application, and are given in Table 6-2 for reference.

For river channels, the Manning coefficient n can be calculated by the following expression:

$$n = \frac{1}{v}\overline{R}^{2/3}\overline{i}^{1/2}$$

where the mean cross-sectional velocity v, the mean hydraulic radius $\overline{R}$ and the mean slope $\overline{i}$ of a channel reach are measured on site.

Table 6-2 Typical values of Manning coefficient n

Surface of channel	Roughness n		
	low	mean	high
1. Earth canal			
Clean, normal	0.020	0.0225	0.025
Rough, with weed	0.027	0.030	0.035
Slightly curved, with weed	0.025	0.030	0.033
Excavated by machine	0.0275	0.030	0.033
Gravel	0.025	0.027	0.030
Fine gravel	0.027	0.030	0.033
Masonry wall, sandy bed	0.030	0.033	0.035
Weedy wall, uneven stone bed	0.030	0.035	0.040
2. Masonry channel			
Clean, straight	0.030	0.033	0.035
Rough, irregular shaped	0.040	0.045	
Smooth, uniform	0.025	0.035	0.040
Well excavated, managed		0.02~0.025	
3. Lining channel			
Mixed (lime, sand, coal ash)	0.014	0.016	
Brick	0.012	0.015	0.017
Regular stone	0.013	0.015	0.017
Stone block with cement	0.017	0.025	0.030
Loose stone block	0.023	0.032	0.035
4. Concrete channel			
Finished, reinforced with cement	0.011	0.012	0.013
Unfinished, reinforced without cementl	0.013	0.014~0.015	0.017
Cement mortar	0.016	0.018	0.021
5. Wooden channel			
Planed	0.012	0.013	0.014
Unplaned	0.013	0.014	0.015

A number of factors have an impact on channel bed slope. If the slopes are too steep or too mild, they cannot meet the requirement to avoid bed deposition and erosion. For sake of minimum excavation and construction difficulty, deep excavation and high filling should be avoided, and efforts should also be made to ensure that the channel bed line is approximately matching with the existing terrain gradient. Therefore, the bed slope i should be considered comprehensively and then determined mainly in accordance with the topographical conditions and the design standard of channel.

6.2.3 Hydraulic calculation of uniform open-channel flow

To distinguish from non-uniform flow, the depth of uniform flow is called the uniform depth or the normal depth, which is denoted by h_0. The cross-sectional area, the hydraulic radius and the Chezy coefficient corresponding to the normal depth are indicated by A_0, R_0 and C_0, respectively. The calculation problems of uniform open-channel flow that are often met in practical engineering can be solved by the basic equation (6-7), which includes the discharge rate Q, the bed slope i, the roughness n and the cross-sectional geometric parameters A and R. For a trapezoidal section, the basic geometric parameters are the normal depth h_0, the base width b and the side slope m. According to Equation (6-7), the relationship between these parameters can be written as the following functional expression

$$Q = f(m, b, h_0, i, n)$$

where there are six variables in the expression, among them the side slope m is usually determined by the soil property or the lining material characteristic of the wall, and the roughness n is normally known in advance. Then, the hydraulic calculation is about how to solve the unknown variable in the basic equation for a set of known variables. The calculation can be undertaken by two types: check and design.

6.2.3.1 Check discharge capacity of a given channel

This is the case: find Q for a given geometry of channel. where b, h_0, m, i and n are known.

From the basic equations of uniform open-channel flow: Equations (6-6) and (6-7),

$$v_0 = C_0 \sqrt{R_0 i} = \frac{1}{n} R_0^{2/3} i^{1/2}$$

$$Q = A_0 v_0 = \frac{1}{n} A_0 R_0^{2/3} i^{1/2}$$

For trapezoidal cross-section, Equation (6-2) becomes

$$A_0 = (b + mh_0) h_0$$

$$\chi_0 = b + 2h_0 \sqrt{1 + m^2}$$

$$R_0 = \frac{A_0}{\chi_0}$$

Substituting A_0 and R_0 into Equation (6-7), we can obtain the flow rate Q.

6.2.3.2 Design a new channel

(1) Given Q, b, h_0, m and n, find i, i.e. design the bed slope of a channel.

Since

$$i = \frac{Q^2}{K_0^2}$$

where $K_0 = A_0 C_0 \sqrt{R_0}$

Based on the Chezy coefficient C_0 from Manning's formula, the bed slope i can be determined directly by the above expression.

(2) Design the size of a channel.

① Given n, i, m, Q, and h_0, find b.

② Given n, i, m, Q, and b, find h_0.

③ Given n, i, m, Q, and the width-depth ratio $\beta = \dfrac{b}{h_0}$, find b and h_0.

Mathematically, the closed equation for each case above certainly has one solution because only one variable is unknown in the six-variable equation. However, Q in Equation (6-7) is an implicit function of b and h_0, so there is no direct solution for both b and h_0, which can be obtained by trial-and-error method.

The trial-on-error method is described as follows: firstly assume some values (generally more than three) of b or h, computing $K = AC\sqrt{R}$ explicitly, then plot the $b-K$ or $h-K$ curve as shown in Fig. 6-7, meanwhile calculating $K_0 = \dfrac{Q}{\sqrt{i}}$ from the given Q and i, and finally find the corresponding b or h to the K_0 value from the $b-K$ or $h-K$ curve, which is the required solution.

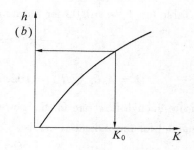

Fig. 6-7 The curve for $K=f(h)$ or $K=f(b)$

This trial-and-error method can apply to all the uniform flows in various cross-sections, so it is a general method.

④ Determine b and h_0 for permissible velocity. In practical engineering, there may be specific requirements for the flow velocity in a channel because too small a velocity results in deposition of sediment or grass growth whilst too large a velocity results in scour of channel bed or wall. Therefore, such a velocity is usually called permissible speed $[v]$. In calculation

$$A_0 = \frac{Q}{[v]}, \quad R_0 = \left(\frac{n[v]}{i^{1/2}}\right)^{3/2}$$

Additionally from the geometrical relationship

$$A_0 = (b + mh_0)h_0, \quad R_0 = \frac{A_0}{\chi_0} = \frac{(b + mh_0)h_0}{b + 2h_0\sqrt{1 + m^2}}$$

Substituting A_0 and R_0 into the above equation and solving them simultaneously, we can obtain the solution of b and h_0.

【Example 6-1】 A trapezoidal water channel in a hydropower station was excavated in clay earth without lining. In the service some grass has grown on both side walls of the channel. The measured data are as follows: the side slope of cross-section is $m = 1.5$, the base breadth $b = 34$ m, the bed slope $i = 1/6500$, and the height from the channel bed to the levee top is 3.2 m (Fig. 6-8). The flow rate required by the power station is 67 m³/s. Now the channel is needed to supply additional water for the local industry. If the free board of the levee is required to be at least 0.5 m, find the maximum water flow rate that can be extracted from the channel for the industrial use?

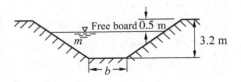

Fig. 6-8

Solution: Considered as a uniform open-channel flow, whilst the free board is 0.5 m, the normal depth will be $h_0 = 3.2 - 0.5 = 2.7$ m. Thus the other cross-sectional parameters are

$$A_0 = (b + mh_0)h_0 = (34 + 1.5 \times 2.7) \times 2.7 = 102.74 \text{ m}^2$$

$$\chi_0 = b + 2h_0\sqrt{1 + m^2} = 34 + 2 \times 2.7 \times \sqrt{1 + 1.5^2} = 43.74 \text{ m}$$

$$R_0 = \frac{A_0}{\chi_0} = \frac{102.74}{43.74} = 2.35 \text{ m}$$

From Table 6-2, $n = 0.03$ for unlining earth channel. Hence the Chezy coefficient

$$C_0 = \frac{1}{n}R_0^{1/6} = \frac{1}{0.03} \times 2.35^{1/6} = 38.4 \text{ m}^{0.5}/\text{s}$$

Then $\quad Q = A_0 C_0 \sqrt{R_0 i} = 102.74 \times 38.4 \times \sqrt{2.35 \times \frac{1}{6500}} = 75.0 \text{ m}^3/\text{s}$

To ensure enough flow rate for the power station, the maximum discharge that can be drawn from the channel for the industrial use is

$$Q = 75.0 - 67 = 8.0 \text{ m}^3/\text{s}$$

【Example 6-2】 In a sandy-soil region, a water channel is excavated with a trapezoidal cross-section lined with stone masonry. The channel side slope $m = 1$, the bed slope $i = 1/800$ and the base width $b = 6$ m. The discharge rate is designed to be 70 m³/s. Determine the levee height. (note: the free board is required to be 0.5 m.)

Solution: In the water channel the levee height is the sum of the flow depth and the free board. So, the question here is in fact to find the normal depth h_0. This is a problem of type ② in channel design. The trial-and-error method should be used, as shown in the following.

According to the surface feature of channel, $n = 0.03$ from Table 6-2. Assuming a number of h values and substituting them into the equation $K = AC\sqrt{R}$ consecutively to calculate the relevant K values, the computing results are listed in the table below:

h /m	A /m²	χ /m	R /m	C /m$^{0.5} \cdot$s^{-1}	K /m³·s^{-1}
2.5	21.25	13.07	1.63	4.5	1 205
3.0	27.00	14.48	1.87	45.5	1 677
3.5	33.25	15.90	2.09	46.5	2 223
4.0	40.00	17.30	2.31	47	2 854

Based on the above table, the $h - K$ curve is plotted as in Fig. 6-9.

From the known condition, the actual discharge conveyance is

$$K_0 = \frac{Q}{\sqrt{i}} = \frac{70}{\sqrt{\frac{1}{800}}} = 1980 \text{ m}^3/\text{s}$$

From the $h - K$ curve, the corresponding depth value to $K_0 (=1980 \text{ m}^3/\text{s})$ is $h_0 = 3.31$ m.

Thus the levee height is

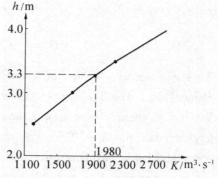

Fig. 6-9

$$h_0 + 0.5 = 3.31 + 0.5 = 3.81 \text{ m}$$

【Example 6-3】 In designing a drainage ditch with a trapezoidal cross-section, it is known that the flow rate $Q = 1.2$ m³/s, the depth $h_0 = 0.7$ m, the bed slope $i = 0.005$, the side slope $m = 1.5$, and the roughness $n = 0.03$. Find the base width b of the ditch.

Solution:

$$A = (b + mh_0)h_0 = 0.7b + 0.735 \text{ m}^2$$

$$\chi = b + 2h_0\sqrt{1 + m^2} = b + 2 \times 2.7 \times \sqrt{1 + 1.5^2} = b + 2.524 \text{ m}$$

$$R = \frac{A}{\chi} = \frac{0.7b + 0.735}{b + 2.524} \text{ m}$$

$$K = AC\sqrt{R} = A\frac{1}{n}R^{2/3} = \frac{1}{0.030}AR^{2/3} \text{ m}^3/\text{s}$$

Assume various b values and compute accordingly the values of A, χ, R and K, as shown in the table below:

h /m	A /m²	χ /m	R /m	$R^{2/3}$ /m$^{2/3}$	K /m³·s^{-1}
1	1.435	3.524	0.407	0.549	26.26
0.5	1.085	3.024	0.359	0.505	18.26
0.4	1.015	2.924	0.347	0.494	16.71
0.42	1.029	2.944	0.350	0.496	17.01

On the other hand, it is requested

$$K = \frac{Q}{\sqrt{i}} = \frac{1.2}{\sqrt{0.005}} = 16.97 \text{ m}^3/\text{s}$$

Hence from the above table

$$b = 0.42 \text{ m}$$

6.2.4 The optimum hydraulic cross-section

Since

$$Q = AC\sqrt{Ri}$$

Fluid Mechanics in Civil Engineering

Letting the Chezy coefficient $C = \dfrac{1}{n} R^{1/6}$ gives

$$Q = \frac{1}{n} A R^{2/3} i^{1/2} = \frac{1}{n} \frac{A^{5/3}}{\chi^{2/3}} i^{1/2}$$

The above equation shows the possible factors that influence channel capacity of uniform open-channel flow. The discharge Q just depends on the shape and size of flow cross-section if the bed slope i and channel roughness n are known respectively from topographic conditions and boundary material properties. Once i, n and A are fixed, Q is reversely proportional to the wetted perimeter χ and has a maximum value as χ is minimum. Put in another way, once i, n and Q are given, A is proportional to χ and will have a minimum as χ is minimum. The cross-section in accordance with this condition is termed the optimum hydraulic cross-section, where the channel capacity will be maximum; in other words, under the same flow condition the excavation would be minimum.

Of various shapes, a circle or a half circle would serve as the optimum hydraulic section because it has the smallest perimeter for a given area. However, in practice for convenience of excavation, it is hard to use such shapes, so trapezoidal or rectangular shapes are often adopted instead. The following discussions will focus on the trapezoidal optimum sections when the side slope m is given.

With a certain m, the trapezoidal section depends on the aspect ratio (width-to-depth ratio) b/h. For a constant flow area A, different aspect ratios have different wetted perimeters and accordingly different flow capacities. From the geometric relationships of trapezoidal cross-section

$$A = (b + mh)h$$
$$\chi = b + 2h \sqrt{1 + m^2}$$

Thus

$$\chi = \frac{A}{h} - mh + 2h \sqrt{1 + m^2} = f(h)$$

Since the optimum hydraulic section has minimum χ value, differentiate χ with respect to h and set the derivative equal to zero

$$\frac{d\chi}{dh} = -\frac{A}{h^2} - m + 2 \sqrt{1 + m^2} = 0$$

Substitute $A = (b + mh)h$ into the above equation and find the width-to-depth ratio for the optimum hydraulic section

$$-\frac{(b + mh)h}{h^2} - m + 2 \sqrt{1 + m^2} = 0$$

$$\frac{b}{h} = 2(\sqrt{1 + m^2} - m) \tag{6-8}$$

If $m = 0$, this will give the width-to-depth ratio for the rectangular optimum hydraulic section

$$\frac{b}{h} = 2$$

which shows that for a rectangular channel the optimum hydraulic section occurs only when the flow depth is one-half of the channel width.

It is worth noting that the optimum hydraulic section is usually narrow and deep, e. g. $b/h = 0.472$ when $m = 2$. Therefore, in practical applications such a cross-section has limitations in use because the soil property, construction or other conditions make such a channel uneconomic. For this reason, the "optimum hydraulic" does not mean the "technologically and economical best". To small-sized channels whose engineering costs are basically dependent on the amount of earthwork and the lining, the optimum hydraulic channel has nearly the best technically economical cross-section, whilst to large-scale channels the rational and economical sections should be determined based on the comprehensive comparison of various factors such as engineering quantity, construction technique and operation management.

6.3 Steady non-uniform open-channel flow

Steady uniform flow is the simplest flow in open channel. In practice, open-channel flows are affected by many factors, such as construction of sluices, dams, bridges and so on or the variation in bed slope, cross-section and channel roughness. Strictly speaking, uniform flow is rarely achieved in practice. Most flows in man-made channels and natural rivers are non-uniform flows.

For non-uniform flow in an open channel, flow parameters such as flow depth, mean cross-sectional velocity, pressures and so on, will change with the distance. The bed line, the free surface line and the total energy line are not parallel to each other, as shown in Fig. 6 – 10. Non-uniform flows in the open channel are classified as either the gradually varied flow or the rapidly varied flow.

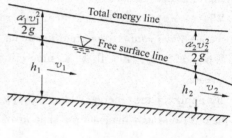

Fig. 6 – 10

6.3.1 Flow regime of open-channel flow

Owing to the existence of free surface, open-channel flows are distinct from pressurized flows. Generally, open-channel flows have three types of flow regime: subcritical, critical and supercritical flow.

To understand the three types of regime, one can observe a simple experiment. If one throws a piece of stone into a large still water, some small wave propagation will occur on the water surface. Centered with the dropping point, the waves spread in all directions at a certain speed v_w, with shape of the waves on the plane like a series of concentric circles, as shown in Fig. 6 – 11a. The propagating speed of such a small wave in still water v_w is called the relative wave speed or wave celerity. If the stone is thrown into a uniform flow, the propagating speed of the wave is the sum of the flow velocity and the wave celerity. When the mean cross-sectional velocity of flow v is smaller than the wave celerity v_w, the small wave will travel upstream at an absolute velocity $(v - v_w)$ and simultaneously downstream at $(v + v_w)$, as shown in Fig. 6-11b, and such a flow is called subcritical flow. When the velocity v is equal to the celerity v_w, the absolute velocity of propagation upstream is zero (Fig. 6-11c), such a flow is termed as critical flow. When the velocity v is larger than the celerity v_w, the wave will travel only downstream at the absolute velocity $(v + v_w)$ without any effect on the upstream flow (Fig. 6-11d), and such a flow is called supercritical flow.

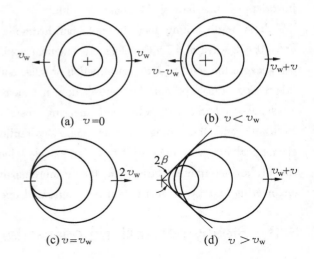

Fig. 6-11

It shows that by comparing the sectional mean velocity v and the small wave celerity v_w, one can estimate whether the disturbed small wave travels upstream or not and thus identify which type of regime the flow belongs to.

When $v < v_w$, the flow is subcritical, where any disturbance wave can propagate both upstream and downstream.

When $v = v_w$, the flow is critical, where despite of spreading downstream at speed $2v_w$, the disturbance wave cannot propagate upstream but forms a stagnation point.

When $v > v_w$, the flow is supercritical, where the disturbance wave propagates only downstream.

By applying the energy and continuity equations to cross-sections 1 and 2, as shown in Fig. 6-12, one can demonstrate that in open channels for any shape of flow cross-section the wave celerity

$$v_w = \sqrt{g\bar{h}} \qquad (6-9)$$

where $\bar{h} = \dfrac{A}{B}$ is the average flow depth of the cross-section, which is virtually the equivalent depth of a rectangular with the width equal to the actual surface width. Herein A is the cross-section area and B is the surface width.

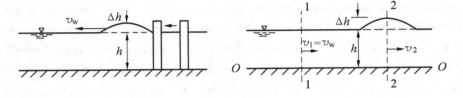

Fig. 6-12

In one-dimensional flow, the absolute velocity of the wave propagation v'_w is the algebraic addition of v and v_w, that is

$$v'_w = v \pm v_w = v \pm \sqrt{g\bar{h}} \qquad (6-10)$$

where the plus sign refers to the celerity along the flow while the minus against the flow.

For critical flow, the mean cross-sectional velocity is equal to the wave celerity

$$v = v_w = \sqrt{g\bar{h}} \qquad (6-11)$$

or written as

$$\frac{v}{\sqrt{g\bar{h}}} = \frac{v_w}{\sqrt{g\bar{h}}} = \sqrt{\frac{2\frac{v^2}{2g}}{\bar{h}}} = 1$$

which demonstrates that the average unit potential energy is twice the average unit kinetic energy if flow is critical.

The quantity of $\dfrac{v}{\sqrt{g\bar{h}}}$ is a dimensionless number, which is known as the Froude number, named after a British scientist, William Froude, and denoted by Fr. It is noted that the square of Froude number is twice the ratio of the average unit kinetic energy to the average unit potential energy. The smaller the depth, the larger the kinetic energy and Fr; or the larger the depth, the smaller the kinetic energy and Fr. For critical flow, $Fr = 1$. Hence, the Froude number can be used to identify the regime of open-channel flows:

$Fr < 1$, subcritical flow;
$Fr = 1$, critical flow;
$Fr > 1$, supercritical flow.

The Froude number is a very important parameter in the analysis of open-channel flow. By dimension analysis, F_I denotes the ratio of inertial force to the gravity G, i. e.

$$\frac{[F_I]}{[G]} = \frac{[ma]}{[mg]} = \frac{\left[\rho L^3 \dfrac{L}{T^2}\right]}{[\rho L^3 g]} = \left[\frac{\dfrac{L^2}{T^2}}{gL}\right] = \left[\frac{v^2}{gL}\right]$$

$$\frac{[F_I]^{1/2}}{[G]} = \left[\frac{v}{\sqrt{gL}}\right]$$

Thus the dimensional expression of square root of the ratio of inertial force to gravity is identical to that of the Froude number, which shows that the physical meaning of the Froude number represents the comparative relationship between the flow inertia and the gravity. When $Fr = 1$ (critical flow), the inertia force is equal to the gravity. When $Fr > 1$ (supercritical flow), the inertia force is greater than the gravity and is dominant in the flow. When $Fr < 1$ (subcritical flow), the inertia action is smaller than the gravity.

6.3.2 Specific energy

Fig. 6-13 shows the cross-section of a subcritical, non-uniform, open-channel flow. The total mechanical energy in unit weight of liquid with reference to the datum $O'-O'$ is

$$E = z_0 + h + \frac{\alpha v^2}{2g}$$

If the datum is set especially through the lowest point on the channel bed, then the total mechanical energy of unit weight of liquid with reference to the new datum $O'-O'$ is

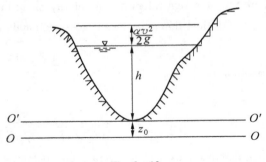

Fig. 6-13

$$E_s = h + \frac{\alpha v^2}{2g} \quad (6-12)$$

where E_s is termed as specific energy or cross-sectional unit energy.

Being different from the total mechanical energy E, which is defined with reference to the same datum plane for all cross-sections and will always decline along the distance due to energy consumption, the specific energy E_s is defined with reference to the datum through the lowest point of each cross-section, so its value may either increase or decrease along the flow. Only in uniform open-channel flow where cross-sectional h and v are constant does E_s remain unchanged with distance, i.e. $\frac{dE_s}{dl} = 0$.

Equation (6-12) can also be written as

$$E_s = h + \frac{\alpha Q^2}{2gA^2} \quad (6-13)$$

Seen from Equation (6-13), when the flow rate Q and the cross-sectional shape and size are constant, E_s is just a function of depth, i.e. $E_s = f(h)$, with which one can plot an $h - E_s$ relationship, termed as the specific energy curve (Fig. 6-14). Different flow rates, cross-sectional shapes or sizes will result in different curves.

Analyzing Equation (6-13), one can see that the specific energy curve is a parabola under the given flow rate and the certain cross-sectional shape and size. The lower curve has an asymptote of the horizontal line whilst the

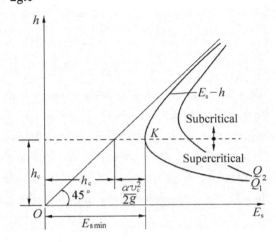

Fig. 6-14

upper curve has an asymptote of the straight line with a slope of 45° to the horizontal axis and passing through the origin. The curve has the minimum value E_{smin} at K that divides the curve into two parts: on the upper part E_s increases with increasing h; on the lower part E_s increases with decreasing h.

Differentiating Equation (6–13) with respect to h gives

$$\frac{dE_s}{dh} = \frac{d}{dh}(h + \frac{\alpha Q^2}{2gA^2}) = 1 - \frac{\alpha Q^2}{gA^3}\frac{dA}{dh} = 1 - \frac{\alpha Q^2 B}{gA^3} = 1 - \frac{\alpha v^2}{g\frac{A}{B}} = 1 - \frac{\alpha v^2}{g\bar{h}}$$

When letting $\alpha = 1.0$, the above equation can be written as

$$\frac{dE_s}{dh} = 1 - Fr^2 \qquad (6-14)$$

Equation (6–14) demonstrates that in open-channel flow the variation of specific energy with depth depends on the cross-sectional Froude number. For subcritical flow $Fr < 1$, thus $\dfrac{dE_s}{dh} > 0$ corresponding to the upper part of the curve, for supercritical flow $Fr > 1$, $\dfrac{dE_s}{dh} < 0$ corresponding to the lower part of the curve, and for critical flow $Fr = 1$, $\dfrac{dE_s}{dh} = 0$ corresponding to the turning (transition) point between the upper part and lower part of the curve, which has the minimum value E_{smin}.

6.3.3 Critical depth

In condition of constant cross-sectional shape and size and flow rate, the depth that has the minimum specific energy E_{smin} is termed as the critical depth, denoted by h_c and shown as in Fig. 6–14. Apparently, h_c should satisfy the following equation:

$$\frac{dE_s}{dh} = 1 - \frac{\alpha Q^2 B}{gA^3} = 0 \qquad (6-15)$$

Usually, the hydraulic quantities related to the critical flow are denoted by a subscript c, so Equation (6–15) can be rewritten as

$$\frac{\alpha Q^2}{g} = \frac{A_c^3}{B_c} \qquad (6-16)$$

Equation (6–15) or (6–16) is the equation of critical flow. When the flow rate and the cross-sectional shape and size are given, the critical depth h_c can be obtained from Equation (6–16).

Generally, the right side of Equation (6–16) is a high-order implicit function about h_c so that directly solving h_c is difficult. Hence, for an arbitrary shape of cross-section the computation of h_c needs to use a trial-and-error method: assume a value of h to calculate the correponding value of $\dfrac{A^3}{B}$, if it is equal to the given $\dfrac{\alpha Q^2}{g}$, then the previously assumed h is the required h_c; otherwise choose another h value to repeat the above process of calculation until the newly

Fluid Mechanics in Civil Engineering

obtained $\frac{A^3}{B}$ is equal to the given $\frac{\alpha Q^2}{g}$. If we are unable to get a satisfactory result after several times of trial-and-error, we can plot an $h \sim \frac{A^3}{B}$ curve, shown in Fig. 6–15, from which we can find the required h_c, which is the vertical coordinate value corresponding with the horizontal coordinate $\frac{A^3}{B} = \frac{\alpha Q^2}{g}$.

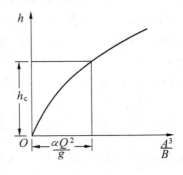

Fig. 6–15

If the channel cross-section is rectangular, Equation (6–16) becomes

$$h_c = \sqrt[3]{\frac{\alpha Q^2}{gb^2}} = \sqrt[3]{\frac{\alpha q^2}{g}} \qquad (6-17)$$

where b is the width of the rectangular channel; q is the flow rate per unit width.

As seen in Equation (6–16), h_c depends on only the flow rate and the cross-sectional shape and size, but it has no relationship with the bed slope i.

The definition of h_c shows that it is the flow depth when the flow is critical. Equation (6–11) also shows that the mean cross-sectional velocity is the same as the critical velocity, given by

$$v_c = \sqrt{g\frac{A}{B}} = \sqrt{g\bar{h}} \qquad (6-18)$$

Through comparing the actual depth h in a channel with the critical depth h_c, or equally comparing the actual velocity v in the channel with the critical velocity v_c, one can also identify different flow regimes:

$h > h_c$, $v < v_c$, $Fr < 1$, subcritical;
$h = h_c$, $v = v_c$, $Fr = 1$, critical;
$h < h_c$, $v > v_c$, $Fr > 1$, supercritical.

6.3.4 Critical bed slope

From the basic equation of uniform open-channel flow $Q = AC\sqrt{Ri}$, one can conclude that the normal depth h_0 depends on the flow rate, cross-sectional shape and size, bed slope and channel roughness. When the flow rate, the cross-sectional section shape and size, and the channel roughness are constant, the normal depth h_0 will change accordingly with the bed slope. Then one can use the basic equation to calculate h_0 for different bed slopes i and plot an $h_0 - i$ curve, shown as Fig. 6–16, which shows that h_0 increases as i decreases and vice versa. If i comes to a

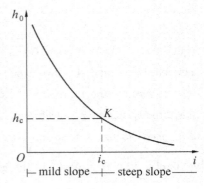

Fig. 6–16

value at which the corresponding h_0 is exactly equal to the critical depth h_c, then the slope of such a value is defined as critical slope i_c.

The uniform flow on the critical slope, on one hand, must satisfy the equation for the critical flow:

$$\frac{\alpha Q^2}{g} = \frac{A_c^3}{B_c}$$

and on the other hand it must satisfy the equation for uniform flow at the same time:

$$Q = A_c C_c \sqrt{R_c i_c}$$

Combining the above two equations simultaneously gives the formula for calculating the critical bed slope

$$i_c = \frac{gA_c}{\alpha C_c^2 R_c B_c} = \frac{g\chi_c}{\alpha C_c^2 B_c} \qquad (6-19)$$

where C_c, R_c, χ_c and B_c are the Chezy coefficient, the hydraulic radius, the wetted perimeter and the surface width corresponding to the critical depth h_c, respectively. For a cross-section with constant roughness and unchanged shape and size, the critical depth varies with the flow rate, and thus the subsequent Chezy coefficient, hydraulic radius, wetted perimeter and surface width are different. Therefore, the critical slope depends on just the flow rate, roughness, section shape and size but not the actual channel bed slope.

Between the actual slope and the critical slope exist three relative relations: $i < i_c$, where the slope is termed as the mild slope; $i > i_c$, the steep slope; $i = i_c$, the critical slope. Also seen from Fig. 6-16, for uniform flow, it must be subcritical flow on the mild slope as $h_0 > h_c$; the flow is supercritical on the steep slope as $h_0 < h_c$; the flow is critical on the critical slope as $h_0 = h_c$. It should be noted that the critical flow is unstable because the flow would quickly deviate from the critical state as long as any of the upstream flow, the roughness and the cross-sectional shape and size varies slightly. Therefore, if the longitudinal gradient of channel is designed to be close to the critical slope, it is difficult in practice to ensure the flow to remain at the expected regime. For this reason, in engineering, the design slope is generally made far from the critical slope, usually having about twice i_c, to ensure the actual flow to be in designed flow regime. Nevertheless, as for the non-uniform flow, unlike the uniform flow, the supercritical flow may happen on the mild slope or the subcritical flow may happen on the steep slope, and the situation may be uncertain.

It is worth noting that the concept of critical, mild and steep slopes is with regard to a certain flow rate passing through a given channel, i. e. the slope types are not unchangeable but can be transformed from one to another with varying flow rates. For instance, the channel slope is mild when the flow is small, but it may become hydraulically steep when the discharge increases greatly and results in decreased i_c. Note that the concepts of mild slopes and steep slopes will be further discussed in the analysis of water surface profiles in non-uniform open-channel flows.

To summarize, the regime of open-channel flow can be determined by the critical velocity v_c, the Froude number Fr, or the critical depth h_c, no matter whether the flow is uniform or non-uniform. Especially to uniform open-channel flow, the critical bed slope i_c can also be used to identify the flow regimes.

【Example 6 – 4】 A channel has the trapezoidal cross-section with the base width $b = 5$ m and the side slope $m = 1.0$. The flow rate is $Q = 8$ m^3/s. Find the critical depth h_c.

Solution: Based on the given section, assume different h values and calculate the corresponding $\dfrac{A^3}{B}$ in a tabular form, and then plot the curve of $h \sim \dfrac{A^3}{B}$. Then, with the given flow condition and the assumption of $\alpha = 1.0$,

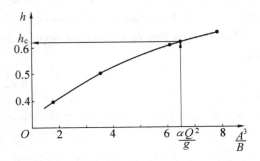

Fig. 6 – 17

$$\frac{\alpha Q^2}{g} = \frac{1.0 \times 8^2}{9.8} = 6.53 \text{ m}^5$$

The calculation table for $h \sim \dfrac{A^3}{B}$

h/m	B/m	A/m^2	$(A^3/B)/\text{m}^5$
0.4	5.8	2.16	1.74
0.5	6.0	2.75	3.47
0.6	6.2	3.36	6.12
0.65	6.3	3.67	7.86

From the above curve, find the depth corresponding to the value of $\dfrac{\alpha Q^2}{g}$, which is the requested critical depth.

$$h_c = 0.61 \text{ m}$$

【Example 6 – 5】 A long straight channel has the rectangular cross-section with the base width $b = 1$ m and the bed roughness $n = 0.014$, and the bed slope is $i = 0.0004$. Under a certain flow rate the normal depth of the channel is $h_0 = 0.6$ m. Estimate the flow regime in the channel.

Solution:

(1) By the critical velocity

According to the given condition, we can obtain the cross-sectional hydraulic radius

$$R = \frac{A}{\chi} = \frac{bh_0}{b + 2h_0} = \frac{1 \times 0.6}{1 + 2 \times 0.6} = 0.273 \text{ m}$$

the Chezy coefficient

$$C_0 = \frac{1}{n} R_0^{1/6} = \frac{1}{0.014} \times 0.273^{1/6} = 57.5 \text{ m}^{0.5}/\text{s}$$

the mean cross-sectional velocity

$$v = C\sqrt{Ri} = 57.5 \times \sqrt{0.273 \times 0.0004} = 0.6 \text{ m/s}$$

and the critical velocity

$$v_c = \sqrt{gh_0} = \sqrt{9.8 \times 0.6} = 2.42 \text{ m/s}$$

Since $v < v_c$, the flow is subcritical.

(2) By the Froude number

Since $Fr = \dfrac{v}{\sqrt{gh_0}} = \dfrac{0.6}{\sqrt{9.8 \times 0.6}} = 0.248 < 1$

The flow is subcritical.

(3) By the critical depth

The critical depth $h_c = \sqrt[3]{\dfrac{\alpha q^2}{g}}$

where the flow rate per unit width $q = vh_0 = 0.6 \times 0.6 = 0.36$ m³/s · m

Substituting it into the equation of h_c gives

$$h_c = \sqrt[3]{\dfrac{1.0 \times 0.36^2}{9.8}} = 0.24 \text{ m}$$

Since the actual depth $h_0 > h_c$, the flow is subcritical.

(4) By the critical bed slope

Since the critical depth $h_c = 0.24$ m, the corresponding quantities of cross-section can be calculated.

$$B_c = b = 1 \text{ m}$$
$$\chi_c = b + 2h_c = 1.48 \text{ m}$$
$$R_c = \dfrac{A_c}{\chi_c} = \dfrac{bh_c}{\chi_c} = 0.16 \text{ m}$$
$$C_c = \dfrac{1}{n} R_c^{1/6} = 52.7 \text{ m}^{0.5}/\text{s}$$

From Equation (6-19), the critical bed slope is

$$i_c = \dfrac{g\chi_c}{\alpha C_c^2 B_c} = \dfrac{9.8 \times 1.48}{1.0 \times 52.7^2 \times 1} = 0.0052$$

Because the actual slope $i = 0.0004$, $i < i_c$, the slope is mild and thus the uniform flow on it would be subcritical.

6.3.5 Hydraulic jump and hydraulic drop

The hydraulic jump and the hydraulic drop are two kinds of local hydraulic phenomena that happen when the flow regimes change (Fig. 6-18). They belong to rapidly varied flows (RVF).

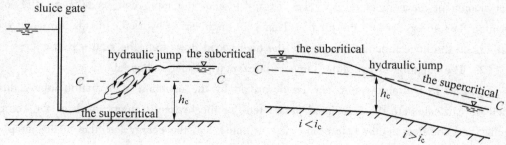

Fig. 6-18

6.3.5.1 Hydraulic jump -Transition from supercritical to subcritical flow

(1) Introduction

When the flow transits from the supercritical to the subcritical flow in an open channel, the hydraulic jump will occur, i. e. in a short distance the flow depth increases rapidly from a depth less than the critical depth to that above the critical depth, with the free surface jumping up abruptly. Generally, such flow jumps often occur in the downstream channel of hydraulic structures such as sluice gates, dams or steep chute, where the coming flows are often forced to be supercritical with small depths and high speeds.

The hydraulic jump zone is sketched in Fig. 6–19, the upper region, which is called "the surface roller", is a water rolling upright violently and whirling with a lot of mixing air bubbles. The lower region under the roller is the main stream with rapidly expanding. The geometric parameters to determine the shape of a typical hydraulic jump are:

The depth of pre-jump h' is the flow depth at cross-section 1 – 1, which is located just before the starting point of the surface roller;

The depth of post-jump h'' is the flow depth at cross-section 2 – 2, which is at the ending point of the surface rolling;

The height of the jump a is $a = h'' - h'$;

The length of the jump l_j is the distance between sections before and after the jump, i. e. cross-sections 1 – 1 and 2 – 2.

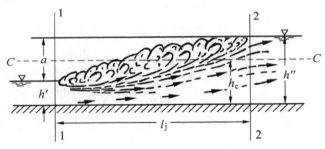

Fig. 6 – 19

Because there is a large number of aeration and rotation in the surface roller where the turbulence and mixing inside the flow are extremely intense, in addition, the velocity distributions of the main stream are being changed constantly; the hydraulic jump could locally dissipate a great amount of mechanical energy. The energy loss in the jump can be 60% ~ 70% of the incoming flow energy (before the jump). Hence in engineering hydraulic jumps are usually used to eliminate the huge kinematic energy of the high speed flows from the spillway structures.

(2) The equation of hydraulic jump in horizontal prismatic channel

The flow in hydraulic jump can be described by the equation of hydraulic jump, through which we can calculate the associated parameters for the hydraulic jump, such as the conjugate depths (i. e. the flow depths before and after a jump) and the energy loss due to the jump, and make an estimate of the jump length.

Because the hydraulic jump is rapidly varied flow, the mechanical energy losses in the jump are so large that they cannot be negligible in the computation. Because of the unknown head losses in hydraulic jumps, the energy equation is not suitable to be applied. Therefore, the momentum equation is appropriate to be applied to the jump section as it does not need the energy losses.

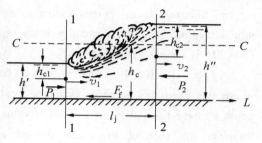

Fig. 6-20

Assuming in a horizontal prismatic channel ($i=0$) a hydraulic jump happens under a flow rate of Q (Fig. 6-20), with the depth h' and mean velocity v_1 before the jump, as well as the depth h'' and mean velocity v_2 after the jump, write the momentum equation for the flow in the jump zone between the pre-jump cross-section 1-1 and the post-jump cross-section 2-2 as

$$P_1 - P_2 - F_f = \rho Q(\beta_2 v_2 - \beta_1 v_1)$$

where P_1 and P_2 denote the hydrodynamic pressure forces on the two end cross-sections, and F_f is the total frictional force of the flow on the solid boundaries.

For simplicity of practical application, three assumptions are made herein on basis of physical features of the hydraulic jump:

① The boundary frictional force are so small for short distance that it is negligible;

② Two sections before and after the jump are of GVF, thus on which the dynamic pressure distribution are hydrostatic;

③ The momentum correction factors for cross-sections 1-1 and 2-2 are unity.

According to the above assumptions, $F_f = 0$, $P_1 = \gamma h_{c1} A_1$, $P_2 = \gamma h_{c2} A_2$, $\beta_1 = \beta_2 = 1.0$, where h_{c1} and h_{c2} denote the depths at the centroid of two cross-sections, A_1 and A_2 are the cross-sectional areas, respectively. Substituting into the momentum equation gives

$$\gamma h_{c1} A_1 - \gamma h_{c2} A_2 = \rho Q \left(\frac{Q}{A_2} - \frac{Q}{A_1} \right)$$

$$\frac{Q^2}{gA_1} + h_{c1} A_1 = \frac{Q^2}{gA_2} + h_{c2} A_2 \qquad (6-20)$$

This is the basic equation for the hydraulic jump in a horizontal prismatic channel. The cross-sectional area and the centroid depth are both the function of the water depth, and the rest quantities are known or constant, so it can also be written as

$$\frac{Q^2}{gA} + h_c A = J(h) \qquad (6-21)$$

where $J(h)$ is termed as the hydraulic jump function, whose function curve is graphically shown in Fig. 6-21.

Mathematically, it can be proved that the depth corresponding to the minimum of the function is also the critical depth h_c for the given discharge in a channel, i.e. $J(h_c) = J(h)_{\min}$. When $h > h_c$, $J(h)$ increases with the depth and when $h < h_c$, $J(h)$ decreases with the depth. So there are two different depths corresponding to an jump function value, or, the hydraulic jump equation (6-20) is simply written as

$$J(h') = J(h'') \qquad (6-22)$$

where the pre-jump depth h' and the post-jump depth h'' have the same jump function value, and therefore the pair of them are termed as the conjugate depths. Fig. 6-21 shows that the smaller the pre-jump depth, the larger the corresponding post-jump depth, and vice versa.

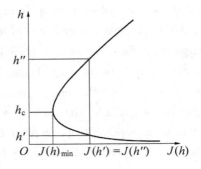

Fig. 6-21

The above derived hydraulic jump equations (6-20) and (6-22) can approximately be applied to the prismatic channels of small bed slopes ($i < 0.05$).

(3) Calculation of hydraulic jump

① Calculation of conjugate depths

The calculation of the conjugate depths is the basis of other jump parameter computations. If one of the conjugate depths is known (h' or h''), the other depth can be obtained by means of applying the jump equation.

For the non-rectangular cross-sections, the hydraulic jump function is complex, so the conjugate depths cannot easily be solved directly. In this case a trial-and-error method or the graphical method can be applied in the solving process.

For the rectangular cross-section, $A = bh$ and the depth at the centroid $h_c = \dfrac{h}{2}$, in addition, $q = \dfrac{Q}{b}$. Substituting into Equation (6-20) and eliminating b gives

$$\frac{q^2}{gh'} + \frac{h'^2}{2} = \frac{q^2}{gh''} + \frac{h''^2}{2}$$

After rearrangement, it gives a quadratic equation

$$h'h''(h' + h'') = \frac{2q^2}{g} \qquad (6-23)$$

Assuming either the pre-jump depth h' or the post-jump depth h'' as the unknown quantity and solving the equation gives

$$h' = \frac{h''}{2}\left(-1 + \sqrt{1 + \frac{8q^2}{gh''^3}}\right) \qquad (6-24)$$

$$h'' = \frac{h'}{2}\left(-1 + \sqrt{1 + \frac{8q^2}{gh'^3}}\right) \qquad (6-25)$$

In the above expressions, since

$$\frac{q^2}{gh''^3} = \frac{v_2^2}{gh''} = Fr_2^2$$

$$\frac{q^2}{gh'^3} = \frac{v_1^2}{gh'} = Fr_1^2$$

The expressions for the conjugate depths can be also written as

$$h' = \frac{h''}{2}(-1 + \sqrt{1 + 8Fr_2^2}) \quad (6-26)$$

$$h'' = \frac{h'}{2}(-1 + \sqrt{1 + 8Fr_1^2}) \quad (6-27)$$

where Fr_1 and Fr_2 are the Froude numbers of the pre-jump section and the post-jump section, respectively.

② Calculation of jump length

In the jump zone of a complete hydraulic jump, the flow is often highly turbulent with large velocity near the bed. Hence, the channel bed needs to be protected from scour damages except solid rock bed. Since the jump length determines the range of the bed that needs protection and reinforcement, its determination is thus of great practical significance. However, the motion of flow in hydraulic jump is so complicated that the relevant theories have not yet been established and its study still relies on practices. In engineering, the empirical expressionsare are frequently used for the calculation of jump length.

The following are some examples of such empirical expressions for jump length in the rectangular channels with horizontal beds:

a. expressed by the post-jump depth

$$l_j = 6.1 h''$$

b. expressed by the jump height

$$l_j = 6.9 (h'' - h')$$

c. expressed by the Froude number and the pre-jump depth

$$l_j = 9.4 (Fr_1 - 1) h'$$

③ Calculation of the head loss of jump

When Fr_1 is very large, the difference of mechanical energy per unit weight of fluid between the sections of pre-jump and post-jump, i.e. the energy eliminated by the jump or the energy dissipation in the jump, ΔE_j, can be used to express the total head loss of the jump. For a horizontal rectangular channel

$$\Delta E_j = E_1 - E_2 = \left(h' + \frac{\alpha_1 v_1^2}{2g}\right) - \left(h'' + \frac{\alpha_2 v_2^2}{2g}\right) \quad (6-28)$$

where approximately $\alpha_1 = \alpha_2 = 1.0$. From Equation (6-23)

$$h'h''(h' + h'') = \frac{2q^2}{g}$$

then

$$\frac{\alpha_1 v_1^2}{2g} = \frac{q^2}{2gh'^2} = \frac{1}{4}\frac{h'}{h''}(h' + h'')$$

$$\frac{\alpha_2 v_2^2}{2g} = \frac{q^2}{2gh''^2} = \frac{1}{4}\frac{h'}{h''}(h' + h'')$$

Substituting into Equation (6-28) and simplifying the resulting equation gives

$$\Delta E_j = \frac{(h'' - h')^3}{4h'h''} \quad (6-29)$$

Equation (6-29) shows that for the given flow rate, the larger the difference between conjugate depths, the greater the energy dissipated by the jump.

【Example 6-6】 A flow through a hydraulic structure has the flow rate per unit width of $q = 15$ m^3/s·m. A hydraulic jump occurs in the downstream channel with the pre-jump depth $h' = 0.8$ m. Find (1) the post-jump depth h''; (2) the jump length l_j; (3) the energy dissipation rate of the jump $\Delta E_j / E_1$.

Solution:

(1) Calculating the Froude number,

$$Fr_1^2 = \frac{q^2}{gh'^3} = \frac{15^2}{9.8 \times 0.8^3} = 44.84$$

$$h'' = \frac{h'}{2}(-1 + \sqrt{1 + 8Fr_1^2}) = \frac{0.8}{2} \times (-1 + \sqrt{1 + 8 \times 44.84^2}) = 7.19 \text{ m}$$

(2) Calculating by $l_j = 6.1 h''$ gives

$$l_j = 6.1 \times 7.19 = 43.84 \text{ m}$$

Calculating by $l_j = 6.9(h'' - h')$ gives

$$l_j = 6.9 \times (7.19 - 0.8) = 44.09 \text{ m}$$

Calculating by $l_j = 9.4(Fr_1 - 1)h'$ gives

$$l_j = 9.4 \times (6.696 - 1) \times 0.8 = 42.84 \text{ m}$$

(3)
$$\Delta E_j = \frac{(h'' - h')^3}{4 h' h''} = \frac{(7.19 - 0.8)^3}{4 \times 0.8 \times 7.19} = 11.34 \text{ m}$$

$$\frac{\Delta E_j}{E_1} = \frac{\Delta E_j}{h' + \frac{q^2}{2gh'^2}} = \frac{11.34}{0.8 + \frac{15^2}{2 \times 9.8 \times 0.8^2}} = 0.61$$

6.3.5.2 Hydraulic drop — the transition flow from subcritical to supercritical

In the hydraulic drop, the flow depth changes from a value higher than the critical depth to a value less than the critical depth, with the local free surface dropping down rapidly. This kind of phenomenon is often seen where the bed slope of channel suddenly changes from a mild to a steep gradient (Fig. 6-22a), where there is a step drop at the end of a mild slope channel (Fig. 6-22b), or where the flow goes into a steep channel from a reservoir (Fig. 6-22c).

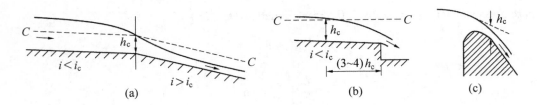

Fig. 6-22

Why do hydraulic drops have to occur in such flow conditions? To illustrate this, consider the following case as an example: flow in a mild slope channel with an end drop sill (Fig. 6-23).

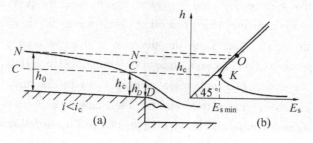

Fig. 6 – 23

Supposing that the bed slope of the channel is extended downstream with no change, there would be a subcritical uniform flow formed in the channel, with a normal depth h_0 and a free surface line $N-N$ parallel to the channel bed. However, if the channel has a sudden step drop at the end cross-section D, the flow in the downstream suddenly loses the channel bed resistance, which will dramatically change the force of the flow on the step where the gravitational force takes a dominant role. The flow attempts to convert the potential energy to kinetic energy, thus resulting in the free surface dropping down as much as possible. The flow at the step consequently becomes a non-uniform accelerated motion, in fact a case of rapidly varied non-uniform flow.

In subcritical flow, the specific energy decreases with the decreasing depth, as the upper curve shown in Fig. 6 – 23b. When the free surface drops at the step, the specific energy gradually declines from point O to K along the upper curve. Under the action of gravity the lowest position that the energy can reach is point K. This is the critical situation for the specific energy development, because if the position gets lower than point K, the flow will become the supercritical flow, i.e. the specific energy increases again, which means that the flow needs to gain extra energy, which is impossible in reality. So, when the flow passes through a drop step, the depth at the end of channel will always be the critical depth h_c, or at least very close to it.

It should be noted that the above analysis is based on the GVF condition. In practice, near the drop step, the free surface falls sharply with the streamlines bending significantly and the flow is no longer gradually varied flow but RVF. The experiments indicate that the actual brink depth, h_D, at the end of the channel is slightly less than h_c, typically $h_D \approx 0.7 h_c$, and that the location of h_c occurs at a place slightly upstream $(3 \sim 4) h_c$ away from the end of channel, D. Generally, in analysis and calculction of surface profiles, the depth at a drop step can approximately be taken as the critical depth h_c and acts accordingly as the control depth.

To summarize, when the boundary condition in open channel has an abrupt change and the flow is changed from subcritical flow into supercritical flow, a hydraulic drop will take place. It is characterized by a transition from subcritical flow with the critical depth occurring at the brink section. This brink section, which has the critical depth, is commonly taken as the control section.

6.3.6 Surface profile of gradually varied flow in prismatic open channel

In gradually varied non-uniform flows the longitudinal free surface line is termed as the water surface profile. The free surface line with increasing depth along the flow is called a back-water curve, whilst the free surface line with decreasing depth along the flow is called a drop-down curve. Because the depth variation along the distance is directly related to many engineering problems, such as the size of flooding area, height of levee, erosion and deposition in a channel, the understanding of water surface profiles is a primary focus in the study of non-uniform open-channel flows. This usually includes two aspects: the qualitative characteristics of the surface profile including its shape, feature and developing trend (rise-up or fall-down); the quantitative variation of the surface profile through computation.

6.3.6.1 Differential equation

In a steady non-uniform open-channel flow, taking cross-sections 1 – 1 and 2 – 2 at a distance of dl where the change in the flow parameters between both sections is assumed to be infinitesimal due to GVF (Fig. 6 – 24), the energy equation is

$$(z + h) + \frac{\alpha v^2}{2g} = (z + dz + h + dh) + \frac{\alpha (v + dv)^2}{2g} + h_{w_{1-2}}$$

Expanding it and neglecting the term $(dv)^2$, the resulting equation is

$$dz + dh + d\left(\frac{\alpha v^2}{2g}\right) + h_{w_{1-2}} = 0$$

For GVF the local head losses can be negligible, so $h_{w_{1-2}} = h_{f_{1-2}}$. Dividing the equation by dl gives

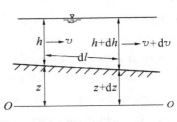

Fig. 6 – 24

$$\frac{dz}{dl} + \frac{dh}{dl} + \frac{d}{dl}\left(\frac{\alpha v^2}{2g}\right) + \frac{h_{w_{1-2}}}{dl} = 0 \tag{a}$$

where

$$\frac{dz}{dl} = -i \tag{b}$$

$$\frac{d}{dl}\left(\frac{\alpha v^2}{2g}\right) = \frac{d}{dl}\left(\frac{\alpha Q^2}{2gA^2}\right) = -\frac{\alpha Q^2}{gA^3}\frac{dA}{dl} \tag{c}$$

For prismatic channels, the area of the cross-section varies with only the depth, i.e. $A = f(h)$, thus

$$\frac{dA}{dl} = \frac{dA}{dh}\frac{dh}{dl} = B\frac{dh}{dl}$$

Then

$$\frac{d}{dl}\left(\frac{\alpha v^2}{2g}\right) = -\frac{\alpha Q^2}{gA^3}B\frac{dh}{dl}$$

$$\frac{h_{f_{1-2}}}{dl} = J \tag{d}$$

For gradually varied flows of open channel, the minor head losses are so small that the total head loss is just the frictional head loss, which is approximately calculated by the equation of uniform flow as

$$J = \frac{Q^2}{A^2 C^2 R} = \frac{Q^2}{K^2}$$

Substituting Equations (b), (c) and (d) into Equation (a), gives

$$-i + \frac{dh}{dl} - \frac{\alpha Q^2}{gA^3} B \frac{dh}{dl} + J = 0 \tag{6-30}$$

or

$$\frac{dh}{dl} = \frac{i - J}{1 - \frac{\alpha Q^2}{gA^3} B} = \frac{i - J}{1 - Fr^2} \tag{6-31}$$

Equation (6-31) is called as the differential equation of gradually varied flow in a prismatic open channel, which is derived for the case of a positive slope ($i > 0$).

6.3.6.2 Water surface profiles

As seen from Equation (6-31), in prismatic open channels, the free surface line in gradually varied flow depends on the numerator and denominator on the right side of the equation, which can be either positive or negative. If the numerator and the denominator have the same sign, $\frac{dh}{dl} > 0$, so the flow depth increases along the flow and the surface profile is a back-water curve; if the numerator and the denominator have different signs, $\frac{dh}{dl} < 0$, so the depth decreases in the direction of flow and the surface profile is a drop-down curve. The depth at the transition point where the signs change, is the depth of flow when both the numerator and denominator are zero. For example, if the depth of a flow h is equal to the normal depth h_0, i.e. $J = i$, the numerator is zero, $i - J = 0$; if $h > h_0$, then the flow speed declines and consequently the less head loss leads to $J < i$, so the numerator is more than zero, $i - J > 0$, vice versa; if $h < h_0$, $i - J < 0$. In the same way, when in a critical flow, the depth h is equal to the critical depth h_c, then $Fr = 1$ and the denominator is zero, so $1 - Fr^2 = 0$; if $h > h_c$, the flow is a subcritical flow, $Fr < 1$ and the denominator is more than zero, so $1 - Fr^2 > 0$; otherwise if $h < h_c$, $1 - Fr^2 < 0$. Therefore, by means of line $N-N$ (the surface profile of uniform flow in Fig. 6-23) and line $C-C$ (the surface profile of critical flow in Fig. 6-23), the flow zone can be divided into subzones, in which the surface curves have distinct variation patterns.

However, since the line $N-N$ location and its relative relation to the line $C-C$ are closely related to the bed slope, in practice, the analysis of surface profile should take into account the bed slope.

(1) Channels of positive slope ($i > 0$)

Such channels are of mild slope ($i < i_c$), steep slope ($i > i_c$), or critical slope ($i = i_c$).

① Channels of mild slope ($i < i_c$)

In these channels, the normal depth h_0 is greater than the critical depth h_c, the flow zone is divided into three subzones: 1, 2 and 3, separated by lines $N-N$ and $C-C$, shown as Fig. 6-25, where the line $N-N$ is above the line $C-C$, because the uniform flow on a mild slope is inevitably subcritical, i.e. $h_0 > h_c$.

Subzone 1 ($h > h_0 > h_c$):

For the flow in this subzone, both $J < i$ and $Fr < 1$, i.e. $i - J > 0$ and $1 - Fr^2 > 0$. Thus the numerator and the denominator on the right side of Equation (6-31) are both positive, so $\dfrac{dh}{dl} > 0$.

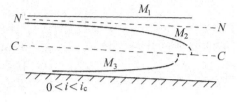

Fig. 6-25

The surface profile is a back-water curve, denoted as M_1 type (the letter M is taken from the initial of mild slope name and the subscript represents the subzone number).

At the upstream end, $h \rightarrow h_0$, $J \rightarrow i$, the numerator on the right side of Equation (6-31) tends to zero but the denominator is a positive number because $h > h_c$. Hence $\dfrac{dh}{dl} \rightarrow 0$, which indicates that the depths are nearly constant along the flow and the surface profile is thus asymptotic to the line $N - N$; for the downstream end, $h \rightarrow \infty$, then $J \rightarrow 0$ and $Fr \rightarrow 0$, so $i - J \rightarrow i$ and $1 - Fr^2 \rightarrow 1$ lead to $\dfrac{dh}{dl} \rightarrow i$, which indicates that at unit distance the increase of depth is equal to the decrease of bed elevation, in other words, the surface profile curve tends to be horizontal.

In summary, an M_1 type profile is a back-water curve that is asymptotic to the line $N - N$ in the upstream and to the horizontal in the downstream. For instance, when a dam is built in a channel it raises the upstream water level above the normal depth, and the water surface profile upstream of the dam is an M_1 type (Fig. 6-26).

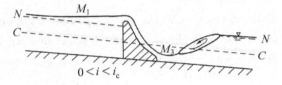

Fig. 6-26

Subzone 2 ($h_0 > h > h_c$):

If a water surface profile occurs in this subzone, the actual depth h is less than the normal depth h_0 but greater than the critical depth h_c, which implies $J > i$, i.e. $i - J < 0$, and since $Fr < 1$, i.e. $1 - Fr^2 > 0$, so $\dfrac{dh}{dl} < 0$, which means that the profile is a drop-down curve, labelled by M_2.

At the upstream end, $h \rightarrow h_0$, similar to the M_1 type, $J \rightarrow i$ and $h > h_c$, thus $\dfrac{dh}{dl} \rightarrow 0$, which results in constant depths or equivalently the $N - N$ asymptote. For the downstream end, $h \rightarrow h_c$ thus $1 - Fr^2 \rightarrow 0$, together with $i - J < 0$, it gives $\dfrac{dh}{dl} \rightarrow -\infty$, which means that the water surface is perpendicular to the line $C - C$. In fact, the flow there is no longer gradually varied flow but a hydraulic drop due to the depth falling rapidly.

Therefore, the profile of type M_2 is a drop-down curve with the line $N - N$ as the upstream asymptote, while dropping in the downstream direction the depth approaches to the critical depth at the brink. An example is the flow in a mild slope channel with a step drop at the end (Fig. 6-

27), in which case the surface profile upstream of the drop step is of an M_2 type, and at the end section flow passes through the critical depth as a hydraulic drop.

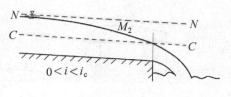

Fig. 6-27

Subzone 3 ($h_0 > h_c > h$)

In this subzone the flow depth h is less than both the normal depth h_0 and the critical depth h_c, $i - J < 0$ and $1 - Fr^2 < 0$, and thus $\frac{dh}{dl} > 0$, which means that the depth increases along the flow and the surface profile is a back-water curve, labelled by M_3.

At the upstream end: the depth is generally determined by the outflow from a hydraulic structure; the downstream depth approaches to the critical depth, i. e. $h \to h_c$ or $1 - Fr^2 \to 0$. Hence, $\frac{dh}{dl} \to \infty$, which implies that the surface is rising rapidly to be perpendicular to the line $C-C$, so a hydraulic jump will in fact occur, as shown in Fig. 6-26.

Therefore, the surface profile of type M_3 is controlled by some outflow at the upstream end, which creates a rising curve in the downstream direction, ending with a depth below the critical depth. For example, the flow discharging from a water retaining structure in a channel of mild slope forms a vena contracta, where the depth is usually less than the critical depth, and the consequent supercritical flow with a continuously decreasing speed, owing to the resistance, results in an increasing depth along the downstream channel, as shown in Fig. 6-26.

② Channels of steep slope ($i > i_c$)

In such channels, since the normal depth h_0 is less than the critical depth h_c, the relative position between the line $N-N$ and the line $C-C$ is different from that in the channels of mild slope. Despite this, the two lines still divide the flow zone into three subzones distinguished by number 1, 2 and 3 (Fig. 6-28).

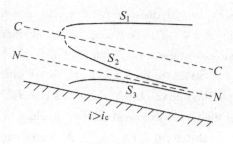

Fig. 6-28

In a similar way as before, the profile patterns for channels with a steep slope can be analyzed by considering three subzones.

Subzone 1 ($h > h_c > h_0$): the surface profile is a back-water curve, termed type S_1, and at the upstream end it approaches the line $C-C$, following a hydraulic jump and at the far end downstream it is asymptotic to a horizontal line. For example, the surface profile upstream of an obstruction built in a channel with a steep slope, is an S_1 profile (Fig. 6-29).

Subzone 2 ($h_c > h > h_0$): the water surface profile is a drop-down curve, marked as type S_2. At the upstream end, it crosses the line $C-C$ where the critual depth occurs and at the far end downstream it is asymptotic to the line $N-N$. For example, water flowing from a mild slope

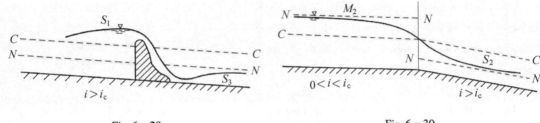

Fig. 6-29   Fig. 6-30

channel into a steep channel will have an M_2 profile in the channel of mild slope but an S_2 type in the channel with a steep slope. The flow at the joint section is the critical depth, which is like a hydraulic drop (Fig. 6-30).

Subzone 3 ($h_c > h_0 > h$): the water surface profile is a back-water curve, marked as type S_3, where the upstream depth is determined by the outflow from a control section (e.g. an obstruction) and downstream it is asymptotic to the line $N-N$. An example is shown in Fig. 6-29.

③ Channels of critical slope ($i = i_c$)

In such channels, the normal depth h_0 is equal to the critical depth h_c, so the line $N-N$ coincides with the line $C-C$ and the flow zone is in fact divided into only two subzones numbered 1 and 3 without 2. The corresponding surface profiles are thus labelled as type C_1 and C_3 respectively, and both back-water curves tend to be horizontal when they approach the line $N-N$ or $C-C$ (Fig. 6-31).

(2) Channels of horizontal slope ($i = 0$)

In horizontal channels, uniform flows do not exist, so only the line $C-C$ exists, and there is no line $N-N$. The flow zone is then divided into two subzones numbered 2 and 3 (Fig. 6-32).

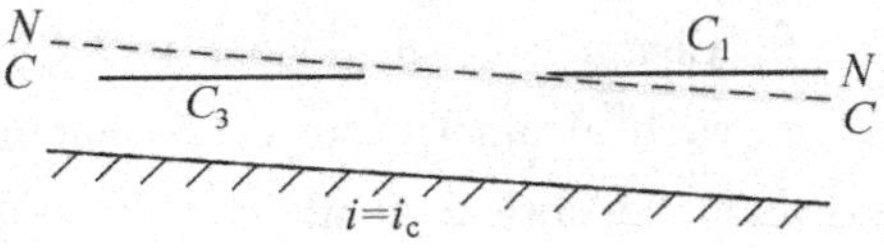

Fig. 6-31

Subzone 2 ($h > h_c$): the water surface profile is a drop-down curve, marked as type H_2. At the upstream end it begins from a horizontal line and at the downstream end the flow is close to the critical depth, where a hydraulic drop forms.

Subzone 3 ($h < h_c$): the water surface profile is a back-water curve, marked as type H_3, with the upstream controlled by the outflow from a hydraulic structure. The downstream depth approaches the line $C-C$ at a large angle, at which point a hydraulic jump will occur.

One example of where these two types of profile occur in a horizontal channel is where the opening height of a sluice gate is less than the critical depth. There will be an H_3 type profile generated immediately downstream of the gate, and if the channel is long enough the H_3 type would lead to a hydraulic jump (Fig. 6-33). If the channel also has a drop step in the end, then there will be a water surface of H_2 type in the channel between the jump and the drop end.

(3) Channels of adverse slope ($i < 0$)

Similar to channels with a horizontal slope, no uniform flow occurs in channels that have an adverse slope. Hence only the line $C-C$ divides the flow zone into two subzones numbered 2 and 3.

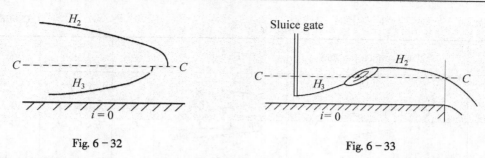

Fig. 6 – 32   Fig. 6 – 33

In Subzone 2, the surface profile is a drop-down curve, marked as A_2 type, and at the upstream end it approaches the horizontal line, whereas at the far end downstream it likes a hydraulic drop when the flow depth is close to the critical depth.

In Subzone 3, the surface profile is a back-water curve marked as A_3 type, with its upstream end controlled by the outflow from a hydraulic structure and its downstream end approaching the horizontal (Fig. 6 – 34). An example of this type of flow is shown in Fig. 6 – 35.

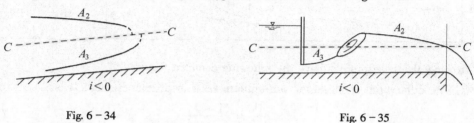

Fig. 6 – 34   Fig. 6 – 35

6.3.6.3 General principle in analysis of water surface profiles

(1) Common features of the surface profiles in prismatic channels

So far, 12 types of possible GVF surface profiles have been identified for prismatic channels, of which M_1, M_2, M_3, S_1 and S_2 are often seen in engineering. All the 12 profiles are summarized in Table 6 – 3.

Table 6 – 3 The summary for surface profiles

	Water surface profile	Example	Type	Depth	Flow	$\dfrac{dh}{dl}$
$i < i_c$			M_1	$h > h_0 > h_c$	GVF	+
			M_2	$h_0 > h > h_c$	GVF	−
			M_3	$h_0 > h_c > h$	RVF	+
$i > i_c$			S_1	$h > h_c > h_0$	GVF	+
			S_2	$h_c > h > h_0$	RVF	−
			S_3	$h_c > h_0 > h$	RVF	+

Continued

	Water surface profile	Example	Type	Depth	Flow	$\dfrac{dh}{dl}$
$i = i_c$			C_1	$h > h_0 = h_c$	GVF	+
			C_3	$h < h_0 = h_c$	RVF	+
$i = 0$			H_2	$h > h_c$	GVF	−
			H_3	$h < h_c$	RVF	+
$i < 0$			A_2	$h > h_c$	GVF	−
			A_3	$h < h_c$	RVF	+

In general, these profiles have the following common features:

① The differential equation for non-uniform GVF in prismatic channels is

$$\frac{dh}{dl} = \frac{i - J}{1 - Fr^2} \qquad (6-31')$$

which is the fundamental equation for the water surface profile analysis and computation. By examining the positive or negative sign of the numerator and denominator for various depths on the right side of the equation, one can qualitatively determine the value of dh/dl and hence the change of surface profile in the direction of flow and its two end limits.

② The line $N - N$ and the line $C - C$ for the given flow rate are not actual water surfaces but two imaginary lines that help in the analysis of water surface profiles. Their relative position depends on the values of normal and critical depths h_0 and h_c, and hence the bed slope type of channel. They are used to divide the flow zone into distinct subzones.

③ The type of surface profile in every subzone is thus uniquely determined, so it is impossible that more than one type of profile would occur in one subzone.

④ Among all types of surface profiles, those in subzones 1 and 3 are all back-water curves while those in subzone 2 are drop-down curves.

⑤ All the surface profiles are asymptotic to $N - N$ when they tend to the normal depths, and approach $C - C$ at a large angle to become RVF when they tend to the critical depths, except for C_1 and C_3.

(2) Steps for drawing water surface profiles

① Specify disturbance sections. The disturbance sections are those where there are sudden changes of bed slope, level, section size or roughness, or at hydraulic structures, as well as the inlet or outlet conditions of the channel. For the cases of small bed slope, the disturbance sections can be taken to be straightly vertical.

② Draw the reference lines $N-N$ and $C-C$. When the type of bed slope has been known, one can determine the relative position of $N-N$ and $C-C$ according to the curve in Fig. 6-16, which shows the relationship between the bed slope and the normal depth; otherwise one needs to calculate the normal depth h_0 and the critical depth h_c respectively and then to plot the lines $N-N$ and $C-C$ accordingly.

③ Find the control sections and the corresponding depths. The so-called control section is the one that is located explicitly with a known depth, i. e. the control depth. Common cases of such control sections and depths are:

a. When the channel is sufficiently long, beyond the influence of non-uniform, there will be uniform flow at the depth h_0, where the surface profile is asymptotic to either from below or above the line $N-N$.

b. When the flow transits from the subcritical to the supercritical flow, the surface profile passes through h_c smoothly by a hydraulic drop, usually located at the junction between sections of slopes from the mild to the steep or at the drop step at the end of channel. When the flow changes from supercritical to subcritical flow, it will also pass through h_c with a hydraulic jump whose position should satisfy the conjugate relationship between the pre-jump and post-jump depths.

c. Since in supercritical flow the disturbance waves can only propagate downstream, the profiles of M_3, S_2, S_3, C_3, H_3 and A_3 in the supercritical flow definitely have their control depths upstream, whilst M_1, M_2, S_1, C_1, H_2, and A_2 must have downstream control depths, because in subcritical flow the disturbance waves can travel upstream as well.

d. The water levels before sluice gates and dams, or the depths at the vena contracta on the downstream spillway of dams, can be considered as the control depths in the downstream direction in subcritical flow or upstream in supercritical flow, respectively.

④ Determine the local variations of surface profiles. Generally, where the flow in a channel is affected by some obstructions such as increasing slope, increasing roughness or dam construction, a back-water profile will occur upstream since the uniform flow will be disturbed; where the flow is accelerated such as in a steepened bed, decreasing roughness or drop step, the uniform flow can no longer be maintained, so a drop-down profile will occur in the upstream channel.

⑤ From the above steps, one can decide which subzone the surface profile will occur in and what type it will belong to. Thus the surface profile can be drawn qualitatively.

【Example 6-7】 There is a sluice gate installed in a mild slope channel with enough distances both upstream and downstream of the sluice gate. At the end of the channel there is a step drop (Fig. 6-36). When the gate discharges the flow with a certain opening, the flow depth before the gate is larger than the normal depth, $h_1 > h_0$, and the contraction depth after the gate is less than the critical depth, $h_2 < h_c$. Qualitatively sketch the surface profile.

Solution: Plot the lines $N-N$ and $C-C$ to divide the flow zone. In the mild channel $h_0 > h_c$, so the line $N-N$ is above the line $C-C$, as shown in Fig. 6-36.

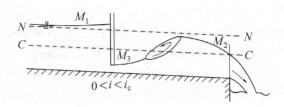

Fig. 6 – 36

(1) The upstream section of the gate

At the far end upstream, the flow is less affected by the gate and is nearly uniform with the free surface asymptotic to $N-N$. Owing to existence of the gate, the water level before the gate rises because for the control depth h_1, $h_1 > h_0 > h_c$. Therefore, the flow is in subzone 1 of mild slope channel, i. e. the surface profile is a back-water curve of M_1 type with an upstream asymptote of $N-N$, as shown in Fig. 6 – 36.

(2) The downstream section of the gate

The flow forms a vena contracta after the gate due to the control gate. The vena contracta cross-section is thus a control section and the flow is supercritical flow with the depth $h_2 < h_c < h_0$. Hence the surface profile after the gate is in subzone 3 of the mild slope channel and thus an M_3-typed backwater curve. If the downstream channel is long enough, a hydraulic jump will occur when the depth increases towards h_c.

(3) The section after the jump

If there was no disturbance in the channel after the jump, the flow would eventually become a uniform flow again and reach the line $N-N$. However, if the length of the downstream channel is relatively short, a drop step at the end of the channel would result in the surface falling. The depth after the jump is between h_c and h_0, so the surface profile will be a drop-down curve of type M_2, the only type in subzone 2 for a mild slope channel. The section at the brink can be treated as a control section with the depth approximately equal to h_c.

All these surface profiles along the channel are illustrated in Fig. 6 – 36.

6.3.7 Computation of surface profiles in steady gradually varied flow

The computation of surface profiles quantitatively to obtain the variation of depth along the channel is based on the basic equation of GVF in open channels. From the qualitative analysis, starting with a known depth, one can apply the energy equation to calculate the depths of the other sections consecutively and thereby can plot the surface profile of the entire channel reach by reach quantitatively.

The direct step method is commonly used in such computations, i. e. the whole channel reach is divided into a number of sub-reaches, which are then substituted in a finite difference form into the differential equation, so that the depths along each section and the related distances between the sections can be solved from the governing equation.

The governing differential equation (6-30) for gradually varied flow of open channel can be rewritten as

$$-i + \frac{dh}{dl} + \frac{d}{dl}\left(\frac{\alpha v^2}{2g}\right) + J = 0$$

Namely,
$$\frac{d}{dl}\left(h + \frac{\alpha v^2}{2g}\right) = i - J$$

Since the sectional specific energy is $E_s = h + \frac{\alpha v^2}{2g}$,

$$\frac{dE_s}{dl} = i - J \qquad (6-32)$$

For a short distance of flow Δl, writing the above differential equation into a finite difference formation with the actual hydraulic slope replaced by the mean hydraulic slope of the sub-reach, gives

$$\Delta l = \frac{dE_s}{i - \overline{J}} = \frac{E_{sd} + E_{su}}{i - \overline{J}} \qquad (6-33)$$

where E_{sd} and E_{su} are the specific energy of the downstream and upstream cross-sections at the two ends of a flow sub-reach, respectively. Equation (6-33) is the basic equation for computing surface profiles by the direct step method. It is suitable for both prismatic and non-prismatic channels.

The hydraulic energy slope can be calculated by the formula of uniform flow, $J = \frac{v^2}{C^2 R}$. For a reach, the averaged hydraulic slope is usually calculated by

$$\overline{J} = \frac{\overline{v}^2}{\overline{C}^2 \overline{R}} \qquad (6-34)$$

or
$$\overline{J} = \frac{1}{2}(J_u + J_d)$$

where $\overline{v}$, $\overline{C}$ and $\overline{R}$ are the mean values of the sectional velocity, Chezy coefficient and hydraulic radius in the flow reach respectively; J_u and J_d denote the hydraulic energy slopes of the upstream and downstream cross-section of the flow reach, respectively.

For a prismatic channel with a certain length, during computation, take the known depth or the control depth (either at the upstream or downstream end of the reach) as the starting depth and assume the depth of an adjacent cross-section according to the qualitative analysis of the surface profile. Then compute the specific energy at the two end cross-sections and the averaged hydraulic energy slope of the flow sub-reach, substitutes them into Equation (6-33), and hence find out the corresponding reach length. Such successive calculations will give the depth at each consecutive cross-section and the distance between them.

If the reach distance is given and the depth at one end is known, then the trial-and-error method is needed for calculating the depth at the other end. Assuming a series of initially guessed depth h and calculating the corresponding distance Δl, one can obtain an $h \sim \Delta l$ curve. On the curve, one can find h corresponding to the given Δl, which is the expected depth.

For a non-prismatic channel, because its sectional geometric parameters vary with the section position, the computation of surface profile has to be undertaken by a trial-and-error method for each sub-reach one by one.

Since the above method is based on substitution of finite difference equation for differential equation, where E_s and J are seen as linearly varying variables, the precision of computation depends on the length of each sub-reach. Therefore, generally the distance of any two adjacent cross-sections cannot be taken too long.

【Example 6 – 8】 A long straight rectangular drainage channel has the roughness $n = 0.025$, the bed slope $i = 0.0002$, the base width $b = 2$ m and the discharge rate $Q = 2.0$ m³/s. At the end of channel the water is discharged into a low-lying river (Fig. 6 – 37). Plot the free surface profile in the channel.

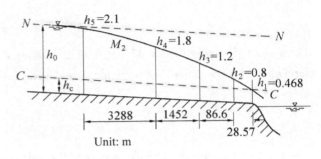

Fig. 6 – 37

Solution:

(1) Determine the slope type of channel, the type of surface profile and the possible range of depth variation.

The normal depth is calculated by Equation (6 – 7), i. e. $h_0 = 2.26$ m, and the critical depth by Equation (6 – 18), $h_c = 0.468$ m. Since $h_0 > h_c$, the slope is mild.

According to the values of h_0 and h_c, draw the lines $N-N$ and $C-C$ in Fig. 6 – 37. At the end of the channel there is a hydraulic drop, and at the brink the depth is h_c. Hence the flow is in subzone 2 of mild slope channel, i. e. it has a surface profile of M_2 type. The depths of the non-uniform flow vary from 2.26 m to 0.468 m along the flow.

(2) Compute the surface profile.

Since it is a subcritical flow in the channel, the downstream end depth h_c is a control depth. Divide the channel into four sub-reaches and start calculation from the downstream towards the upstream. Letting $h_1 = h_c = 0.468$ m, $h_2 = 0.8$ m, $h_3 = 1.2$ m, $h_4 = 1.8$ m and $h_5 = 2.1$ m, from Equations (6 – 33), (6 – 34) and $E_s = h + \dfrac{\alpha v^2}{2g}$, the computation is in a tabulated form below:

	h/m	A/m²	v/m·s⁻¹	$\bar{v}$/m·s⁻¹	$(v^2/2g)$/m	E_s/m	ΔE_s/m
1	0.468	0.934	2.14	1.695	0.234	0.7	-0.180
2	0.8	1.6	1.25	1.64	0.08	0.88	-0.355
3	1.2	2.4	0.833	0.694	0.035	1.235	-0.581
4	1.8	3.6	0.556	0.516	0.016	1.816	-0.296
5	2.1	4.2	0.476		0.012	2.112	

Continued

	R/m	$\bar{R}/m$	$C/m^{0.5} \cdot s^{-1}$	$\bar{C}/m^{0.5} \cdot s^{-1}$	$\bar{J}$	$i-\bar{J}$	$\Delta l/m$	$\sum \Delta l/m$
1	0.32	0.38	33.07	34.0	0.006 5	-0.006 3	28.57	28.6
2	0.44		34.94					
		0.493		35.55	0.004 3	-0.004 1	86.59	115
3	0.545		36.15					
		0.594		36.66	0.000 6	-0.000 4	1 452	1 567
4	0.643		37.16					
		0.660		37.32	0.000 29	-0.000 09	3 288	4 855
5	0.677		37.48					

According to the calculated results, plot the surface profile curve in an appropriate scale vertically and horizontally, as shown in Fig. 6-37.

6.4 Weir flow and underflow of sluice gates

When an obstacle or contral structure exists in an open channel, any subcritical flow will be changed, typically leading to the upstream water surface level rising and the downstream surface falling. Flows over an obstacle are often called weir flow (Fig. 6-38a), and the obstacle or structure is called a weir. The overflow on a weir has a continuous free surface with the notably bending streamlines, which is rapidly varied flow (RVF). If there is a sluice gate above the block and the flow is controlled by the sluice gate through the opening between the top of block and the bottom edge of the gate, the flow over the block is called the underflow of sluice gate (Fig. 6-38b). Because the flow is controlled by the gate, the water surface between the upstream and the downstream of the gate is not continuous. The flow over weir and the flow under sluice gate are two different kinds of flow due to the difference in their different boundary conditions, so their flow features and discharge are different.

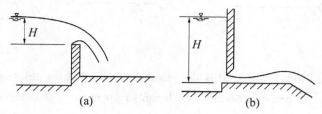

Fig. 6-38

From the hydraulics perspective, if the gate opening height is e and the water head before the sluice gate is H, then the flow under the gate is either a weir flow or an underflow of sluice gate depending on the relative opening value e/H. The general criteria used to distinguish the type of flow are:

When the sluice gate is over a flat weir, $\dfrac{e}{H} \leqslant 0.65$, the flow is underflow of the sluice gate;

otherwise $\frac{e}{H} \geqslant 0.65$, the flow is simply weir flow.

When the sluice gate is over a curved weir, $\frac{e}{H} \leqslant 0.75$, the flow is underflow of the sluice gate; otherwise $\frac{e}{H} \geqslant 0.75$, the flow is simply weir flow.

Weirs and sluices are widely used in engineering. In drainage or diversion systems, they are used to control the water level or the flow rate. In addition, the flow over a weir is used to measure the flow rate in either natural rivers or laboratories.

The hydraulic computations about the flow over weirs or under sluice gates are mainly about the calculation of the flow capacity, and hence corcern about the estimation of appropriate discharge cofficients and the design of the discharge sections.

6.4.1 Types and basic formula of weir flow

6.4.1.1 Weir types

In weir flow, when the flow approaches the crest, owing to contraction of the streamlines, the flow velocities gradually increase with the falling free surface. The upstream cross-section where the water surface does not drop significantly is marked as the pre-weir cross-section. The observations demonstrate that such a section is about $(3-5)H$ away from the weir (Fig. 6-39b). The quantities describing the weir flow are shown in Fig. 6-39.

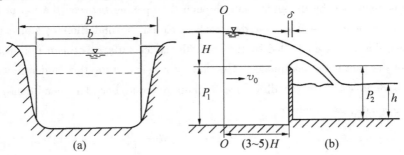

Fig. 6-39

Notes: b is the width of weir, B is the channel width, δ is the thickness of weir crest, P_1 and P_2 are the upstream and downstream weir heights respectively, H is the water head above the weir, h is the water depth downstream of the weir, and v_0 is the approach velocity which is the mean velocity in cross-section $O-O$ upstream of the weir.

According to the experimental data, the pattern of flow over a weir crest varies with the ratio of the thickness δ to the head H. In terms of the ratio δ/H, the flow over a weir is commonly classified into three types: the sharp-crested weir flow, the practical weir flow and the broad-crested weir flow.

(1) The sharp-crested weir flow $\frac{\delta}{H} \leqslant 0.67$ (Fig. 6-40)

In this type of weir flow the bottom flow in front of weir is blocked by the weir wall and thus the streamlines are bent upward due to inertia. When the lower fringe of the nappe falls down back to the crest level, it is about 0.67H away from the upstream wall of the weir. Hence, when $\dfrac{\delta}{H} \leqslant 0.67$, the shape of the nappe over the crest will not be affected by weir thickness, the lower fringe of the nappe contacts the crest linearly, and the free

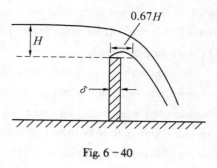

Fig. 6-40

surface is like a single drop-down curve. Since the overflow nappe is stable and the distribution of its pressure and velocity is steady as well, the sharp-crested weir is primarily used as a device for flow rate measurement.

(2) The practical weir flow $0.67 \leqslant \dfrac{\delta}{H} \leqslant 2.5$ (Fig. 6-41)

Because the wall of a practical weir is thicker, the lower fringe of nappe contacts the crest surface, which will then affect the overflow due to the falling nappe being restricted by the crest surface. Although the thickness of the weir has some impact on the flow, due to gravity the flow over the crest will fall freely with the free surface continuously declining. The profile of a practical weir has two common types: a curved line type (Fig. 6-41a) and the polygonal line type (Fig. 6-41b). A large or medium hydraulic spillway generally takes the curved line type in order to achieve a large discharge, whilst the small spillway often takes the polygonal line type for simplicity.

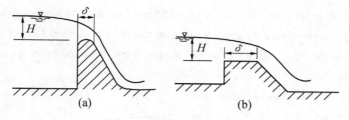

Fig. 6-41

(3) The broad-crested weir flow $2.5 \leqslant \dfrac{\delta}{H} \leqslant 10$ (Fig. 6-42)

The thickness of a broad-crested weir is large enough so that it will significantly affect the overflow, such that the flow into the crest zone is markedly constricted by the crest, thus resulting in a decreasing cross-section of flow and an increasing velocity. Consequently, the increasing kinetic energy results in an inevitable decrease in potential energy. In addition, local head losses will exist around the crest. Hence at the inlet a falling free surface fall is formed; afterwards

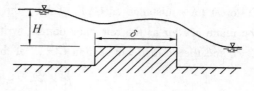

Fig. 6-42

Fluid Mechanics in Civil Engineering

owing to the support from the crest top, part of the free surface will be nearly parallel to the top of weir. If the downstream water level is low, the outflow of the weir will have a second drop-down, as shown in Fig. 6-42.

In some cases, although there is no lump block in a channel, the same effect may be produced due to a contraction of channel width. The flow then still behaves in a similar manner to broad-crested weir flow, for example, the flow meets with a suddenly narrowed session in a channel. Such a flow is called as the ridge-free broad-crested weir flow or side contracted weir flow, which can be calculated as broad-crested weir flow. Some practical examples are the flow between bridge piers, the flow at the inlet of a culvert or tunnel, or through a flat-bed sluice (Fig. 6-43).

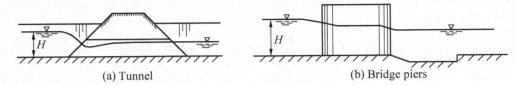

(a) Tunnel (b) Bridge piers

Fig. 6-43

When the weir width increases up to $\delta > 10 H$, the frictional head loss cannot be ignored, and then the flow is no longer weir flow, but open-channel flow. The above method is just an approximate criterion for identifying different types of weir flow. For example, for the same weir, it may become a practical weir when the head on the crest is large, or it may become a broad-crested weir when the head is smaller.

6.4.1.2 Fundamental formula for weir flow

Despite different types, weir flows have similar features and the same fundamental formula for discharge calculation, because they are the same in the form of motion (subcritical outflow over a barrier) and are under the influence of an acting force (predominant gravity and negligible frictional resistance).

Take the flow over a broad-crested weir as an example, shown in Fig. 6-44. The water surface declines initially over the crest. If the downstream water level is low, the depth h_{co} at the vena contracta near the inlet is less than the critical depth h_c, and the flow nearby remains supercritical and parallel to the crest for a distance as GVF. Taking the top plane of weir as the reference datum, write the energy equation between the upstream section 1-1 and the downstream section $C'-C'$ at the vena contracta,

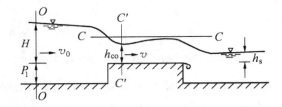

Fig. 6-44

$$H + \frac{\alpha_0 v_0^2}{2g} = h_{co} + \frac{\alpha v^2}{2g} + \zeta \frac{\alpha v^2}{2g}$$

where $H + \dfrac{\alpha_0 v_0^2}{2g}$ is termed as the total head of weir, which includes the approaching velocity head $\dfrac{\alpha_0 v_0^2}{2g}$ and the static head H, denoted by H_0. h_{co} depends on H_0 and can be expressed as $h_{co} = kH_0$, where k is a factor related to the entrance shape and the variation of flow cross-section. ζ is the minor loss coefficient of the inlet. Substituting H_0 and kH_0 into the equation gives the velocity,

$$v = \dfrac{1}{\sqrt{\alpha + \zeta}} \sqrt{1-k}\sqrt{2gH_0} = \varphi\sqrt{1-k}\sqrt{2gH_0}$$

Thus the flow rate is $\quad Q = vkH_0 b = \varphi k\sqrt{1-k}\, b\sqrt{2g}H_0^{3/2}$

where φ is the velocity factor, $\varphi = \dfrac{1}{\sqrt{\alpha + \zeta}}$.

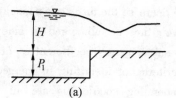 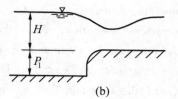

(a)　　　　　　　　　　(b)

Fig. 6-45

Letting $m = \varphi k\sqrt{1-k}$, m is termed as the discharge coefficient, which is an overall coefficient depending on the acting water head and the boundary conditions of weir, the variation of m with the weir type, entrance shape and weir height is complex. The value of m is commonly determined by experiments or empirical formulae.

For broad-crested weir with right angle inlet (Fig. 6-45a),

when $0 \leqslant \dfrac{P_1}{H} \leqslant 3.0$

$$m = 0.32 + 0.01\dfrac{3 - \dfrac{P_1}{H}}{0.46 + 0.75\dfrac{P_1}{H}} \tag{6-35}$$

when $\dfrac{P_1}{H} > 3.0$

$$m = 0.32$$

For broad-crested weir with round edge inlet (Fig. 6-45b),

when $0 \leqslant \dfrac{P_1}{H} \leqslant 3.0$

$$m = 0.36 + 0.01\dfrac{3 - \dfrac{P_1}{H}}{1.2 + 1.5\dfrac{P_1}{H}} \tag{6-36}$$

when $\dfrac{P_1}{H} > 3.0$

$$m = 0.36$$

Taking m into the flow rate expression gives

$$Q = mb\sqrt{2g}H_0^{3/2} \quad (6-37)$$

which is the fundamental formula for weir flow. It is suitable for the sharp-crested, practical and broad-crested weirs in a rectangular channel. Equation (6-37) shows that the weir flow rate is proportional to H_0 to the power of $2/3$, i.e. $Q \propto H_0^{2/3}$.

In practical application, sometimes when the downstream water level is high or the downstream height of the weir is small, the outflow from a weir is obstructed by the downstream water so that the flow rate of weir will be reduced. Such a weir is called a submerged or drowned outflow; otherwise, the weir flow is a free outflow. In other cases, the weir type flow can be formed at a narrowed section of a channel, such as the flow through a channel section with bridge piers or gate piers, which would cause side contraction to the flow and decrease the discharge capacity. Such a flow is called as side-contracted weir flow. When the weir flow is submerged or side contracted, the reduction of flow rate can be evaluated by the submergence coefficient σ_s ($\sigma_s < 1$) or the coefficient of side contraction ε ($\varepsilon < 1$). The criteria for identifing submerged flow and side-contracted flow and the determination of corresponding coefficients are all based on either experiment data or empirical formulae. The details can be seen in relevant references. For example, the submergence coefficient of broad-crested weir is given in Table 6-4, where h_s denotes the downstream depth above the weir as shown in Fig. 6-44, while the coefficient of side contraction for the single broad-crested weir can be calculated by the following empirical formula

$$\varepsilon = 1 - \frac{a}{\sqrt[3]{0.2 + \frac{P_1}{H}}} \sqrt[4]{\frac{b}{B}\left(1 - \frac{b}{B}\right)} \quad (6-38)$$

where a is the shape factor of pier; $a = 0.19$ for a rectangular pier and $a = 0.10$ for a round pier. When the weir flow is free outflow without side contraction, $\sigma_s = 1$ and $\varepsilon = 1$.

In conclusion, the computational formula for the discharge capacity of weir can actually be written as

$$Q = \sigma_s \varepsilon mb\sqrt{2g}H_0^{3/2} \quad (6-39)$$

Table 6-4 The values of submergence coefficient (σ_s)

$\frac{h_s}{H_0}$	0.80	0.81	0.82	0.83	0.84	0.85	0.86	0.87	0.88	0.89
σ_s	1.00	0.995	0.99	0.98	0.97	0.96	0.95	0.93	0.90	0.87
$\frac{h_s}{H_0}$	0.90	0.91	0.92	0.93	0.94	0.95	0.96	0.97	0.98	
σ_s	0.84	0.82	0.78	0.74	0.70	0.65	0.59	0.50	0.40	

【Example 6-9】 A broad-crested weir, which has right angle inlet and no side contraction, has the width $b = 2$ m, the upstream weir height $P_1 = 0.5$ m and the upstream head above the weir $H = 1.8$ m. Assuming the weir flow is free outflow, find the flow rate over the weir.

Solution:

$\dfrac{P_1}{H} = \dfrac{0.5}{1.8} = 0.278 < 3$, and from Equation (6-35) the discharge coefficient is

$$m = 0.32 + 0.01 \dfrac{3 - \dfrac{P_1}{H}}{0.46 + 0.75 \dfrac{P_1}{H}} = 0.361$$

Firstly assuming the approaching velocity head is negligible, i.e. $\dfrac{\alpha_0 v_0^2}{2g} \approx 0$, $H_0 = H$. Applying Equation (6-37) gives the flow rate over the weir in the conditions of no side contraction and free outflow

$$Q = mb\sqrt{2g}H_0^{3/2}$$
$$= 0.361 \times 2 \times \sqrt{2 \times 9.8} \times 1.8^{3/2}$$
$$= 7.719 \text{ m}^3/\text{s}$$

Then, compute the first approximate approaching velocity v_0 and the total head above the weir H_0 for $Q = 7.719 \text{ m}^3/\text{s}$

$$v_0 = \dfrac{Q}{b(H + P_1)} = \dfrac{7.719}{2 \times (1.8 + 0.5)} = 1.678 \text{ m/s}$$

$$H_0 = H + \dfrac{\alpha v_0^2}{2g} = 1.8 + \dfrac{1.678^2}{2 \times 9.8} = 1.944 \text{ m}$$

Furthermore, use $H_0 = 1.944$ m to compute the flow rate Q'

$$Q' = 0.361 \times 2 \times \sqrt{2 \times 9.8} \times 1.944^{3/2} = 8.66 \text{ m}^3/\text{s}$$

Once again, use $Q' = 8.66 \text{ m}^3/\text{s}$ to compute the second approximate values of v_0' and H_0'

$$v'_0 = \dfrac{8.66}{2 \times (1.8 + 0.5)} = 1.883 \text{ m/s}$$

$$H'_0 = 1.8 + \dfrac{1.883^2}{2 \times 9.8} = 1.981 \text{ m}$$

Using $H_0' = 1.981$ m, compute the flow rate Q''

$$Q'' = 0.361 \times 2 \times \sqrt{2 \times 9.8} \times 1.981^{3/2} = 8.909 \text{ m}^3/\text{s}$$

So far the relative error between Q' and Q'' is

$$\dfrac{8.909 - 8.66}{8.909} = 2.7\%$$

If it satisfies the accuracy requirement, stop the computation. The flow rate over the weir is then $8.91 \text{ m}^3/\text{s}$.

【Example 6-10】 A rectangular broad-crested weir has a rectangular side pier, the weir width $b = 2$ m, weir heights $P_1 = P_2 = 1$ m. The upstream head over the weir is $H = 1.8$ m, the upstream channel width is $B = 3$ m, and the downstream water depth is $h_s = 2.8$ m. Find the discharge over the weir (assuming the approaching velocity v_0 is negligible).

Solution:

Firstly, check whether the outflow of the weir is free or submerged. Since v_0 is negligible, $H_0 = H$,

$$\frac{h}{H_0} = \frac{2.8 - 1}{2} = 0.9 > 0.8$$

From Table 6-4 it is known that the weir flow is of submerged outflow and the submergence coefficient is found to be $\sigma_s = 0.84$. Also, since $B > b$, the side contraction should be considered, and the corresponding coefficient is calculated by Equation (6-38)

$$\varepsilon = 1 - \frac{a}{\sqrt[3]{0.2 + \frac{P_1}{H}}} \sqrt[4]{\frac{b}{B}} \left(1 - \frac{b}{B}\right)$$

$$= 1 - \frac{0.19}{\sqrt[3]{0.2 + \frac{1}{2}}} \sqrt[4]{\frac{2}{3}} \left(1 - \frac{2}{3}\right)$$

$$= 0.936$$

The discharge coefficient m can be calculated by Equation (6-35)

$$m = 0.32 + 0.01 \frac{3 - \frac{P_1}{H}}{0.46 + 0.75 \frac{P_1}{H}}$$

$$= 0.32 + 0.01 \times \frac{3 - \frac{1}{2}}{0.46 + 0.75 \times \frac{1}{2}}$$

$$= 0.35$$

The flow rate is then calculated by Equation (6-39)

$$Q = \sigma_s \varepsilon m b \sqrt{2g} H_0^{3/2}$$

$$= 0.936 \times 0.84 \times 0.35 \times 2 \times \sqrt{2 \times 9.8} \times 2^{3/2}$$

$$= 6.89 \text{ m}^3/\text{s}$$

6.4.2 Fundamental formula of underflow of a sluice gate

Sluices in practical engineering are of many types. Their gates are often broad-crested weir (ridge-free broad-crested weirs included) or curved practical weirs. The forms of the gate are mainly of two types: vertical plate gate and radial gate. When the gate is open to some extent, as long as the flow is controlled by the sluice gate opening, the outflow is known as the underflow. The outflow from the sluice gate would continue to converge due to inertia. If the gate opening is e, at the place of $(0.5 \sim 1.0) e$ downstream from the gate there will be a vena contracta at section $C'-C'$, in which the flow is supercritical and has a minimum depth. After a certain distance, the supercritical flow then meets possibly the subcritical flow downstream, which will produce a hydraulic jump (Fig. 6-46).

The main purpose for such a flow is to find the discharge for different openings e under a certain total head H_0, or determine the gate opening width b. Different types of gate, or different shapes of bottom weir, obviously have different impacts on outflow convergence and energy losses and thus discharge capacities.

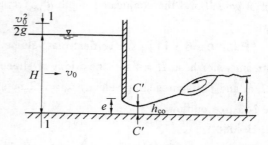

Fig. 6-46

When the downstream water level is lower and the outflow after gate has sufficient convergence, applying the energy equation for cross-section 1-1 before gate and the vena contracta cross-section $C'-C'$ after gate gives

$$Q = \mu be \sqrt{2gH_0} \qquad (6-40)$$

where μ denotes the discharge coefficient of sluice gate, which contains such factors as the kinetic energy correction coefficient of section $C'-C'$, the minor loss coefficient of gate, the vertical contraction coefficient of sluice opening and the relative gate opening. Equation (6-40) is the fundamental formula for free underflow of gate, and shows that the outflow discharge of gate is proportional to the total head H_0 with the power of $1/2$, i.e. $Q \propto H_0^{1/2}$.

As an overall factor reflecting the energy losses and the contraction extent, the discharge coefficient μ depends on the form of bottom weir, gate type and relative opening e/H. For instance, the coefficient of vertical plate gate is calculated by the following empirical formula,

$$\mu = 0.60 - 0.176 \frac{e}{H} \qquad (6-41)$$

For the sluice gate with side piers or gate piers, the computation is not necessarily needed to take the impact of side contraction into account separately. The experimental results show that for underflow of sluice gate, the side piers or gate piers have little effect on discharge.

When the downstream level is higher than the conjugated depth corresponding to the vena contracta depth, the jump zone downstream may be pushed towards the gate, and the depth at the contraction cross-section will increase to be even higher than the critical depth, and the flow then becomes subcritical. The discharge will decrease and the water level upstream rises.

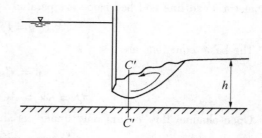

Fig. 6-47

The flow in such a flow situation is called a submerged flow (Fig. 6-47), in which the flow rate can be computed by Equation (6-40) with including an additional submergence coefficient σ_s. This is

$$Q = \sigma_s \mu be \sqrt{2gH_0} \qquad (6-42)$$

where the submergence coefficient σ_s depends on the downstream water level h_s, the water head in front of gate H and the conjugated depth h''_{co}. σ_s is usually determined by empirical formulae or charts.

【Example 6-11】 A vertical plate sluice gate is on a flat channel. It is known that the upstream water head $H = 4$ m, the width of sluice opening $b = 5$ m, the height of gate opening $e = 1$ m and the approaching velocity $v_0 = 1.2$ m/s. Determine the flow rate Q if the underflow of gate is a free outflow.

Solution:

By Equation (6-41) the discharge coefficient is calculated

$$\mu = 0.60 - 0.176 \frac{e}{H} = 0.60 - 0.176 \times \frac{1}{4} = 0.556$$

The total head upstream $H_0 = H + \frac{\alpha v_0^2}{2g} = 4 + \frac{1 \times 1.2^2}{2 \times 9.8} = 4.07$ m

Thus, the flow rate is

$$Q = \mu b e \sqrt{2gH_0}$$
$$= 0.556 \times 5 \times 1 \times \sqrt{2 \times 9.8 \times 4.07} = 24.83 \text{ m}^3/\text{s}$$

Chapter summary

This chapter has described the characteristics of open-channel flow and the hydraulic calculation of steady uniform and non-uniform flow. Compared with the pressurized flow, open-channel flow has a free surface so that the depth, cross-sectional area and flow velocity vary with the distance.

1. The geometric features of open-channel, including bed slope, cross-sectional shape and size, and cross-section changes with distance, have an important impact on the regime and the discharge of open-channel flow.

2. In uniform open-channel flow, the depth and velocity are constant, the total head line, piezometric head line and bed slope are parallel, that is

$$J = J_e = i$$

The basic equations are

$$v = C\sqrt{RJ}$$
$$Q = AV = AC\sqrt{RJ} = K\sqrt{i}$$

Open-channel flow occurs only under certain conditions: In steady flow, any non-uniform flow of open channel always tends to become uniform when the conditions are met. The depth of uniform flow is called the normal depth, expressed by h_0.

3. Theoretical analysis shows that for the same flow capacity, an optimum hydraulic cross-section has the smallest area; or that for a constant area of flow, the optimum hydraulic cross-section has the largest flow capacity. An optimum hydraulic cross-section is not a technically the most economic optimum section.

4. Non-pressurized uniform pipe flow is uniform open-channel flow, so the conditions,

hydraulic characteristics and basic formula of uniform open-channel flow apply.

5. Due to the existence of disturbance, uniform flow of open channel becomes non-uniform flow, in which the properties of flow change along distance. Non-uniform flow has three flow regimes: subcritical flow, critical flow, and supercritical flow, which are dependent on the Froude number Fr:

$$Fr < 1 \quad \text{Subcritical flow}$$
$$Fr = 1 \quad \text{Critical flow}$$
$$Fr > 1 \quad \text{Supercritical flow}$$

6. For different flow regimes, the specific energy varies with depth. The depth that corresponds a minimum specific energy is called the critical depth, h_c, which depends on the flow rate and cross-sectional shape and size. For a rectangular channel,

$$h_c = \sqrt[3]{\frac{\alpha q^2}{g}}$$

The flow regime can also be identified by critical depth (h_c):

$$h > h_c \quad \text{Subcritical flow}$$
$$h = h_c \quad \text{Critical flow}$$
$$h < h_c \quad \text{Supercritical flow}$$

7. At a certain flow rate, when the normal depth (h_0) of uniform flow is equal to the critical depth h_c, the corresponding channel slope is called the critical bed slope, i_c. The actual slope, i, is used to compare with the critical slope, i_c, to decide the specific slope type: mild slope, $i < i_c$; critical slope, $i = i_c$; steep slope, $i > i_c$. This can also be used to identify the flow regime of water surface profiles in open channel flow.

8. The hydraulic jump and the hydraulic drop (water fall) are local hydraulic phenomena that affect the flow regime as they indicate a transition, typically via the critical flow. Both phenomena are referred to as rapidly varied flow. Since the hydraulic jump consumes a large amount of mechanical energy, it is often used to eliminate the tremendous kinetic energy of high speed flow in the downstream of a structure such as a spillway.

9. Gradually varied flow in a prismatic channel can be described by

$$\frac{dh}{dl} = \frac{i - J}{1 - Fr^2}$$

which is the theoretical basis for a qualitative analysis on water surface profiles. For steady gradually varied flow in prismatic channels, there are 12 possible types of water surface profile, among which M_1, M_2, M_3 and S_2 types are commonly seen in engineering.

10. Based on the qualitative analysis, the water surface profile of gradually varied flow can quantitatively be obtained by the direct step method, which starts from a control section, and is based on,

$$\Delta l = \frac{dE_s}{i - \overline{J}} = \frac{E_{sd} + E_{su}}{i - \overline{J}}$$

11. The basic formula for discharge calculation of weir flow is

$$Q = \sigma_s \varepsilon m b \sqrt{2g} H_0^{\frac{3}{2}}$$

and the basic formula for calculating the flow under a sluice gate is

$$Q = \sigma_s \mu b e \sqrt{2g} H_0^{\frac{1}{2}}$$

where the coefficients are obtained from empirical formulae.

Multiple-choice questions (one option)

6-1 The driving force of open-channel flow is _____.
(A) inertial force (B) pressure force (C) gravity (D) resistance force

6-2 In a prismatic channel _____.
(A) the longitudinal axis can be curvilinear
(B) the channel bed slope can change with distance
(C) the shape and size of cross-section can be changed with distance
(D) The flow area of cross-section only changes with the depth of flow

6-3 Uniform flow of open channel may occur in _____.
(A) the prismatic channel with positive slope
(B) a horizontal prismatic channel
(C) the channel of positive slope having structures
(D) the channel with varying roughness along the channel

6-4 In a pipe of diameter D, the hydraulic radius of half full pipe flow is _____.
(A) D (B) $0.5D$ (C) $0.125D$ (D) $0.25D$

6-5 Uniform flow of open channel is _____.
(A) turbulent and rapidly varied flow (B) unsteady flow
(C) steady gradually varied flow (D) unsteady laminar and turbulent flow

6-6 For the given steady flow in a certain channel, when the channel slope increases, the normal depth of flow will be _____.
(A) increased (B) decreased (C) unchanged (D) undetermined

6-7 For the given steady flow in a certain channel, when the channel slope increases, the critical depth of flow will be _____.
(A) increased (B) decreased (C) unchanged (D) undetermined

6-8 The aspect ratio of optimal hydraulic rectangular cross-section equals _____.
(A) 1 (B) 1.5 (C) 1.73 (D) 2

6-9 Optimum hydraulic cross-section is defined as _____.
(A) cross-section of the minimum cost
(B) cross-section of the maximum sectional area of a certain flow rate
(C) cross-section with the minimum roughness
(D) cross-section of the minimum wetted perimeter for a certain sectional area

6-10 For rapidly varied flow _____.
(A) $Fr > 1$ (B) $h > h_c$ (C) $i > i_c$ (D) $dE_s/dh > 0$

6-11 For gradually varied flow _____.

(A) $Fr > 1$ (B) $h > h_c$ (C) $i > i_c$ (D) $dE_s/dh > 0$

6-12 When rapidly varied flow meets gradually varied flow in a channel, there will occur _____.

(A) hydraulic drop (water fall)   (B) hydraulic jump
(C) continuous and gradual transition   (D) all the above

6-13 In a hydraulic jump, in cross-sections before and after the jump _____ are the same.

(A) the depths   (B) the flow velocities
(C) the mechanical energies   (D) the function values of hydraulic jump

6-14 In a channel of mild slope, it is impossible to have _____.

(A) uniform flow   (B) the water surface profile of S_2 type
(C) the water surface profile of M_3 type   (D) hydraulic jump

6-15 Usually in weir flow _____.

(A) the upstream flow is rapidly varied flow
(B) the flow at the entrance of overflow is gradually varied flow
(C) the flow is formed over an obstacle
(D) the water surface profile upstream and downstream the weir is discontinuous

6-16 For weir flow, _____ have a minimal impact on the flow.

(A) thin-walled weirs   (B) practical weirs
(C) broad-crested weirs   (D) culverts and bridge piers

6-17 In weir and gate outflow, relative to free outflow, the flow rate of submerged outflow is _____.

(A) larger   (B) smaller   (C) the same   (D) all above possible

Problems

6-1 Uniform flow in a trapezoidal earth canal: the bottom width $b = 2.4$ m, side slope $m = 1.5$, Manning's roughness $n = 0.025$ and bed slope $i = 0.001$. If the flow depth is 1.2 m, determine the flow velocity and discharge capacity of the channel.

6-2 The circular spillway tunnel of a reservoir has a diameter of $d = 8$ m, the bottom slope $i = 0.002$, and roughness $n = 0.014$. The flow is unpressurized uniform flow. If the water depth h is 6.2 m, determine the flood discharge Q.

6-3 In a trapezoidal concrete channel with uniform flow: the discharge $Q = 35$ m³/s, base width $b = 8.2$ m, side slope $m = 1.5$, Manning's roughness $n = 0.012$, and bed slope $i = 0.00012$; determine the height of the embankment (assume the freeboard is 0.5 m).

6-4 In a trapezoidal earth irrigation canal designed for uniform flow: the bed slope $i = 0.002$, side slope $m = 1.5$, Manning's roughness $n = 0.025$, and design flow at a flow rate of $Q = 4.2$ m³/s. If the depth of flow is $h = 0.95$ m, determine the bottom width b of channel.

6-5 A rectangular concrete channel with masonry surface (rough) requires to carry a flow of $Q = 9.7$ m³/s. If the bed slope $i = 0.001$, determine the dimension of the optimum hydraulic cross-section.

6-6 Construction of a trapezoidal canal requires to carry a flow of $Q = $ m³/s. The channel has the bed slope $i = 0.0022$, side slope $m = 1.0$ and Manning's roughness $n = 0.003$. Design the cross-sectional dimension of channel according to the maximum permit velocity of 0.8 m/s.

6-7 A rectangular aqueduct has the width $b = 1.8$ m, the length $l = 116$ m, the entrance bottom elevation is 52.06 m, and the wall surface is cement plaster ($n = 0.011$). If the design flow of $Q = 7.65$ m³/s and the depth of flow $h = 1.7$ m are required, find the bed slope i for the aqueduct and the bottom elevation at the exit.

6-8 A trapezoidal channel carries water at $Q = 10$ m³/s, the bottom slope $i = 1/3000$, and the side slope coefficient $m = 1.0$. The channel is masonry lining on the surface with cement mortar. Determine the cross-sectional dimension of channel for an aspect ratio $b/h = 5$.

6-9 A rectangular cross-section channel has the bottom width $b = 3$ m, roughness $n = 0.022$, and bottom slope $i = 0.0005$. If the flow rate of $Q = 4.8$ m³/s is required, determine:

(1) the wave speed;

(2) the Froude number;

(3) the flow regime using different methods.

6-10 If an 8m wide rectangular channel carries water at a flow rate of $Q = 17.25$ m³/s, determine the critical depth of flow.

6-11 A trapezoidal earth canal has the bottom width $b = 12$ m, side slope $m = 1.5$, and roughness $n = 0.025$. If the flow rate of $Q = 18$ m³/s is required, determine the critical depth and the critical bed slope.

6-12 A hydraulic jump occurs in a horizontal prismatic rectangular channel, where the discharge per unit width is $q = 0.351$ m³/(s·m), the upstream depth of hydraulic jump $h' = 0.0528$ m, determine the downstream depth of hydraulic jump, h''.

6-13 In a prismatic rectangular channel, the upstream depth of a hydraulic jump $h' = 0.2$ m, and the downstream depth of the jump, $h'' = 1.4$ m, determine the discharge per unit width of channel q.

6-14 A hydraulic water jump occurs in a flat prismatic rectangular channel, with the upstream Froude number $Fr_1 = \sqrt{3}$, what is the relationship between the jump depths h' (upstream) and h'' (downstream)?

6-15 Analyze the possible shape of the water surface profiles in Fig. 6-48.

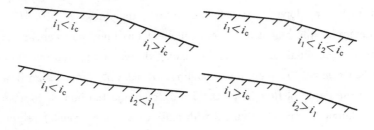

Fig. 6-48

6-16 As shown in Fig. 6-49, the prismatic channels are long enough, and the bed slope $i_1 < i_c$, $i_2 > i_3 > i_c$ and the opening (e) of the gate is smaller than the critical water depth h_c. Sketch the water surface profiles with the identification of the profile type.

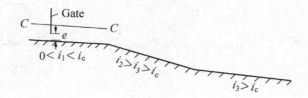

Fig. 6-49

6-17 A long straight rectangular channel is used to deliver water to the lower water basin, as shown in Fig. 6-50. If the channel bottom width $b = 1$ m, bed slope $i = 0.0004$, roughness $n = 0.014$ and the normal depth $h_0 = 0.5$ m, determine:

(1) the flow rate of channel;

(2) the flow depth of the exit section at the end of the channel;

(3) the schematic profile of water surface in the channel.

6-18 Fig. 6-51 shows the flat rectangular channel with a control gate, which has a flow rate of $Q = 12.7$ m³/s, the depth of contraction section $h = 0.8$ m, channel width $b = 3.8$ m, and the roughness $n = 0.012$. The downstream of the flat channel is connected with a steep channel; if the critical depth h_{co} is expected at the connecting point, how far should the connecting point be away from the contraction section?

6-19 A broad-crested weir with right angle inlet has no side contraction. The weir width $b = 4.0$ m, the height of the weir $P_1 = P_2 = 0.60$ m, and the acting water head of weir $H = 1.2$ m. If the downstream water depth $h = 0.80$ m, determine the flow rate Q.

6-20 If the downstream water depth is $h = 1.70$ m in the above problem (6-19), what is the flow rate over the weir?

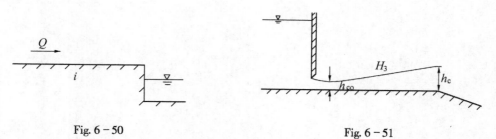

Fig. 6-50   Fig. 6-51

6-21 A broad-crested weir with round inlet has no side contraction. The flow rate $Q = 12$ m³/s, weir width $b = 1.8$ m, weir height $P_1 = P_2 = 0.80$ m, and the downstream water depth $h = 1.73$ m. Determine the upstream water head H.

6-22 For a weir of flat top with the dimension shown in Fig. 6-52 (units in m), determine:

(1) the type of weir;

(2) the flow rate over the weir.

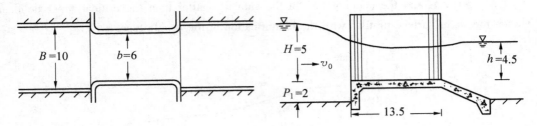

Fig. 6 - 52

6 - 23 In a vertical sluice gate on a flat base, the depth of flow in front of the gate is $H = 4$ m, the width of gate $b = 5$ m, the opening of gate $e = 1$ m, and the approaching velocity of gate is 1.2 m/s. Determine the discharge with free outflow.

Chapter 7 Seepage flow

7.1 The phenomenon of seepage and the seepage model

7.1.1 Seepage phenomenon

Seepage flow is the flow of liquid in a porous medium such as soil and rock. Seepage flow occurs when where there is any water flowing through earth dams, the surrounding aquifers of wells and underground water-collecting tunnels, channel and river banks, and some foundations below certain engineering structures. Seepage theory is not only applied in such fields as water conservancy, chemical engineering, petroleum, geology, mining engineering, but is also widely used in many aspects in civil engineering as follows:

(1) Hydraulic calculation of water catchment structures such as wells and corridors.

Wells and catchment corridors are commonly used for well point drainage in the construction of various water supply pipelines and buildings. The main aim of most calculations is to determine the flow rate of seepage and to select the dimensions of the catchment structures.

(2) Water storage and irrigation projects

The groundwater level increases with the rising water level in reservoirs, rivers and channels nearby. On the one hand, a large leakage via a permeable layer will result in the decrease of both water storage and water level; on the other hand, a large leakage will result in swamping and salinization of nearby farmlands and soil, which could influence the stability of embankments and gate dams, or destroy crops.

(3) Water intake projects

Water intake structures from reservoirs or rivers are usually constructed on wet soil, so some engineered anti-permeable and anti-floating measures are often required. Understanding the seepage and geological characteristics of the base of a river bed are therefore essential, as these will help engineers take appropriate precaustions in construction.

(4) Water-retaining earth dams

As shown in Fig. 7-1, the soil in an earth dam plays an important role in retaining water, as the upstream water will infiltrate into the dam body and exit from the downstream side. The free surface line of any seepage in the dam is called the seepage line. The part of the dam under the seepage line is immersed into water, resulting in a decrease of soil strength. Moreover, serious seepage will affect the stability of earth dam, and even cause the collapse of the entire dam.

(5) Impervious water-retaining structures on permeable foundation

As shown in Fig. 7-2, when a concrete gated dam is used for retaining water, the effect of the difference between the upstream and downstream water levels, causes infiltration into the subsoil layer beneath the structure; some water will move downstream, and finally seeps from the base of the dam into the river downstream. As a result of the hydrodynamic pressure, the upward pressure generated by the water on the base of the gated dam, called the uplift pressure, may be significant for the stability of the whole structure.

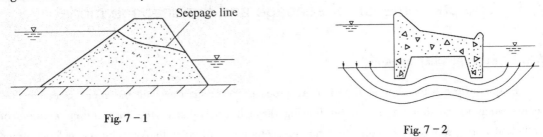

Fig. 7-1

Fig. 7-2

As the flow resistance in the pores between soil or sand particles is large, the seepage velocity is usually small. This means that the velocity head can be ignored and the total head can be replaced by the piezometric head. Hydraulic calculation of seepage includes the following aspects:

① Determine the seepage discharge. For example, calculate the inflow discharge of seepage into wells, catchment corridors and other structures.

② Determine the position of seepage line, such as those in earth dams and near pumping wells.

③ Determine seepage pressures, including the uplift pressure acting on the base of dams or structures.

④ Estimate the destructive effect of seepage on soil. When the seepage velocity is large, small particles in soil can be moved away through the pores, which is called "shifting-sand" phenomenon. With the amount of shifting-sand increases, the pores in the soil become larger, resulting in damage to the soil structure. Moreover, if the seepage flow is so large that it will fully fill ditches, wellheads or other structures in a short period of time, this will then delay construction plans and even cause significant adverse consequences, such as collapse. This phenomenon is called "piping". To prevent its occurrence, some technical measures are usually necessary.

7.1.2 State of water in soil

Water in soil can stay in the form of water vapour, attached water, film water, capillary water and gravitational water.

Water vapour exists in soil in a form of vapour. As its amount is small, the water vapour can usually be ignored in the seepage.

Because the soil particles are negatively charged making water molecules polarized, both attached and film water are absorbed in the soil. Attached water is weakly absorbed in the outer layers while film water is strongly attached in the inner layers. Considering that these two types of water are hard to remove and their quantity is small, they are always neglected in the analysis of seepage.

Capillary water can move through the pores of soil due to the effect of surface tension. Except in certain special situations, this water is not considered either.

Gravitational water is the one that moves in soil pores under gravity. When the water content of the soil is high, most of the water in the soil exists in the form of gravitational water.

This chapter will study the law of motion of only gravitational water in soil.

7.1.3 The characteristics of soil seepage

Soil properties have a significant impact on seepage. The larger the porosity of soil is and the more uniform the particle distribution, the higher the permeability of the soil.

The porosity of soil is the ratio of the void volume within a soil body, ω, to the total volume of the soil body W:

$$n = \frac{\omega}{W} \qquad (7-1)$$

Particle uniformity can be expressed by a non-uniform coefficient K_{60}:

$$K_{60} = \frac{d_{60}}{d_{10}} \qquad (7-2)$$

where d_{60} is the sieve diameter through which 60 percent of soil particles in weight can pass, and d_{10} is the sieve diameter through which 10 percent of particles in weight can pass.

Generally, $K_{60} \geqslant 1$. The larger K_{60} is, the less uniform the particle distribution is. For the soil that consists of homogeneous particles, $K_{60} = 1$.

Based on the structure of soil and characteristics of seepage, soil can be classified into two types, namely, homogeneous and heterogeneous. Homogeneous soil has the same permeability or hydraulic conductivity everywhere in all dimensions; otherwise the soil is heterogeneous soil.

The layered structure or columnar structure of natural soil makes the permeability different at different directions in the soil. Accordingly, the concepts of isotropic soil and anisotropic soil are used. An isotropic soil has the same permeability or hydraulic conductivity in all directions at the same point; otherwise a soil is an isotropic soil. For example, the soil consisting of equal-sized sphere particles arranged regularly is isotropic soil.

This chapter only focuses on the motion law of seepage in homogeneous isotropic soil.

7.1.4 Seepage models

Particles in a soil are significant different in shape and size, and the distributions of soil pores are also extremely irregular, so the real seepage flow is very complicated. It is difficult to determine the actual velocity of seepage at any position, either by theoretical analysis or by experiments, which in fact is not necessary in practical engineering. Using an engineering statistical method, seepage movement can be described by the average value of flow, i. e. an idealized seepage model is used, thus simplifying the real seepage flow.

In the simplified model, the main flow direction of seepage is considered rather than the irregular path of seepage flow, and also the existence of soil particle is neglected. So the water in seepage flow is assumed to flow through the whole soil body, and thus the flow is continuous. In essence, the seepage flow in the partically void space of soil is regarded as a movement of continuum in continuous space medium.

Based on this model, the hydraulic concepts and methods described previously, such as cross-section, streamline, and average velocity, can be applied to the study of seepage flow.

In the seepage model, the seepage velocity of a small cross-section can be defined as

$$u = \frac{\Delta Q}{\Delta A} \qquad (7-3)$$

where ΔQ is the seepage discharge through the small cross-section, and ΔA is the area of cross-section that consists of soil particles and their void, which is larger than the actual cross-section of flow.

So the velocity in the model is much smaller than the actual velocity. However, because the seepage velocity is small, the kinetic energy is often neglectable, so is the impact caused by the difference on engineering application.

To meet the requirement of engineering, several principles should be followed when the simplified model is used for real situations:

① The discharge in the seepage model should be same as that in the practical situation;

② For a certain section, the pressures of fluid in the model and real situation should equal;

③ The head losses should be same.

Based on the concept of the seepage model, seepage can be classified as either steady or unsteady seepage flow, uniform or non-uniform seepage flow, rapidly varied or gradually varied seepage flow, and pressurized seepage or unconfined seepage flow.

7.2 The basic law of seepage flow

7.2.1 Darcy's law

To solve the basic problems of seepage in engineering practice, in 1852—1855, a French hydraulic engineer, named Henry Darcy, generalized the results of his experiments into the Darcy law, which is named after his name.

The experiment device is shown in Fig. 7-3. The main device is a cylinder with an opening on the top, which holds homogeneous sand. An intake pipe and an overflow pipe are connected to the top of the cylinder, which maintains a constant water level. The side wall of the cylinder is connect with two piezometric tubes, l apart from each other. Water enters into the cylinder from the top, infiltrates through the sand, and exits from the filter board D. The seepage discharge is measured by container *C*. Because the water level at the upper part of the cylinder is constant, the seepage is steady seepage flow, and the water surfaces in the piezometric tubes are constant. Darcy observed that the two piezometric tubes with different installation heights have different water levels, which shows there is a head loss in seepage flow.

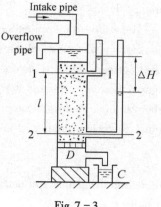

Fig. 7-3

Through further experiments, Darcy noticed that for cylinders in different dimensions and types of soil particles, seepage discharge *Q* is proportional to cross-sectional area *A* and energy gradient *J*, and associated with the permeability of soil. Mathematically, it can be expressed:

$$Q = kAJ = kA\frac{\Delta H}{l} \qquad (7-4)$$

where *k* is the coefficient of soil permeability.

The average seepage velocity in the cross-section of the cylinder is

$$v = \frac{Q}{A} = kJ$$

Darcy's tests were conducted in the equal-diameter cylinder with homogeneous sand, so the seepage was uniform seepage flow. Therefore, the seepage velocity of every point equals the average velocity of section. The seepage velocity equation can then be expressed as

$$u = kJ \qquad (7-5)$$

Equation (7-5) is the Darcy formula of seepage, also called Darcy's law. It suggests that seepage velocity is directly proportional to energy gradient and associated with the coefficient of soil permeability. So Darcy's law is also called the linear law of seepage.

Afterwards, through much more practice and research, Darcy's law has been extended to apply to other common situations, such as non-uniform, unsteady seepage, in which the expression is then written as

$$u = kJ = -k\frac{dH}{dl}$$

where *u* is the seepage velocity at a certain point, *H* is the total head, and *l* is the length of seepage path.

7.2.2 The limitations of Darcy's law

Darcy's law was obtained by the experiment of sand soil, and was later extended to use for clay and rock with cracks, etc. However, further study shows that seepage flow does not follow Darcy's law in certain situations. Thus, to solve problems in practice, the limitations of Darcy's law have to be considered.

According to Darcy's law, the seepage head loss is directly proportional to the velocity, the same as the law of head loss in laminar flow, which means that Darcy's law can only be applied to laminar flow.

Regarding the application criteria of Darcy's law, some researchers use the size of particle diameter, whilst most researchers think that it is more appropriate to use the Reynolds number (Re) instead. Studies have shown that the critical Reynolds number (Re_k) is not a constant from laminar flow to turbulent flow, but that it changes with the particle diameter, the porosity and other factors. Pavlovskii proposed that when $Re < Re_k$, the seepage flow is laminar flow. Re is the actual Reynolds number given by

$$Re = \frac{1}{0.75n + 0.23} \frac{vd}{\nu} \qquad (7-6)$$

where n is the porosity of soil, and d is the effective diameter of soil, which is usually used as d_{10} in unit cm.

Re_k is the critical Reynolds number. Typically,

$$Re_k = 7 - 9 \qquad (7-7)$$

For non-laminar flow, the law of flow can be described as follows

$$v = kJ^{\frac{1}{m}} \qquad (7-8)$$

when $m = 1$, it is laminar flow; when $m = 2$, it is fully turbulent flow; when $1 < m < 2$, it is in a transition region of flow.

It is worth noting that all the laws above are only applied to the situations where the soil structure is not damaged by the seepage, which means that the seepage occurs in the soil with a stable structure. If the seepage causes the movement of soil particles, such as shifting-sand or piping, then the laws do not apply anymore.

7.2.3 The coefficient of permeability

The coefficient of permeability (k) is a comprehensive index that describes the characteristics of soil seepage. Its value has a great impact on the seepage calculation. Many factors influence the coefficient, such as the shape and size of soil particles, the structure and porosity of soil and the temperature of the water. So it is difficult to precisely determine the coefficient of permeability. Usually, the following methods are used to determine k in practical engineering:

(1) Empirical method

In preliminary estimation, because of the lack of actual data, some related specifications and data from already finished projects can be consulted, and then empirical formulae or laws that are obtained from the parameters such as the shape and size of soil particles, the porosity of soil, and the temperature of water, are used to determine the value of k.

(2) Experimental measurement

Use the measuring device of seepage as shown in Fig. 7-3 in laboratory, where

$$Q = kA \frac{\Delta H}{l}$$

then

$$k = \frac{Ql}{A \times \Delta H}$$

The value of k can be obtained by the experimental data.

Although the device and measurement in this method are simple, it is still difficult to reflect the real conditions because not all natural soil is truly homogeneous and it is hard to guarantee that there is no disturbance to the structure of the soil in the process of taking samples and in experiment. Thus, several repeated tests are needed to obtain a mean value.

(3) Field measurement

Drill a well or dig a hole in the field of seepage, fill water into or pump water out from the hole/well, measure the discharge and water head, and then calculate the value of k from theoretical formulae. This method can measure the mean coefficient of seepage in a large field of seepage. The data obtained are reliable, but it requires a large scale of field and is costly in terms of manpower and material resources.

In the approximate calculation, the value of k can be obtained from Table 7 − 1.

Table 7 − 1   The reference values of seepage coefficient

Types	Coefficient of seepage $k/\text{cm} \cdot \text{s}^{-1}$
clay	$< 6 \times 10^{-6}$
loam	$6 \times 10^{-6} - 1 \times 10^{-4}$
loess	$3 \times 10^{-4} - 6 \times 10^{-4}$
fine sand	$1 \times 10^{-3} - 6 \times 10^{-3}$
coarse sand	$2 \times 10^{-2} - 6 \times 10^{-2}$
pebble	$1 \times 10^{-1} - 6 \times 10^{-1}$

7.3   Dupuit's formula of steady gradually varied seepage flow

7.3.1   The velocity distribution in steady uniform and non-uniform seepage flows

Based on the seepage model, seepage flow can be classified into uniform and non-uniform seepage flow by means of open-channel flow descriptors. The seepage in which the hydraulic parameters such as velocity and pressure are constant is called uniform seepage; otherwise it is non-uniform seepage flow. In non-uniform seepage, it is called non-uniform and gradually varied seepage if the streamlines are approximately parallel with each other. Otherwise it is called non-uniform and rapidly varied seepage. As seepage flow is subject to Darcy's law, uniform or non-uniform seepage flow has some different chracteristics from open-channel flow.

According to Darcy's law, the energy gradient J can also be expressed as $J = -\dfrac{\mathrm{d}H}{\mathrm{d}l}$. Because the seepage velocity is usually small, the velocity head can be ignored. Thus the total head can be replaced by the piezometric head.

7.3.1.1 The velocity distribution of cross-section in steady uniform seepage flow

If the seepage flow with bed slope i is uniform, as shown in Fig. 7-4, then the depth of water h, the mean velocity v of cross-section and the gradient of flow surface profile are all constant along the flow, the same as uniform flow in open channels. The point velocity of any small streamtube in isotropic soil equals the mean velocity of cross-section, that is

$$v = u = -k\frac{dH}{dl} = kJ = ki \tag{7-9}$$

As shown in Fig. 7-4, the rectanglar velocity distribution at any cross-section remains the same.

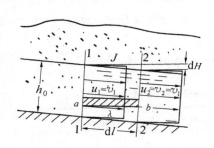

Fig. 7-4

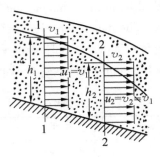

Fig. 7-5

7.3.1.2 The velocity distribution of cross-section in steady gradually varied seepage flow

For the non-uniform and gradually varied seepage flow, the depth of water h, the mean velocity v of cross-section and the gradient of flow surface profile will change along the flow, as shown in Fig. 7-5. Take cross-sections 1-1 and 2-2 for analysis. As the velocity head is small enough to be neglected, the piezometric head becomes the total head. So $H_1 = z_1 + \frac{p_1}{\gamma}$ can represent the total head at section 1-1. Similarly, $H_2 = z_2 + \frac{p_2}{\gamma}$ can be used as the total head at section 2-2. Because the pressure distribution of water in the cross-section of gradually varied flow follows the law of hydrostatic pressure distribution, that is, $z + \frac{p}{\gamma}$ = constant, and then $dH = H_2 - H_1$ = constant. dH between cross-sections of any small streamtube is $dH = H_2 - H_1$. Given the distance dl of a streamtube, the energy gradient over the streamtube is $J = -\frac{dH}{dl} = \frac{H_1 - H_2}{dl}$. Because the streamlines of gradually varied flow are approximately parallel, the length of any small streamtube between sections 1-1 and 2-2 is dl, which also means that the point velocity u of the streamtube is constant and equals $-k\frac{dH}{dl}$. Thus, the mean velocity distribution of every cross-section is rectangular, so the values of mean and point velocity are the same. But for different cross-sections, the height and length of the rectangle, which represents the water depth and the mean velocity respectively, are not the same for different energy gradient $J = -\frac{dH}{dl}$.

The basic formula for the non-uniform and gradually varied seepage flow is

$$v = u = -k\frac{dH}{dl} = kJ \qquad (7-10)$$

That is Dupuit's formula, which was derived by the French scholar J. Dupuit in 1857. This formula only applies to gradually varied seepage flow, not rapidly varied seepage flow. In uniform seepage the velocity at any point in a seepage field is the same, but in gradually varied seepage the velocity of points only in the same cross-section is constant.

7.3.2 The basic differential equation and the seepage curve of steady gradually varied seepage flow

The free surface of gravitational water in unconfined seepage is called seepage surface. The seepage surface in planar problems is seepage curve. To obtain the seepage curve in engineering, the differential equation of gradually varied seepage can be derived from Dupuit's formula, and the seepage curve can then be obtained by integration.

As shown in Fig. 7-6, for the gradually varied seepage with bed slope i over an impermeable layer, the water depth of initial section 1-1 is h, and the water depth of section 2-2 is $h + dh$ through flow path dl. Thus, the piezometric head difference between 1-1 and 2-2 is

$$-dH = (idl + h) - (h + dh) = idl - dh$$

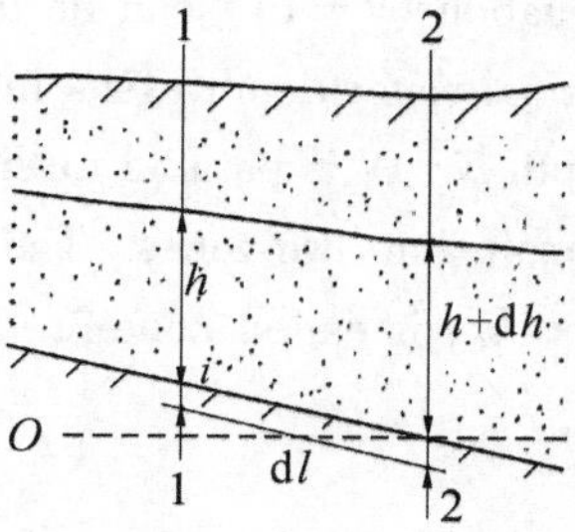

According to Dupuit's formula, the cross-sectional mean velocity of gradually varied seepage is

Fig. 7-6

$$v = k\left(-\frac{dH}{dl}\right) = k\left(i - \frac{dh}{dl}\right) \qquad (7-11)$$

The discharge is

$$Q = kA\left(i - \frac{dh}{dl}\right) \qquad (7-12)$$

Equation (7-12) is the basic differential equation of steady gradually varied seepage flow. It can be used to calculate and analyze the seepage curve of gradually varied seepage.

The normal depth and critical depth of water are very important for the analysis of the water surface profile in open channels. As the velocity head can be neglected in seepage, the unit specific energy of cross-section is $E_s = h + \frac{\alpha v^2}{2g} = h$. Thus, the specific energy curve becomes a straight line, so the concepts including the critical depth of water, critical bed slope, mild or steep slope, and subcritical or supercritical flow do not exist in seepage. Slope in seepage only has three types: positive slope, horizontal slope and adverse slope. The actual water depth is used to compare with only the normal water depth of uniform flow.

Because cross-sections of seepage are usually very wide, they are often regarded as rectangles. So the area of cross-section $A_0 = bh$, and the normal water depth of uniform seepage in planar problems can be calculated as follows

Fluid Mechanics in Civil Engineering

$$Q = vA_0 = kJA_0 = kibh_0$$

Hence

$$h_0 = \frac{Q}{kib} \qquad (7-13)$$

where h_0 is the normal water depth, and b is the width of the rectangles.

Seepage curves in three types of slope are described in the following sections.

7.3.2.1 Seepage curve in positive slope ($i > 0$)

Uniform flow may occur in positive slope. The seepage discharge equals the corresponding discharge of uniform flow:

$$Q = kibh_0 = k\left(i - \frac{dh}{dl}\right)bh$$

Then

$$\frac{dh}{dl} = i\left(1 - \frac{h_0}{h}\right) \qquad (7-14)$$

Equation (7-14) can be used to analyze the type of seepage curve in positive slope. Because normal water depth exists in the underground channel of positive slope, the line of normal water depth $N-N$ is parallel to the bed slope, as shown in Fig. 7-7, and line $N-N$ divides the flow domain into two zones. The part where $h > h_0$ is called Zone 1, whilst the other zone, where $h < h_0$, is called Zone 2.

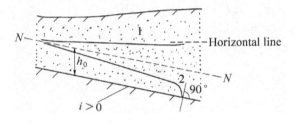

Fig. 7-7

In Zone 1, $h > h_0$, $\frac{dh}{dl} > 0$, and the seepage curve is a backwater curve. Towards the upstream end of the curve, $h \to h_0$, $\frac{dh}{dl} \to 0$, and line $N-N$ is the asymptote of seepage curve in the upstream. Towards the downstream end of the curve, $h \to \infty$, $\frac{dh}{dl} \to i$, and the horizontal line is the asymptote in the downstream.

In Zone 2, $h < h_0$, $\frac{dh}{dl} < 0$, and the seepage curve is a falling curve. To the upstream end of the curve, $h \to h_0$, $\frac{dh}{dl} \to 0$, and line $N-N$ is still the asymptote of the curve at the upstream end. At the downstream end of the curve, $h \to 0$, $\frac{dh}{dl} \to -\infty$. Thus, the seepage curve in the downstream has a tendency to be orthogonal to the channel bottom.

The seepage curve can be obtained by integrating Equation (7-14).

Given $\eta = \dfrac{h}{h_0}$, then $dh = h_0 d\eta$, put into Equation (7-14):

$$\frac{h_0 d\eta}{dl} = i\left(1 - \frac{1}{\eta}\right)$$

$$dl = \frac{h_0}{i}\left(1 + \frac{1}{\eta - 1}\right) d\eta \qquad (7-15)$$

Given the distance l between two sections, the water depth at the upstream section is h_1, $\dfrac{h_1}{h_0} = \eta_1$, the water depth at the downstream section is h_2, and $\dfrac{h_2}{h_0} = \eta_2$, so the integration becomes

$$\int_0^l dl = \int_{\eta_1}^{\eta_2} \frac{h_0}{i}\left(1 + \frac{1}{\eta - 1}\right) d\eta$$

then

$$l = \frac{h_0}{i}\left(\eta_2 - \eta_1 + \ln\frac{\eta_2 - 1}{\eta_1 - 1}\right)$$

$$= \frac{h_0}{i}\left(\eta_2 - \eta_1 + 2.3\lg\frac{\eta_2 - 1}{\eta_1 - 1}\right) \qquad (7-16)$$

This is the seepage curve equation of planar seepage in positive slope.

7.3.2.2 Seepage curve in horizontal slope ($i = 0$)

Put $i = 0$ into $Q = k\left(i - \dfrac{dh}{dl}\right) bh$. Then:

$$\frac{dh}{dl} = -\frac{Q}{kbh} < 0$$

The seepage curve must be a falling curve, shown in Fig. 7-8. The upstream part of the curve depends on the boundary conditions. In the limiting case, $h \to \infty$, $\dfrac{dh}{dl} \to 0$, so a horizontal line is the asymptote of seepage curve. In the downstream part of the curve, $h \to 0$, $\dfrac{dh}{dl} \to -\infty$, so the seepage curve in the downstream has a tendency to be orthogonal to the channel bottom.

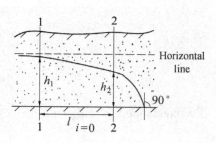

Fig. 7-8

Integrating the equation $\dfrac{dh}{dl} = -\dfrac{Q}{kbh}$, and given $q = \dfrac{Q}{b}$, then

$$\frac{q}{k} dl = -h dh$$

The integral formula is $\int_0^l \dfrac{q}{k} dl = \int_{h_1}^{h_2} -h dh$. Thus,

$$\frac{ql}{k} = \frac{1}{2}(h_1^2 - h_2^2) \qquad (7-17)$$

This is the seepage curve equation of planar seepage in horizontal slope.

7.3.2.3 Seepage curve in adverse slope ($i<0$)

According to Equation (7-12), $Q = k\left(i - \dfrac{dh}{dl}\right)bh$, then

$$\frac{dh}{dl} = i - \frac{Q}{kbh} < 0$$

Thus, the seepage curve must be a falling curve, shown in Fig. 7-9. In the upstream part of the curve, $h \to \infty$, $\dfrac{dh}{dl} \to i$, so the horizontal line is the asymptote of the curve. In the downstream part of the curve, $h \to 0$, $\dfrac{dh}{dl} \to -\infty$, and then the seepage curve is orthogonal to the channel bottom.

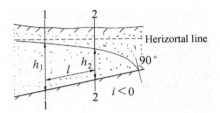

Fig. 7-9

Letting $i' = |i|$ and imaging a uniform seepage with equal width in bed slope i', its discharge is equal to that of gradually varied seepage with adverse slope i. Thus,

$$Q = ki'h_0'b$$

where h_0' is the normal water depth of the imaginary uniform seepage. Then,

$$\frac{dh}{dl} = i - \frac{Q}{kbh} = i - \frac{ki'h_0'b}{kbh} = i - \frac{i'h_0'}{h}$$

$$= i'\left(\frac{i}{i'} - \frac{h_0'}{h}\right) = -i'\left(1 + \frac{h_0'}{h}\right)$$

If $\eta' = \dfrac{h}{h_0'}$, then $dh = h_0' d\eta'$. Thus,

$$\frac{h_0' d\eta'}{dl} = -i'\left(1 + \frac{1}{\eta'}\right)$$

Then,

$$\frac{i'}{h_0'}dl = -\frac{d\eta'}{1 + \dfrac{1}{\eta'}} = -\frac{\eta' d\eta'}{1 + \eta'}$$

By integration,

$$\int_0^l \frac{i'}{h_0'}dl = \int_{\eta_1'}^{\eta_2'}\left(-\frac{\eta'}{1+\eta'}\right)d\eta'$$

Then,

$$\frac{i'}{h_0'}l = \eta_1' - \eta_2' + \ln\frac{1+\eta_2'}{1+\eta_1'}$$

Rewriting it in common logarithm,

$$\frac{i'}{h_0'}l = \eta_1' - \eta_2' + 2.3\lg\frac{1+\eta_2'}{1+\eta_1'} \qquad (7-18)$$

This is the seepage curve equation of planar seepage flow in adverse slope.

【Example 7-1】 Shown in Fig. 7-10, between the channel and the river is a permeable layer, the bed slope of the impermeable layer $i = 0.02$, the coefficient of seepage $k = 0.005$ cm/s, the distance between the channel and river is 180 m, the water depth of channel is $h_1 = 1.0$ m, and the water depth of the river (exit of seepage) is $h_2 = 1.9$ m. Assuming that the seepage is planar seepage, calculate the seepage discharge per unit width and the seepage curve.

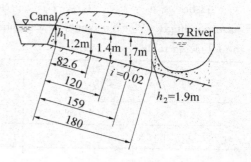

Fig. 7-10

Solution: As $h_1 < h_2$, the seepage curve is a rising curve.

From Equation (7-16),

$$l = \frac{h_0}{i}\left(\eta_2 - \eta_1 + 2.3\lg\frac{\eta_2 - 1}{\eta_1 - 1}\right).$$

Thus,

$$il - h_2 + h_1 = 2.3 h_0 \lg\frac{h_2 - h_0}{h_1 - h_0}$$

For the given data,

$$h_0 \lg\frac{1.9 - h_0}{1.0 - h_0} = \frac{1}{2.3} \times (0.02 \times 180 - 1.9 + 1.0) = 1.172$$

Now calculate by a trial-and-error method. Assuming a value of h_0, then calculate $h_0 \lg\frac{1.9 - h_0}{1.0 - h_0}$. Repeat the process until the value of $h_0 \lg\frac{1.9 - h_0}{1.0 - h_0}$ approaches 1.172. The process of calculation is shown in Table 7-2:

Table 7-2 Calculation of h_0

h_0/m	$h_0 \lg\dfrac{1.9 - h_0}{1.0 - h_0}$
0.92	1.001
0.94	1.131
0.96	1.315

By interpolation, $h_0 = 0.9451$ m. Thus, the unit discharge is

$$q = kih_0 = 5 \times 10^{-5} \times 0.02 \times 0.9451 = 9.45 \times 10^{-7} \text{ m}^3/(\text{s} \cdot \text{m})$$

From $l = \dfrac{h_0}{i}\left(\eta_2 - \eta_1 + 2.3\lg\dfrac{\eta_2 - 1}{\eta_1 - 1}\right)$:

$$l = \frac{0.9451}{0.02}\left(\frac{h_2}{0.9451} - \frac{1}{0.9451} + 2.3\lg\frac{\dfrac{h_2}{0.9451} - 1}{\dfrac{1}{0.9451} - 1}\right)$$

Assuming $h_2 = 1.2$, 1.4, 1.7 m respectively, we can obtain l being 82.6, 120, 159 m correspondingly. Plot the points in a graph to obtain the seepage curve. The results are shown in the following table:

h_1/m	$\eta_1 = \dfrac{h_1}{h_0}$	h_2/m	$\eta_2 = \dfrac{h_2}{h_0}$	l/m
1.0	1.058	1.2	1.27	82.6
		1.4	1.48	120.0
		1.7	1.8	159.0

7.4 Seepage calculation of wells and catchment corridors

Wells and catchment corridors are usually applied for groundwater exploitation to meet the needs for domestic water, agricultural irrigation and industrial production. Pumping water from wells and corridors will result in groundwater level recession nearby. So wells and corridors are often also used for water drainage in wet-soil construction.

7.4.1 Catchment corridors

A rectangular catchment corridor is shown in Fig. 7-11. The water depth is h, the distance from the corridor is l, and the water depth is H. Where the groundwater level in the region beyond the distance of l is not affected, l is called the impact distance of the corridor. Because the corridor is built on a horizontal impermeable layer, the bed slope $i = 0$. From Equation (7-17),

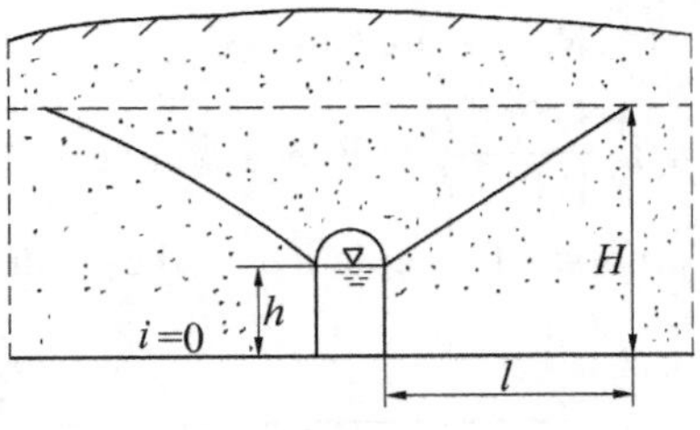

$$\frac{q}{k}l = \frac{1}{2}(H^2 - h^2) \qquad (7-19)$$

Fig. 7-11

where q is the seepage discharge per unit width in one side.

7.4.2 Fully penetrating open wells

The groundwater with a free surface is called unconfined groundwater or phreatic water. The corresponding well is called an open well. According to the positions of both the well and bottom impermeable layer, open wells can be classified into a fully penetrating well or a partially penetrating well. When the bottom of a well has already reached the impermeable layer, the well is a fully penetrating well. Otherwise, it is a partially penetrating well. The cross-section of well is usually circular.

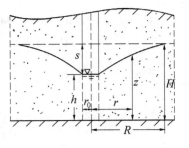

Fig. 7-12

Assume that the aquifer thickness of a fully penetrating well is H, and the radius of the well is r_0. Pumping water from a well will decrease the groundwater level nearby, which will form a funnel-shaped surface with a series of seepage lines symmetrically with respect to the central line of the well. Supposing that the extraction discharge is constant, the volume of the aquifer is large enough and the soil structure is stable, then the seepage will be steady uniform flow and the water depth h of well will keep constant.

For the surface of a concentric cylinder of radius r with the well, its area is $A = 2\pi rz$, where z is the aquifer thickness of the cross-section, and r is measured away from the well axis. For gradually varied seepage, the piezometric heads at every point of the cylindrical surface are the same. The energy gradient of any small streamtube is $J = \dfrac{dz}{dr}$. From Dupuit's formula, the cylindrical surface discharge of the gradually varied seepage is

$$Q = Av = 2\pi rz \cdot k \dfrac{dz}{dr}$$

By separating the variables

$$2\pi z dz = \dfrac{Q}{k} \dfrac{dr}{r}$$

Integrating it

$$\int_h^H 2\pi z dz = \int_{r_0}^R \dfrac{Q}{k} \dfrac{dr}{r}$$

where R is the radius of influence of the well, beyond which the groundwater level is not affected. Then

$$\pi(H^2 - h^2) = \dfrac{Q}{k} \ln \dfrac{R}{r_0}$$

Thus

$$Q = 1.366 \dfrac{k(H^2 - h^2)}{\lg \dfrac{R}{r_0}} \qquad (7-20)$$

This is the Dupuit's outflow formula of fully penetrating open well. When it is a water injection well, the outflow is negative, which means that the equation still can be applied.

When Q is constant, the difference between the water level inside the well and the original groundwater level, $s = H - h$, is also constant. Equation $(7-20)$ becomes

$$Q = 2.732 \dfrac{kHs}{\lg \dfrac{R}{r_0}} \left(1 - \dfrac{s}{2H}\right)$$

When $\dfrac{s}{2H} \ll 1$, it can be simplified as

$$Q = 2.732 \dfrac{kHs}{\lg \dfrac{R}{r_0}} \qquad (7-21)$$

Because it is easy to measure the value of s from h, Equation $(7-21)$ is commonly used in engineering.

Equation (7 – 21) clearly shows that Q is in direct proportion to k, H and s, which have significant impact on Q, but the influence of R on Q is relatively small.

The radius of influence (R) is usually measured by water pumping or injecting tests. In any preliminary calculation, or when the requirements are not high, R can be determined empirically. For fine sand, $R = 100 \sim 200$ m, for medium sand, $R = 250 \sim 500$ m, and for coarse sand, $R = 700 \sim 1000$ m. It also can be obtained by the empirical formula:

$$R = 3000s \sqrt{k} \tag{7-22}$$

where the units of s and R are m, and the unit of k is m/s. It is worth noting that because the radius of influence is an approximate parameter and is affected by Q and s, the values of R obtained by different methods vary considerably. However, the errors on the values of radius of influence would not have much impact on the calculation of discharge because the discharge is inversely proportional to the logarithm of the radius.

The seepage of a partially penetrating well is more complicated because the bottom of well is permeable, where Dupuit's formula does not apply anymore. Its outflow discharge is often obtained by empirical formulae, which is not covered in this chapter.

【Example 7 – 2】 An in-situ water-injection method is used to measure the coefficient of seepage k of a phreatic and water-bearing sandy ground. The hole of $d = 30$ cm is drilled and directly reaches the impermeable layer, as shown in Fig. 7 – 13. When the water injection discharge is $Q = 0.002$ m³/s, the water depth in the well is $h = 10$ m. If the thickness of the natural aquifer before water injection is $H = 8$ m, then calculate the value of k.

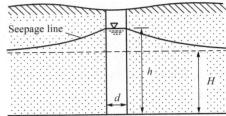

Fig. 7 – 13

Solution: According to Dupuit's Formula for fully penetrating open well, the coefficient of seepage is

$$k = -\frac{Q}{1.366(H^2 - h^2)} \lg \frac{2R}{d}$$

where R is the radius of influence, approximately 500 m by in-situ measurement. Then,

$$k = -\frac{0.002}{1.366(8^2 - 10^2)} \times \lg \frac{2 \times 500}{30 \times 10^{-1}}$$

$$= 6.19 \times \frac{10^{-5} \text{m}}{\text{s}} = 0.0062 \text{ cm/s}$$

7.4.3 Fully penetrating artesian wells

When an aquifer is between two impermeable layers, the pressure on the aquifer is larger than the atmospheric pressure. This type of underground water is called confined water. The corresponding well is called a confined well or an artesian well. Fig. 7 – 14 shows a fully penetrating artesian well.

Supposing that the thickness of aquifer is constant, represented by t; the bed slope is $i = 0$; the water depth in the well is l; the piezometric head of the confined water is H; the well radius is r_0; and the radius of influence is R. Take the same method as used in the fully penetrating open well. For the concentric cylindrical surface with radius r, Dupuit's formula for gradually varied seepage flow becomes

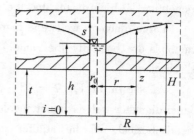

Fig. 7-14

$$Q = 2\pi r \cdot t \cdot k \frac{dz}{dr}$$

By separating the variables

$$dz = \frac{Q}{2\pi kt} \frac{dr}{r}$$

By integrating it

$$z = \frac{Q}{2\pi kt}\ln r + C$$

For the boundary condition, when $r = r_0, z = h$, then

$$C = h - \frac{Q}{2\pi kt}\ln r_0$$

Thus,

$$z = \frac{Q}{2\pi kt}\ln r + h - \frac{Q}{2\pi kt}\ln r_0$$

$$z - h = \frac{Q}{2\pi kt}\ln \frac{r}{r_0} = 0.37 \frac{Q}{kt} \lg \frac{r}{r_0}$$

Note that $r = R$, $z = H$. Then the outflow discharge of fully penetrating artesian well is

$$Q = 2.73 \frac{kt(H - h)}{\lg \frac{R}{r_0}} = 2.73 \frac{kts}{\lg \frac{R}{r_0}} \qquad (7-23)$$

where $s = H - h$ is the value of descending water level in the well.

If the discharge of outflow Q, the affected radius R, the well radius r_0, the coefficient of seepage k and the aquifer thickness t are all known, then from Equation (7-23)

$$s = \frac{Q \lg \frac{R}{r_0}}{2.73 kt} \qquad (7-24)$$

Equation (7-24) is based on the hypothesis $h > t$. For the case of $h < t$, the discharge formula of steady gradually varied seepage flow can also be derived, as seen in the problems at the end of this chapter.

7.4.4 The drainage of large-diameter well and foundation ditch

Large-diameter wells and foundation ditches are intake structures to collect shallow groundwater. The diameter of large-diameter well is usually $2 \sim 10$ m. This kind of well is often partially penetrated, which means that the well has water inflow both the sidewall and the bottom.

Large-diameter wells and foundation ditches have similar characteristics and their calculation methods are basically the same. Herein only take the large-diameter well as an example for discussion.

As shown in Fig. 7 - 15, the sidewall of the large-diameter well is impermeable and the bottom is hemispherical. If the aquifer underneath is deep enough, the water supply of the well is through the bottom.

The flow section is a hemisphere surface, which is concentric with the bottom of the well. Given that the radius is r, and the corresponding water depth is z, then

$$Q = \frac{1}{2} \cdot 4 \pi r^2 \cdot k \frac{dh}{dr}$$

By separating the variables and integrating it

$$Q \int_{r_0}^{R} \frac{dr}{r^2} = 2 \pi k \int_{H-s}^{H} dh$$

Then,

$$Q \left(\frac{1}{r_0} - \frac{1}{R} \right) = 2 \pi k s$$

As $R \gg r_0$, $\frac{r_0}{R} = 0$. Then,

$$Q = 2 \pi k r_0 s \tag{7-25}$$

This is the discharge formula of large-diameter well with hemispherical bottom. If necessary, the water yield (quantity) should be obtained from the relationship of measured Q and s.

7.4.5 Well group

Whether groundwater intake or excavation of foundation ditches is used to reduce groundwater level, usually several wells are made in one region for pumping. The group of wells is called a well group, as shown in Fig. 7 - 16.

Unless the distance between wells is too far, the seepage between any two wells will affect each other, which will make the seepage flow and seepage surface of a well group very complicated. So this section only focuses on a well group under simple conditions.

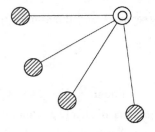

Fig. 7 - 16

Assume that each well is fully penetrated, its size and the discharge of pumping are unchanged, and the distance between wells is not too far. According to the superposition principle of potential flow, when several wells work at the same time, the potential function at any point equals the sum of potential function of each well working independently at the same point. Thus, the seepage discharge of well group can be obtained.

By introducing the potential function, φ, for a fully penetrating open well,

$$\varphi = \frac{1}{2}kz^2$$

$$d\varphi = kzdz$$

Then,

$$Q = v \cdot A = 2\pi rz \cdot k\frac{dz}{dr} = 2\pi r\frac{d\varphi}{dr}$$

For fully penetrating artesian wells $(h > t)$,

$$\varphi = ktz$$

$$d\varphi = ktdz$$

Then,

$$Q = 2\pi rt \cdot k\frac{dz}{dr} = 2\pi r\frac{d\varphi}{dr}$$

Thus, the potential function of fully penetrating well can be expressed as:

$$\varphi = \frac{Q}{2\pi}\ln r + C \tag{7-26}$$

According to the superposition principle of potential flow, the value of φ at any point when n wells work at the same time equals the combined values of φ_i when each well works separately. Then,

$$\varphi = \sum_{i=1}^{n}\varphi_i = \sum_{i=1}^{n}\frac{Q_i}{2\pi}\ln r_i + \sum_{i=1}^{n}C_i = \sum_{i=1}^{n}\frac{Q_i}{2\pi}\ln r_i + C$$

where r_i is the distance of the well i away from the given point, and C is a constant depending on the boundary conditions.

Because the outflow of each well is the same,

$$Q_1 = Q_2 = \cdots = Q_n = \frac{Q}{n}$$

where Q is the total outflow of well group. Then,

$$\varphi = \frac{Q}{2\pi} \cdot \frac{1}{n}\sum_{i=1}^{n}\ln r_i + C$$

$$= \frac{Q}{2\pi} \cdot \frac{1}{n}\ln\prod_{i=1}^{n}r_i + C$$

Supposing that the radius of influence for the well group is the same as that for each well, and the potential function at R is φ_R, then

$$\varphi_R - \varphi = \frac{Q}{2\pi}\left(\ln R - \frac{1}{n}\ln\prod_{i=1}^{n}r_i\right) \tag{7-27}$$

For open wells, $\varphi_R = \frac{1}{2}kH^2, \varphi = \frac{1}{2}kh^2$. Then,

$$Q = 1.366 \frac{k(H^2 - h^2)}{\lg R - \frac{1}{n}\lg \prod_{i=1}^{n} r_i} \qquad (7-28)$$

For artesian wells,

$$\varphi_R = kHt, \varphi = kht$$
$$\varphi_R - \varphi = kt(H - h)$$

Then,

$$Q = 2.73 \frac{kts}{\lg R - \frac{1}{n}\lg \prod_{i=1}^{n} r_i} \qquad (7-29)$$

where h is the water depth at any point, and s is the decrease of water level.

【Example 7-3】 In the construction of a foundation-engineering project, to reduce the groundwater level, six fully penetrating open wells with radius $r_0 = 0.1$ m are set around the rectangular foundation ditch, as shown in Fig. 7-17. The coefficient of seepage of the soil layer is $k = 0.05$ cm/s, the effected radius of the well group is $R = 600$ m, and the thickness of the aquifer is $H = 15$ m. If the pumping discharge of each well is the same, and when the decrease of groundwater level at the central point G in the foundation ditch is $s = 6$ m, what is the pumping discharge of each well?

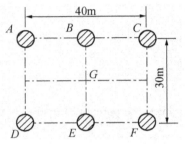

Fig. 7-17

Solution: The distance of each well to the central point G is

$$r_A = r_C = r_D = r_F = \sqrt{20^2 + 15^2} = 25 \text{ m}$$
$$r_B = r_E = 15 \text{ m}$$

Supposing the pumping discharge of each well is Q_w, then Equation (7-28) for the fully penetrating open well becomes

$$Q_w = \frac{1}{6} \frac{1.366 \times k(H^2 - h^2)}{\lg R - \frac{1}{6}\lg(r_A r_B r_C r_D r_E r_F)}$$

$$= \frac{1}{6} \times \frac{1.366 \times 0.0005 \times (15^2 - 9^2)}{\lg 600 - \frac{1}{6}\lg(25^4 \times 15^2)}$$

$$= 0.0113 \text{ m}^3/\text{s}$$

7.5 Graphical solution by drawing flow net

The seepage discussed previously is steady gradually varied seepage flow, which can be solved by Dupuit's formula. However, in steady rapidly varied seepage flow, streamlines either bend significantly or expand and contract sharply, so Dupuit's formula does not suit anymore.

According to the definition of potential flow, the seepage that follows Darcy's law is potential flow. For the planar seepage of imcompressible fluid, its flow field can be described by the Laplace equation. When the boundary conditions are simple, analytical method can be used to solve the Laplace equation; when the boundary conditions are complicated, graphic method or test method can be used. This section will introduce a graphical method, called flow net method.

For the planar seepage of imcompressible fluid, equipotential lines are always orthogonal to the streamlines everywhere, forming an orthogonal grid called a flow net. Take the seepage in a permeable foundation as an example to explain the solution of planar seepage by flow net.

7.5.1 Drawing of flow net for the planar confined seepage

As shown in Fig. 7-18, the flow net of seepage in the permeable foundation can be drawn in the following steps:

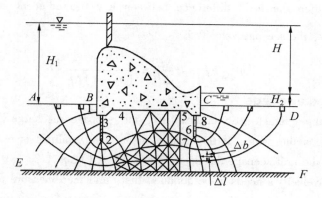

Fig. 7-18

(1) Firstly, according to the seepage boundary conditions, determine the streamlines and the equipotential lines on the boundary. For example, the upstream permeable boundary AB is an equipotential line, because the value of piezometric head at any point on the line is the same, i.e. the velocity potential is the same. Similarly, the downstream permeable boundary CD is another equipotential line. The underground borderline of the structure, i.e. $B-1-2-3-4-5-6-7-8-C$, is a streamline. The bottom boundary EF is also a boundary streamline.

(2) Flow net is characterized by a set of orthogonal square grids. When the net is preliminarily drawn, streamlines and equipotential lines can be sketched roughly according to the trend of boundary line. They should all be smooth curves without having abrupt turning points.

(3) In general, the flow net drawn preliminarily does not always meet the requirements. To check whether the net is correct or not, diagonal lines of grid are often drawn in the net. If every diagonal line is orthogonal and equals, and approximately forms a curved square grid, then the net is correct. However, because the boundary shape is irregular, at the boundary with abrupt change, triangle, pentagon or other irregular shapes will inevitably occur. This is because the net cannot be subdivided into an infinitely small net, but this will not necessarily affect the application of the entire flow net.

(4) The shape of a flow net only depends on the boundary conditions, so it is not affected by

the upstream or downstream water level. However, local modification will affect the whole flow net. Thus, the flow net needs to be drawn and modified carefully, sometime repeatedly, in order to obtain the required accuracy.

7.5.2 Seepage calculation by flow net

7.5.2.1 Seepage velocity

As shown in Fig. 7 – 18, to calculate the seepage velocity in a grid, the mean streamline strength of the grid, Δl, needs to be measured so that the head difference ΔH in the grid can be obtained. Based on the characteristic of flow net, the head difference between any two adjacent equipotential lines is constant. If the number of equipotential lines in the flow net is m (including boundary equipotential lines), and the water level difference between the upstream and downstream is H, then the head difference between any two adjacent equipotential lines is $\Delta H = \dfrac{H}{m-1}$. Thus, the seepage velocity is

$$u = kJ = k\frac{\Delta H}{\Delta l} = \frac{kH}{(m-1)\Delta l} \qquad (7-30)$$

7.5.2.2 Seepage discharge

According to the characteristic of flow net, the seepage discharge Δq between any two adjacent streamlines is the same. If the total number of streamlines is n (including boundary streamlines), then the unit seepage discharge is $q = (n-1)\Delta q$.

To obtain Δq, velocity u needs to be obtained first, and then measure the width of grid Δb, i.e. the average spacing of two adjacent streamlines. So $\Delta q = u\Delta b$, therefore, $q = (n-1)\Delta q$ becomes

$$q = \frac{kH(n-1)}{(m-1)\Delta l}\Delta b \qquad (7-31)$$

If $\Delta b \approx \Delta l$, then $q = \dfrac{kH(n-1)}{m-1}$.

7.5.2.3 Seepage pressure

As shown in Fig. 7 – 19, set a rectangular coordinate system arbitrarily in the seepage region, and choose the impermeable foundation as the transverse axis and the base at level $O-O$.

The total head at point N:

$$H_N = z_N + \frac{p_N}{\gamma}$$

Then the pressure of fluid at point N is

$$\frac{p_N}{\gamma} = H_N - z_N$$

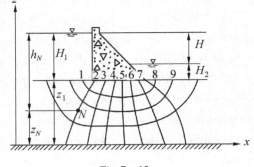

Fig. 7 – 19

The upstream riverbed boundary is equipotential line 1. The total head on line 1 is

$$H_{up} = z_1 + \frac{p_1}{\gamma} = z_1 + H_1$$

Supposing that the seepage head loss from the upstream riverbed to point N is h_{f_1-N}, then

$$H_N = H_{up} - h_{f_1-N} = z_1 + H_1 - h_{f_1-N}$$

Thus,

$$\frac{p_N}{\gamma} = z_1 + H_1 - h_{f_1-N} - z_N = h_N - h_{f_1-N} \tag{7-32}$$

where h_N is the depth of point N below the upstream water surface.

If the water level difference between upstream and downstream is H, there are m equipotential lines and point N is on the second line, then $h_{f_1-N} = \frac{H}{m-1} = \Delta H$, $\frac{p_N}{\gamma} = h_N - \Delta H$, and the pressure of fluid at the ith equipotential line is $\frac{p_i}{\gamma} = h_N - (i-1)\Delta H$.

Equation (7-32) illustrates that the pressure of water at an arbitrary point N in the seepage region equals the depth at the same point, which is the elevation of the upstream water surface minus the head loss from the infiltration point to the point.

7.5.2.4 The uplift on the base of structures (total seepage pressure)

In engineering, one of the most important parameters to calculate is the total seepage pressure on the base of structures, which may cause problems to the stability of the structure.

Taking Fig. 7-19 as an example: calculate the pressure on the base of the structure at first, and then calculate the pressure of the fluid at the intersection point between the equipotential lines and the base, i.e. 1, 2, 3, 4, 5, 6, 7, 8, 9. The depths at these points, calculated from the upstream water surface, are the same although the head losses from the upstream boundary to these points are different. Obviously, the head losses at points 2 to 9 are $\Delta H, 2\Delta H, 3\Delta H, \ldots, 8\Delta H$. Then the corresponding pressure of fluid can be obtained. Using these values, now draw the graph of pressure distribution under the structure. According to the area of the graph, the uplift on the base of foundation per unit width can be obtained. If the pressure in water height is used and the area of the pressure distribution graph is A_p, then the uplift on the base of foundation per unit width is

$$P = \gamma A_p$$

【Example 7-4】 As shown in Fig. 7-18, the upstream water depth is 8.1 m, and the downstream water depth is 3.0 m. The width of the dam is 24 m, the base of the dam is 1.0 m below the river bed, and the key wall between the upstream and downstream is 1.0 m high. The length of sheet piling, upstream and downstream, are 5 m and 3 m, respectively. The seepage coefficient of the sand foundation is 5×10^{-2} cm/s. Determine: ① the seepage discharge of the dam foundation; ② the seepage pressure at points 3 & 5 on the base of the dam; ③ if the grid of the seepage exit at downstream, which is adjacent to the base of the dam, is 6 m long, then what are the seepage gradient and the seepage velocity?

Solution: (1) The difference between the upstream and downstream water level is
$$H = 8.1 - 3 = 5.1 \text{ m}$$
In the flow net, the number of equipotential lines is $m=18$, i.e. the number of segments is $m-1 = 17$; the number of streamlines $n = 5$, i.e. the sections $n-1 = 4$. Then the seepage discharge of the dam foundation is
$$Q = qB = B\frac{kH(n-1)}{(m-1)} = 24 \times \frac{5 \times 10^{-4} \times 5.1 \times 4}{17} = 14.4 \text{ L/s}$$

(2) The water head difference between any two adjacent equipotential lines is
$$\Delta H = \frac{H}{m-1} = \frac{5.1}{17} = 0.3 \text{ m}$$
Thus, the seepage pressure at point 3 is
$$\frac{p_3}{\gamma} = H_1 + z_3 - 7.5\Delta = 8.1 + 1.0 + 1.0 - 7.5 \times 0.3 = 7.85 \text{ m (76.93 kPa)}$$
where z_3 represents the vertical distance from point 3 to the river bed.

Similarly, the seepage pressure at point 5 is
$$\frac{p_5}{\gamma} = H_1 + z_5 - 11\Delta H = 8.1 + 1.0 - 11 \times 0.3 = 5.8 \text{ m (56.84 kPa)}$$

(3) The seepage gradient in the grid is
$$J = \frac{\Delta H}{\Delta l} = \frac{0.3}{6} = 0.05$$
Thus the corresponding seepage velocity is
$$u = kJ = 5 \times 10^{-4} \times 0.05 = 2.5 \times 10^{-5} \text{ m/s}$$

Chapter summary

This chapter has introduced the basic concepts and principles of seepage and its calculation in engineering. Graphical solution by a flow net is also introduced.

1. In hydraulics, gravitational water in porous medium is the object of study in seepage. The seepage model is a simplified model for describing actual seepage flow, and some principles have to be followed when the model is used for actual seepage problems. Based on the seepage model, the common hydraulic concepts and methods can be applied in the study of seepage.

2. Low velocity is the hydraulic characteristic of seepage. Thus, the kinetic energy usually can be ignored, and the total head equals the piezometric head:
$$H = z + \frac{p}{\gamma}$$

3. Darcy's law is the basic principle of seepage, which can only be applied for laminar seepage flow.

For uniform seepage, the velocity of point equals the mean cross-sectional velocity. Then,
$$u = v = kJ$$

For non-uniform seepage, the velocity of point is

$$u = kJ = -k\frac{\mathrm{d}H}{\mathrm{d}l}$$

Dupuit's formula is the basic formula for steady gradually varied seepage flow. In the flow cross-section of gradually varied seepage, the velocity of point equals the mean velocity of cross-section. Then,

$$u = v = -k\frac{\mathrm{d}H}{\mathrm{d}l}$$

4. Based on Dupuit's formula, the water surface curve equation of steady gradually varied seepage can be derived, so can the seepage calculation formula for wells and catchment corridors.

5. For the planar seepage problems under complicated conditions, the method of flow net can be used to calculate the seepage velocity, discharge and pressure of fluid.

Review questions

7-1 What are the factors of influencing the coefficient of seepage?

7-2 Why should we introduce the seepage model?

7-3 What conditions should be met when the seepage model is applied for actual seepage study?

7-4 What are the similarities and differences between Darcy's law and Dupuit's formula?

7-5 What hypotheses are introduced in the derivation of discharge formula for fully penetrating wells?

7-6 Is a seepage curve streamline or equipotential line? Why?

7-7 Why would triangle or pentagon exist sometimes in flow net? Do they have any influence on seepage?

7-8 Try to analyze the difference between the uplift and hydrostatic pressure forces of seepage on impermeable base?

Multiple-choice questions (one option)

7-1 Compared to the actual seepage, the seepage model has _____.
  (A) the same discharge   (B) the same velocity
  (C) different pressure of point (D) different resistance

7-2 Among the seepage coefficients of different soils (k_1 for clay; k_2 for silt; k_3 for coarse sand), _____.
  (A) $k_1 > k_2 > k_3$ (B) $k_1 < k_2 < k_3$
  (C) $k_1 > k_3 > k_2$ (D) $k_2 > k_1 > k_3$

7-3 Darcy's formula in seepage is only applied for _____.
  (A) uniform seepage   (B) gradually varied seepage
  (C) turbulent seepage   (D) laminar seepage

7-4 In the flow cross-section of gradually varied seepage, the seepage velocity at every point comforms to _____ distribution.

(A) linear (B) parabolic (C) uniform (D) logarithmic

7-5 The seepage curve of gradually varied seepage flow changes _____ along the flow.
(A) increasingly (B) horizontally
(C) decreasingly (D) all the above are possible

7-6 For fully penetrating wells, the incorrect statement is _____.
(A) the larger the radius of influence is, the larger the outflow
(B) the outflow is directly proportional to the coefficient of seepage
(C) the larger the radius of well is, the larger the outflow
(D) the outflow is directly proportional to the decrease of well water level

7-7 Based on the flow net, _____ in planar seepage region can be calculated.
(A) the velocity and discharge (B) the hydrodynamic pressure of water
(C) the uplift pressure force (D) all the above

Problems

7-1 In the apparatus of Darcy's test (shown as Fig. 7-3), the diameter of the cylinder is 20 cm, the interval of the two piezometers is 40 cm, the difference of the corresponding piezometric heads is 20 cm, and the measured discharge is 0.002 L/s. Calculate the coefficient of seepage k.

7-2 Fig. 7-20 shows a cylindrical water-filter. Its diameter $d = 1.2$ m, the height of the filter is 1.2 m, and the coefficient of seepage is $k = 0.0001$ m/s. Calculate the seepage discharge Q if $H = 0.6$ m.

7-3 In a flow cross-section of gradually varied seepage, the slope gradient of the seepage curve is 0.005, and the coefficient of seepage is 0.00004 m/s. Calculate the seepage velocity at arbitrary points in the cross-section and the mean seepage velocity of the cross-section.

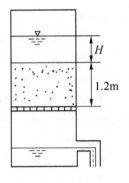

Fig. 7-20

7-4 Above a horizontal impermeable ground is a seepage layer, which has the width of 800 m and the seepage coefficient of $k = 0.0003$ m/s. The water depths are 8 m and 6 m separately measured in two observation wells, which are 1000 m apart from each other along the seepage path. Calculate the seepage discharge Q.

7-5 For fully penetrating artesian wells, when $H > t$ and $h < t$, show the seepage discharge $Q = 1.366 \times \dfrac{k(2Ht - t^2 - h^2)}{\lg \dfrac{R}{r_0}}$.

7-6 In the seepage with positive slope, the bed slope $i = 0.0025$, the water depths are 3 m and 4 m respectively at the two sections, which are 500 m apart from each other, and the coefficient of seepage in the soil is $k = 0.0005$ m/s. Determine the seepage discharge per unit width.

7-7 A fully penetrating open well over a horizontal impermeable layer has the radius of $r_0 = 10$ cm, the aquifer thickness is 8 m, the radius of influence is 500 m, and the coefficient of seepage is $k = 0.00001$ m/s. Determine the outflow Q when the decrease of water level reaches 6 m.

7-8 As shown in Fig. 7-21, the well group consists of 6 fully penetrating open wells. $a = 50$ m, $b = 40$ m, the total discharge of the well group is $Q = 3$ L/s, and the outflow for each well of radius 0.2 m is the same. The aquifer thickness $H = 12$ m, the seepage coefficient $k = 0.0001$ m/s, and the radius of influence $R = 700$ m. Determine the value of water level decrease s at the central point of the well group G.

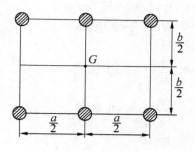

Fig. 7-21

7-9 The radius of a fully penetrating artesian well is $r_0 = 0.1$ m. The aquifer thickness is $t = 5$ m. Drill a hole at 10 m away from the well axis. Before pumping, the depth of groundwater is measured as $H = 12$ m. When the outflow reaches 10 L/s, the decrease of water level in the well reaches 2 m, and the water level of the observation hole drops 1 m, what are the seepage coefficient of the aquifer k and the radius of influence R?

7-10 Lowering the groundwater level is required in a construction site. Fig. 7-22 shows eight wells set uniformly in a circle with radius $r = 10$ m. The radius of each well is $r = 10$ cm. If the aquifer thickness $H = 15$ m, the seepage coefficient $k = 0.0005$ m/s, and the affected radius of the well group $R = 500$ m, to make the water level decrease of central point O reach 4 m, what is the outflow of each well?

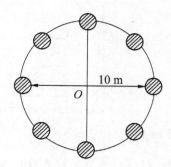

Fig. 7-22

Answers to selected problems

Chapter 1:

1-1 $\nu = 1.003 \times 10^{-6}$ m^2/s

1-2 $dp = 1.003 \times 10^7$ Pa

1-3 [L]

1-4 $\left.\dfrac{du}{dy}\right|_{\frac{y}{H}=0.25} = 1.06 \dfrac{u_m}{H}$

$\left.\dfrac{du}{dy}\right|_{\frac{y}{H}=0.5} = 0.84 \dfrac{u_m}{H}$

1-5 $\tau = 1150$ N/m^2

1-6 $\mu = 0.105$ Pa·s

1-7 $z = g$

1-8

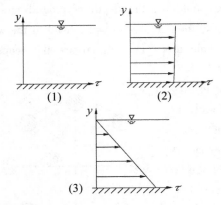

Chapter 2:

2-1 $p_A - p_C = -22.5$ kPa, $p_C - p_D = 1482.5$ kPa, $p_O - p_B = 0$

2-2 30 220 N/m^2

2-3 3 m

2-4 (1) $z_A = 5$ m, $z_B = 3$ m, $\dfrac{p_0}{\gamma_{abs}} = 7$ m (water),

$\dfrac{p_A}{\gamma_{abs}} = 7$ m(water), $\dfrac{p_B}{\gamma_{abs}} = 9$ m(water)

(2) $z_A = 0$, $z_B = -2$ m, the same pressure as (1)

2-5 $p_c = 107.4$ kN/m^2, $h = 0.96$m (water)

2-6 3 m (long), $h_m = 0.235$ m (Hg)

2-7 $p_{1abs} = 63.7$ kN/m^2, $p_{1rel} = -34.3$ kN/m^2,

$\dfrac{p_1}{\gamma} = -3.5$ m(water),

$\dfrac{p_{v1}}{\gamma} = 3.5$ m(water), $z_1 + \dfrac{p_1}{\gamma} = -3.0$ m;

$p_{2abs} = 67.6$ kN/m^2, $p_{2rel} = -30.4$ kN/m^2,

$\dfrac{p_2}{\gamma} = -3.1$ m(Hg),

$\dfrac{p_{v2}}{\gamma} = 3.1$ m(Hg),

$z_2 + \dfrac{p_2}{\gamma} = -3.0$ m(Hg)

2-8 $h_1 = 2.91$ m, $h_2 = 2.09$ m; $h = 4.12$ m

2-9 (1) $\Delta p = 1.225$ kN/m^2;
(2) $\Delta l = 0.312$ m

2-10 $p_{Aabs} = 93.1$ kN/m^2, $p_{Arel} = -4.9$ kN/m^2
$p_v = 4.9$ kN/m^2, $h_v = 0.5$ m (water)

2-11 $p_M = 7.51$ kN/m^2

2-12 $p_{Aabs} = 103.1$ kN/m^2, $h_1 = 0.52$ m (water)

2-13 $\gamma_{oil} = 8.02$ kN/m^3, $G_{oil} = 2.49$ kN

2-14 $h_1 = 1.26$ m (oil), $h_2 = 6$ m (water), $h = 0.807$ m (Hg)

2-15 $p_B - p_A = 0.52$ kN/m^2, $p_{vA} = 5886$ kN/m^2, $p_{vB} = 5366$ kN/m^2

2-18 $p = 45.3$ kN, 1.6 m from B, 2.304 m from A

2-19 $p = 773$ kN, $y_D = 0.501$ m (from top)

2-20 (1) $p_x = 3312.4$ kN (to right), $p_z = 862.4$ kN (downward);
(2) $p_x = 176.4$ kN (to left), $p_z = 88.2$ kN (downward); $p_{x(all)} = 3136$ kN (to right), $p_z = 950.6$ kN (downward)

2-21 (1) $T_1 \geqslant 144.5$ kN;
(2) $T \geqslant 66.15$ kN

2-23 (1) $p_1 = p_2 = p_3 = 26.13$ kN;
(2) $y_1 = 1.54$ m, $y_2 = 2.81$ m, $y_3 = 3.645$ m

2-26 $p_x = 352.8$ kN, $p_z = 46.18$ kN (downward)

2-27 $x = 3$ m

2-28 $T = 41.05$ kN

2-29 $\geqslant 8$ mm

Answers to selected problems

Chapter 3:

3-1 $a = \sqrt{(1+x+t)^2 + (1+y-t)^2}$

3-2 $a|_{x=8} = 0.03$ m/s², $a_x|_{x=8} = a_y|_{x=8} = 0.0214$ m/s²

3-3 $y = C$, straight streamlines parallal to x

3-4 $x^2 + y^2 = C$, streamlines are the same as pathlines, which are concentric

3-7 $Q = 0.49$ L/s, $v_1 = 6.25$ cm/s, $v_2 = 25$ cm/s

3-8 $Q_{CD} = 7.85$ L/s, $Q_{BC} = 29.35$ L/s, $Q_{AB} = 79.35$ L/s, $v_{AB} = 1.617$ m/s, $v_{BC} = 1.66$ m/s

3-9 $Q = 0.447$ m³/s

3-10 $Q_3 = 0.8$ m³/s, $Q_2 = 1.6$ m³/s, $Q_1 = 2.4$ m³/s, $v_3 = 3.2$ m/s, $v_2 = 6.4$ m/s, $v_1 = 9.6$ m/s

3-11 $d = 300$ mm, $v = 1.178$ m/s

3-12 $v_1 = 18.05$ m/s, $v_2 = 22.25$ m/s

3-13 $v = 11.31$ m/s, $\delta = 0.0212$ m

3-14 $Q = 0.212$ L/s, $v = 0.075$ m/s

3-15 Flow C. S. 1 → C. S. 2, $h_{w1-2} = 1.74$ m

3-16 $Q = 0.139$ m³/s, $v_1 = 1.107$ m/s, $v_2 = 4.43$ m/s

3-17 $Q = 0.058$ m³/s

3-18 $d = 0.8$ m

3-19 $v = 8.283$ m/s

3-20 $v_2 = 2.74$ m/s, $Q = 0.86$ L/s

3-21 $Q = 61.6$ L/s

3-22 $\frac{p_2}{\gamma} = 1.545$ m

3-23 $Q = 0.903$ m³/s, $p_1 = 7668$ N/m²

3-24 $Q = 0.55$ m³/s

3-25 $h_v = -3.752$ m (water column)

3-26 $\frac{A_1}{A_2} = \frac{v_2}{v_1} \leq \sqrt{\frac{H}{h+b}}$

3-27 $R_x = 384.1$ kN →

3-28 $V \geq 0.51$ m³

3-29 (1) $Q = 0.053$ m³/s;
      (2) $T = 1272.3$ N

3-30 $R_x = 2028$ N, $R_y = 1323.5$ N, $R = 2421.7$ N, $\theta = 33.13°$, upright

3-31 $R'_x = 10.28$ kN, $R'_y = 6.79$ kN, $R' = 12.32$ kN, $\theta = 33.5°$, downleft

3-32 $P = 98.35$ kN, →

3-33 $R = 51.2$ kN, $P = 91.9$ kN > R

3-34 $R_x = 2987.1$ kN

3-35 $R_x = 0.051$ kN, to left, $R_y = 0$

Chapter 4:

4-1 $Re_1 > Re_2$, $Re_1 : Re_2 = 2$

4-2 (1) $Re = 1914 < 2000$, Laminar flow;
     (2) $Re = 4786 > 2000$, Turbulent flow;
     (3) $Q = 51.3$ cm³/s;
     (4) $v = 320$ cm/s

4-3 (1) $Re = 416\,700 > 2000$, Tubulent flow;
     (2) $v = 18$ cm/s

4-4 $\tau_0 = 3.92$ N/m²; $h_f = 0.8$ m

4-5 $v_* = 0.1854$ m/s; $\tau_0 = 34.37$ N/m²

4-6 $\mu = 0.0332$ N·s/m²

4-7 $\gamma = \gamma_0/\sqrt{2}$

4-8 $\delta_o = 9.5 \times 10^{-2}$ mm

4-9 0.087 mm; 0.051 mm

4-10 $\gamma Q = 165.7$ kN/h, $\rho Q = 16.9$ t/h, $Q = 18.77$ m³/h

4-11 $\lambda = 0.04$

4-12 $\lambda \approx 0.021$, $h_f = 7.15$ m, $\Delta p = 70.08$ kN/m²

4-13 $\lambda = 0.038$, $Q = 0.1245$ m³/s

4-14 (1) $R_e = 1078$, laminar flow;
      (2) $\lambda = 0.0594$;
      (3) $h_f = 0.0103$ m;
      (4) $h_f = 0.0191$ m

4-15 (1) $h_f = 0.65$ m;
      (2) $h_f = 0.0373$ mm;
      (3) $h_f = 1.987$ m

4-16 $C = 44.27$; $n = 0.0109$

4-17 $\zeta = 0.32$

4-18 $Q = 0.372$ L/s

4-19 $F_D = 720$ N

4-20 $R = 4.15$ m

Chapter 5:

5-1 $\mu = 0.62$

5-2 $t = 948.5$ s

5-3 $\Delta H = 0.2$ m

5-4 $Q = 0.3878$ m³/s

5-5 $Q_H = 1.22$ L/s; $Q_P = 1.61$ L/s; $\frac{p_v}{\gamma} = 1.51$ m (water)

5-6 $\mu = 0.94$

5-7 $d = 1.2$ m

5-8 $\Delta H = 1.07$ m; $Q = 3.57$ L/s

5-9 $t = 937.6$ s

5-10 $Q = 3.41$ L/s; $\Delta H = 0.072$ m

5-11 $H = 2.597$ m

5-12 $Q = 9.09$ L/s; $\frac{p_c}{\gamma} = -1.7565$ m (Hg)

5-13 $H = 32.177$ m; $P_w = 11.837$ kW

5-14 $Q = 132.42$ L/s

5-15 $H = 16.84$ m

5-16 $H = 21.61$ m

5-17 $Q_2 = 169.7$ L/s; $Q_3 = 467.65$ L/s

5-18 $Q' = \sqrt{\dfrac{n}{n+3}}\, Q$

5-19 $Q_A = 53.83$ L/s; $Q_B = Q_C = 66.17$ L/s; $h_{MN} = 7.35$ m

5-20 $Q_1 = Q_2 = Q_4 = Q_5 = Q/2$; $Q_3 = 0$; Q_1, Q_4 decrease; Q_2, Q_5 increase; Q_3 flow downward

Chapter 6:

6-1   6.66 m³/s

6-2   240.78 m³/s

6-3   2.85 m

6-4   3.17 m

6-5 $h = 1.7$ m; $b = 3.4$ m

6-6 $h = 0.5$ m; $b = 2.0$ m

6-7 $i = 0.0026$; $z = 51.76$ m

6-8 $h = 1.46$ m; $b = 7.3$ m

6-9   GVF $h_0 = 1.8$ m; $h_c = 0.64$ m; $Fr = 0.21$

6-10   1.07 m

6-11 $h = 0.6$ m; $i_c = 0.0074$

6-12   0.665 m

6-13   1.48 m³/(s·m)

6-14   2

6-17 $Q = 0.283$ m³/s; $h = 0.20$ m

6-18   132 m

6-19   8.97 m³/s

6-20   8.25 m³/s

6-21   2.52 m

6-22   14.1 m³/s

6-23   24.8 m³/s

Chapter 7:

7-1 $k = 0.0001273$ m/s

7-2 $Q = 0.05655$ L/s

7-3 $v = u = 2 \times 10^{-7}$ m/s

7-4 $Q = 3.36$ L/s

7-5   Hint: write differential equation for segments: H-t and t-h, followed by integration.

7-6 $q = 8.69 \times 10^{-5}$ m³/(s·m)

7-7 $Q = 0.2216$ L/s

7-8 $s = 1.35$ m

7-9 $k = 1.465$ mm/s; $R = 1000$ m

7-10 $Q = 4.44$ L/s

References

[1] DOUGLAS J F, GASIOREK J M, SWAFFIELD J A, et al. Fluid Mechanics. 5th ed. New Jersey: Prentice Hall, 2006.
[2] HAMILL L. Understanding Hydraulics. 3th ed. London: Palgrave Macmillan, 2011.
[3] MASSEY B S, SMITH J W. Mechanics of Fluids. 7th ed. Northamptonshire: Nelson Thornes, 1998.
[4] MUNSON B R, YOUNG D F, OKIISHI T H. Fundamentals of Fluid Mechanics. 5th ed. New York: Wiley, 2005.
[5] PRASUHN A L. Fundamentals of Fluid Mechanics. New Jersey: Prentice Hall, 1980.
[6] STREETER V L, WYLIE E B, BEDFORD K W. Fluid Mechanics. 9th ed. New York: McGraw-Hill, 1998.
[7] 尹小玲，于布. 水力学. 3版. 广州：华南理工大学出版社，2014.